EROSION AND SEDIMENTATION, SECOND EDITION

Praise for the First Edition:

"... (presents) the mechanics of sediment motion alongside those subjects in fluid mechanics that are fundamental to understanding sediment transport. The interweaving of the two subjects is carried out particularly well ... Each topic is covered clearly, with many carefully designed figures, examples, and exercises ... an excellent primer on both fundamental concepts of sediment-transport theory and methods for practical applications ... well written, and nicely illustrated, and it will serve as either a handbook for workers in the field or a textbook for beginning students of the subject. Julien has done a truly admirable job in making ... (a) difficult subject much more accessible to beginning Earth scientists and engineers."
JONATHAN NELSON, *Journal of Hydraulic Engineering*

"... this well-written text can be equally useful to undergraduates, graduates, geologists and geophysicists ..."
CHRISTOPHER KENDALL, *The Leading Edge*

"... a welcome addition ... logically planned out, well written with clear explanations"
R. L. SOULSBY, *Journal of Fluid Mechanics*

"... clearly and concisely written, covers not only the theory but also measurement methods and also provides many worked examples. It will serve well its primary purpose as a textbook for post-graduate courses on erosion and sedimentation ... thoroughly recommended."
IAN R. CALDER, *Hydrology and Earth System Sciences*

"... an excellent and accessible treatment of the 'classic' engineering approach to erosion and sedi-mentation ... a worthy primer for those requiring mastery over the essential technicalities of hydraulic analysis."
NICHOLAS J. CLIFFORD, *Progress in Physical Geography*

The new edition of Pierre Julien's acclaimed textbook brings the subject of sedimentation and erosion completely up-to-date. The structure of the first edition is essentially unchanged, but all the chapters have been updated, with several chapters reworked and expanded significantly. Examples of the new additions include the Modified Einstein Procedure, sediment transport by size fractions, sediment transport of sediment mixtures, the concept of added mass, and new solutions to the Einstein Integrals. Many new examples and exercises have been added. *Sedimentation and Erosion* is an essential textbook for students in civil and environmental engineering as well as the geosciences, and also as a handbook for researchers and professionals in engineering, the geosciences, and the water sciences.

PIERRE Y. JULIEN is Professor of Civil and Environmental Engineering at Colorado State University. He has co-authored more than 400 scientific publications including two textbooks, 20 lec-ture manuals and book chapters, 145 refereed journal articles, 130 professional presentations, 150 conference papers, and 100 technical reports. As a Professional Engineer, he has completed projects and assignments for at least 50 different agencies and consulting firms world-wide. He is member of 12 professional societies and served as Editor of the ASCE *Journal of Hydraulic Engineering*. He has delivered numerous keynote addresses at international conferences. In 2004, he received the H. A. Einstein Award for his outstanding contributions to sedimentation and river mechanics. He is also the author of the textbook *River Mechanics* (2002, Cambridge University Press).

EROSION AND SEDIMENTATION

2nd Edition

PIERRE Y. JULIEN
Colorado State University

CAMBRIDGE
UNIVERSITY PRESS

CAMBRIDGE UNIVERSITY PRESS
Cambridge, New York, Melbourne, Madrid, Cape Town, Singapore,
São Paulo, Delhi, Dubai, Tokyo

Cambridge University Press
The Edinburgh Building, Cambridge CB2 8RU, UK

Published in the United Kingdom by Cambridge University Press, UK

www.cambridge.org
Information on this title: www.cambridge.org/9780521830386

First published 1995
Second edition 2010

Printed in the United Kingdom at the University Press, Cambridge

A catalog record for this publication is available from the British Library

Library of Congress Cataloging in Publication data
Julien, Pierre Y.
Erosion and sedimentation / Pierre Y. Julien. – 2nd ed.
p.   cm.
Includes bibliographical references and index.
ISBN 978-0-521-83038-6 – ISBN 978-0-521-53737-7 (pbk.)
1. Erosion.   2. Sedimentation and deposition.   I. Title.
QE571.J85 2010
551.3′53–dc22   2010017821

ISBN 978-0-521-83038-6 Hardback
ISBN 978-0-521-53737-7 Paperback

To Helga and Patrick

Contents

Preface

This textbook has been prepared for graduate students and professionals keeping up with recent technological developments in the field of sedimentation engineering. This text is not a voluminous encyclopedia; it is rather a concise digest to be found in the briefcases of students and professionals. It scrutinizes selected methods that meet learning objectives, underlining theory and field applications.

The material can be covered within a regular forty-five-hour graduate-level course at most academic institutions. Colorado State University offers several graduate courses in hydraulics, sedimentation and river engineering, and stream rehabilitation. Two advanced courses CIVE716 Erosion and Sedimentation and CIVE717 River Mechanics are offered in sequence. This book has been prepared for the first course and the author's companion book entitled *River Mechanics*, at Cambridge University Press, has been tailored to the needs of the second course. The prerequisites include undergraduate knowledge of fluid mechanics and basic understanding of partial differential equations.

The chapters of this book contain a variety of exercises, examples, problems, computer problems, and case studies. Each type illustrates a specific aspect of the profession from theoretical derivations through exercises and derivations, to practical engineering solutions through the analysis of simple examples and complex case studies. Most problems can be solved with a few algebraic equations; a few require the use of a computer. Problems and equations marked with a single diamond (♦) are important; those with a double diamond (♦♦) are considered most important. The answers to some problems are provided to check calculations. Tests and homework assignments can include problems without answers.

Recent technological developments in engineering encourage the use of computers for quantitative analyses of erosion, transport, and sedimentation. Numerous algorithms in this text can be easily programmed. No specific computer code or language is emphasized or required. Instead of using old software packages, the textbook promotes student creativity in developing their own tools and programs with the best software available at any given time.

This second edition has been significantly revised and improved. Specific goals with the revisions have been twofold: (1) clarify and reinforce the most important concepts and principles through added explanations, figures, examples, and new homework problems; and (2) expand and update the technical content in the light of recent developments in the literature and engineering practice. Throughout the revisions, the main objectives have been to keep this text: (1) concise and effective when learning new concepts and derivations; and (2) helpful as a reference for future use in engineering practice. A major concern throughout the revisions has been to keep this textbook light and affordable.

I am grateful to my own teachers and professors at Laval University, and particularly to my advisors M. Frenette, J. L. Verrette, Y. Ouellet and B. Michel. I am also grateful to my mentors at Colorado State University, and specifically to D. B. Simons, H. Rouse, E. V. Richardson, and my esteemed colleagues. Many graduate students offered great suggestions for improvement to this textbook. Jenifer Davis diligently typed successive drafts of the manuscript and Jean Parent prepared professional figures. Finally, it has been a great pleasure to collaborate with the Cambridge University Press production staff.

Symbols

Symbols

a, a_x	acceleration
a	thickness of the bed layer
A	surface area
A_T, A_U, A_G, A_B	gross, upland, gully, and bank erosion
A_t	basin drainage area
B	constant of the resistance formula
B	bulking factor
c_Δ	bedform celerity
c_0	integration constant
c_{Bd}, c_{cl}	coefficients
C	Chézy coefficient
C_a	near-bed sediment concentration
$Co = l_c / \sqrt{l_a\, l_b}$	Corey particle shape factor
$C_v, C_w, C_{ppm}, C_{mg/l}$	sediment concentration by volume, weight, in parts per million, and milligrams per liter
$C_t, C_\forall, C_f$	time-, spatial-, and flux-averaged concentration
C_D, C_E	drag and expansion coefficients
$\hat{C}$	cropping-management factor of the USLE
d_{10}, d_{50}	grain size with 10%, or 50%, of the material finer by weight
d_s	sediment size
d_*	dimensionless particle diameter
D	molecular diffusion coefficient
e	void ratio
e_B	Bagnold coefficient
E	specific energy
E_b	near-bed particle pick-up rate
f	Darcy–Weisbach friction factor $E = 2d_s / h$

F	force
Fr	Froude number
Fr_d	densimetric Froude number
$F_1, F_2, J_1, J_2, I_1, I_2$	components of the Einstein integrals
g	gravitational acceleration
G	specific gravity
Gr	gradation coefficient
h	flow depth
h_c, h_n	critical and normal flow depth
H	Bernoulli sum
He	Hedstrom number
i	rainfall intensity
I	universal soil-loss equation rainfall intensity
k_s, k'_s	boundary and grain roughness height
K	consolidation coefficient
$\hat{K}$	soil erodibility factor of the USLE
K_d	dispersion coefficient
ℓ	liter
ℓ_a, ℓ_b, ℓ_c	particle dimensions
$\ell_1, \ell_2, \ell_3, \ell_4$	moment arms for particle stability analysis
ℓ_m	mixing length
l, L	lengths
L_b, L_s, L_t	bedload, suspended load, total load
$\hat{L}$	field length factor of the USLE
m, M	mass
M, N	particle stability coefficients
M'_D, M''_D	moments
n	Manning n
$\vec{n}$	vector normal to a surface
p	pressure
p_0	porosity
P	wetted perimeter
$\hat{P}$	conservation practice factor
q	unit discharge
q_b, q_s, q_t	unit sediment discharge (bed, suspended, total)
Q	total discharge
Q_b, Q_s, Q_t	sediment discharge (bed, suspended, total)
r	radial coordinate
R	radius of a sphere
$\hat{R}$	rainfall erosivity factor of the USLE
R_T	sampler lowering rate

Re, Re_B, Re_d, Re_p	Reynolds numbers
$Re_* = \frac{u_* d_s}{v}$	grain shear Reynolds number
$R_h = A/P$	hydraulic radius
Ro	Rouse number
Sh	Shen–Hung parameter
$S_p = (\ell_b \ell_c / \ell_a^2)^{1/3}$	sphericity of a particle
S_0, S_f, S_w	bed, friction, and water surface slopes
$\hat{S}$	slope steepness factor of the USLE
SF	particle stability factor
S_{DR}	sediment-delivery ratio
t, T	time
t_d, t_t, t_v	dispersion, transversal, and vertical time scales
T	sediment transport parameter
T_c	consolidation time
T_E	trap efficiency
T_R	life expectancy of a reservoir
T_w	wave period
$T°_C, T°_F$	temperature in degrees Celsius and Fahrenheit
u, v_x, v_y, v_z	velocity
u^*	shear velocity
V	depth-averaged flow velocity
$\forall$	volume
W	channel width
x, y, z	coordinates
X_C	settling distance
X_D	total rate of energy dissipation
X_r	runoff length
Y	sediment yield
$\hat{z}$	upward vertical direction
z_b	bed elevation
z_0	elevation where the velocity is zero
Z	dependent variable

Greek symbols

α_e	energy correction factor
$\beta, \delta, \lambda, \theta$	angles of the particle stability analysis
β_m	momentum correction factor
β_s	ratio of sediment to momentum exchange coefficient
γ	specific weight

Γ	circulation
δ	laminar sublayer thickness
∂	partial derivative
∇^2	Laplacian
Δ, Λ	dune height and wavelength
Δp_i	sediment size fraction
ε	turbulent mixing coefficients
ε_m	eddy viscosity
ζ	turbulent-dispersive parameter
η_0, η_1	particle stability number
θ	angular coordinate
Θ	angle
ϑ	mixing stability parameter
κ	von Kármán constant
λ, δ, β	angles of particle stability analysis
λ	linear concentration
μ	dynamic viscosity
υ	kinematic viscosity
$\xi = z/z_m$	normalized depth
ρ, ρ_s	mass density of water and sediment
Π	dimensionless parameter
Π_W	wake strength
σ	normal stress components
σ	standard deviation for sediment diffusion
σ_g	gradation coefficient
σ_t	mixing width
τ_{yx}	shear stress in x direction from gradient in y
τ_0, τ_c	bed shear stress and critical shear stress
τ^*	Shields parameter
τ_y, τ_d	yield and dispersive stresses
ϕ	angle of repose
Φ, Ψ	potential and stream functions
χ	rate of work done per unit mass
χ_D	dissipation function
$\odot, \ominus, \oslash, \otimes$	modes of deformation
ω	settling velocity
ω_f	flocculated settling velocity
$\vec{\omega}$	vorticity
Ω_e, Ω_g	elastic energy and gravitation potential
$\forall$	volume

Superscripts and diacriticals

$a, b, \hat{a}, \hat{b}$	coefficient and exponent of resistance formula
$(\hat{v}), \bar{v}, v^+$	fluctuating, average and time velocity
F'_D, F''_D	surface and form drag
τ', τ''	grain and form resistance
$\tilde{E}, \tilde{H}$	integrated value
$\dot{m}, \dot{C}_v$	point source
$\tilde{u} = v_x/u_*$	relative velocity

Subscripts

a, C_a, v_a	bed layer characteristics
a_x, a_y, a_z	x, y, z acceleration components
C_0, C_1, C_2	concentration at different times
$d_i, \tau_{ci}, \tau^*_{ci}, \Delta p_i$	characteristics of size fraction i
d_s, ρ_s, γ_s	sediment properties
E_{min}	minimum specific energy
f_b, f_w	bed and wall friction factor
f_p, h_p, T_p	upper-regime plane bed parameters
F_D, C_D	drag force and coefficient
F_H, F_V	horizontal and vertical forces
F_W, F_B, F_D, F_L	weight, buoyant, drag, and lift forces
h_c, h_n	critical and normal flow depth
h_d, V_d, f_d, α_d	densimetric values
L_b, L_s	bedload, suspended load
L_m, L_u	measured, unmeasured load
L_w, L_{bm}	washload, bed material load
M_w, M_s, M_T	mass of water, solids, and total
p_r	relative pressure
q_{bv}, q_{bm}, q_{bw}	unit sediment discharge by volume, mass, and weight
S_0, S_f, S_w	bed slope, friction slope, and water surface slope
T°_F, T°_C	temperature in $^\circ$F and $^\circ$C
u_∞, p_∞	velocity and pressure infinitely far
v_a	near-bed particle velocity
v_r, v_p	reference velocity and velocity against a particle
W, W_L, W_p, W_s	water content, liquid, plastic, and shrinkage limits
X_d, X_t, X_v	lengths for dispersion, transversal, and vertical mixing
ρ_{md}, γ_{md}	dry specific mass and weight of a mixture
ρ_m, v_m, μ_m	properties of a water–sediment mixture
τ_c, τ^*_c	critical shear stress and critical Shields parameter
τ_0, τ_b, τ_s	boundary shear stress, at the bed, and on the side slope
τ^*, d^*	dimensionless sedimentation parameters

1

Introduction

Erosion and sedimentation refer to the motion of solid particles, called sediment. The natural processes of erosion, transportation and sedimentation, sketched in Figure 1.1, have been active throughout geological time and have shaped the present landscape of our world. Today, they can cause severe engineering and environmental problems.

Human activities usually accelerate the processes of erosion, transport, and sedimentation. For instance, soil erodibility is enhanced by plowing and tillage. The protective canopy is weakened by grubbing, cutting, or burning of existing vegetation. Besides producing harmful sediment, erosion may cause serious on-site damage to agricultural land by reducing the productivity of fertile soils. Under some circumstances, the erosion rate can be 100 to 1,000 times greater than the geological erosion rate of 0.1 ton/acre year (25 ton/km^2 year).

Severe erosion can occur during the construction of roads and highways when protective vegetation is removed and steep cut and fill slopes are left unprotected. Such erosion can cause local scour problems along with serious sedimentation downstream. Approximately 85% of the 571,000 bridges in the United States are built over waterways. The majority of these bridges span rivers and streams that are continuously adjusting their beds and banks. Bridges on more active streams can be expected to experience scour problems as a result of stream realignment. Local scour at bridge piers and erosion of abutments are the most common causes of bridge failure during floods.

Mining operations may introduce large volumes of sediment directly into natural streams. Mine dumps and spoil banks often continue to erode by natural rainfall for many years after mining operations have ceased. For example, some drainage and flood problems in the Sacramento Valley, California, as well as problems of construction and maintenance of navigation channels, can be traced directly to mining activities that took place more than a century ago at the time of the gold

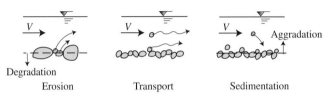

Figure 1.1. Processes of erosion, transport, and sedimentation

rush. Gravel stream mining can cause severe channel instabilities such as upstream headcutting, which may trigger instability problems at highway bridges.

Stream and river control works may have a serious local influence on channel erosion. Channel straightening, which increases slope and flow velocity, may initiate channel and bank erosion. If the bed of a main stream is lowered, the beds of tributary streams are also lowered. In many instances, such bed degradation is beneficial because it restores the flood-carrying capacity of channels.

Sediment transport affects water quality and its suitability for human consumption or use in various enterprises. Numerous industries cannot tolerate even the smallest amount of sediment in the water that is necessary for certain manufacturing processes, and the public pays a large price for the removal of sediments from the water it consumes every day.

Dam construction influences channel stability in two ways. It traps the incoming sediment, and it changes the natural flow and sediment load downstream. As a net result, degradation occurs below dams and aggradation might increase the risk of flooding upstream of the reservoir. Severe problems of abrasion of turbines, dredging, and stream instability and possible failure are often associated with reservoir and dam construction. Damage can be observed downstream from dam failure sites. In recent years, dam removal has become increasingly popular. The redistribution of sediment, at times contaminated, after dam removal will foster new research developments.

Sediment not only is the major water pollutant, but also serves as a catalyst, carrier, and storage agent of other forms of pollution. Sediment alone degrades water quality for municipal supply, recreation, industrial consumption and cooling, hydroelectric facilities, and aquatic life. In addition, chemicals and waste are assimilated onto and into sediment particles. Ion exchange occurs between solutes and sediments. Thus, sediment particles have become a source of increased concern as carriers and storage agents of pesticides, residues, adsorbed phosphorus, nitrogen, and other organic compounds, heavy metals, actinides and radioactive waste, as well as pathogenic bacteria and viruses.

The problems associated with sediment deposition are varied. Sediments deposited in stream channels reduce flood-carrying capacity, resulting in more frequent overflows and greater floodwater damage to adjacent properties. The

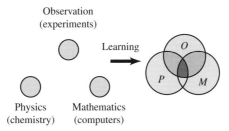

Figure 1.2. Learning erosion and sedimentation

deposition of sediments in irrigation and drainage canals, in navigation channels and floodways, in reservoirs and harbors, on streets and highways, and in buildings not only creates a nuisance but inflicts a high public cost in maintenance, removal, or in reduced services. Sedimentation is of vital concern in the conservation, development, and utilization of our soil and water resources.

As sketched in Figure 1.2, learning sedimentation involves three major elements: (1) observation; (2) physics and; (3) mathematics; these elements may seem disjointed at first. In the classroom, we develop the ability to: (1) make good observations; (2) understand the mechanics of the problem; and (3) find appropriate engineering solutions. Competence is developed through the skills in these three areas. The course CIVE716 aims at developing observational skills through a field trip and the analysis of laboratory and field data. Physical understanding is promoted through the in-depth analysis of the equations governing sediment transport. The mathematical skills are also developed through multiple exercises, calculation examples, homework, and computer problems. Altogether, these skills are displayed in numerous case studies illustrating solutions to sedimentation engineering problems.

This book rests on Newtonian mechanics and integrates concepts from fluid mechanics and sediment transport theory. Chapter 2 outlines the physical properties of sediments and dimensional analysis. Chapter 3 presents the fundamental principles of fluid mechanics applied to sediment-laden flows. Chapter 4 explains the concept of lift force and describes the motion of single particles in inviscid fluids. Chapter 5 analyzes viscous fluids and explains the concept of drag force. Applications of the concept of turbulence to sediment-laden flows are summarized in Chapter 6. Chapter 7 extends the analysis of the beginning of motion of single particles to complex three-dimensional cases with applications to stable channel design. The complex topics of bedform configurations and resistance to flow are reviewed in Chapter 8. The general topic of sediment transport is divided into three chapters: bedload in Chapter 9, suspended load in Chapter 10, and total load in Chapter 11. Sedimentation is covered in Chapter 12 with emphasis on reservoirs.

2

Physical properties and dimensional analysis

The processes of erosion, transport, and deposition of sediment particles introduced in Chapter 1 relate to the interaction between solid particles and the surrounding fluid. This chapter describes physical properties of water and solid particles in terms of dimensions and units (Section 2.1), physical properties of water (Section 2.2) and of sediment (Section 2.3). The method of dimensional analysis (Section 2.4) is then applied to representative erosion and sedimentation problems.

2.1 Dimensions and units

The physical properties of fluids and solids are usually expressed in terms of the following fundamental dimensions: mass (M), length (L), time (T), and temperature ($T°$). The fundamental dimensions are measurable parameters which can be quantified in fundamental units.

In the SI system of units, the basic *units* of mass, length, time, and temperature are the kilogram (kg), the meter (m), the second (s), and the degree Kelvin (°K), respectively. Alternatively, the Celsius scale (°C) is commonly preferred with the freezing point of water at 0°C, and the boiling point at 100°C.

A newton (N) is defined as the force required to accelerate one kilogram at one meter per second squared. Knowing that the acceleration due to gravity at the Earth's surface g is 9.81 m/s², the weight of a kilogram is obtained from Newton's second law: $F = \text{mass} \times g = 1 \text{ kg} \times 9.81 \text{ m/s}^2 = 9.81\text{N}$. The unit of work (or energy) is the joule (J) which equals the product of one newton times one meter. The unit of power is a watt (W) which is a joule per second. Prefixes are used to indicate multiples or fractions of units by powers of 10.

$$
\left.\begin{array}{l} \mu(\text{micro}) = 10^{-6} \\ \text{m(milli)} = 10^{-3} \\ \text{c(centi)} = 10^{-2} \end{array}\right\} \quad \left\{\begin{array}{l} \text{k(kilo)} = 10^{3} \\ \text{M(mega)} = 10^{6} \\ \text{G(giga)} = 10^{9} \end{array}\right.
$$

Table 2.1. *Geometric, kinematic, dynamic, and dimensionless variables*

Variable	Symbol	Fundamental dimensions	SI Units
Geometric variables (L)			
length	L, x, h, d_s	L	m
area	A	L^2	m^2
volume	$\forall$	L^3	m^3
Kinematic variables (L, T)			
velocity	V, v_x, u, u_*	LT^{-1}	m/s
acceleration	a, a_x, g	LT^{-2}	m/s^2
kinematic viscosity	v	L^2T^{-1}	m^2/s
unit discharge	q	L^2T^{-1}	m^2/s
discharge	Q	L^3T^{-1}	m^3/s
Dynamic variables (M, L, T)			
mass	m	M	1 kg
force	$F = ma, mg$	MLT^{-2}	1 kg m/s^2 = 1 Newton
pressure	$p = F/A$	$ML^{-1}T^{-2}$	1 N/m^2 = 1 Pascal
shear stress	$\tau, \tau_{xy}, \tau_o, \tau_c$	$ML^{-1}T^{-2}$	1 N/m^2 = 1 Pascal
work or energy	$E = F \cdot d$	ML^2T^{-2}	1 Nm = 1 Joule
power	$P = E/t$	ML^2T^{-3}	1 Nm/s = 1 Watt
mass density	ρ, ρ_s	ML^{-3}	kg/m^3
specific weight	$\gamma, \gamma_s = \rho_s g$	$ML^{-2}T^{-2}$	N/m^3
dynamic viscosity	$\mu = \rho v$	$ML^{-1}T^{-1}$	1kg/ms = 1Ns/m^2 = 1 Pas
Dimensionless variables ($-$)			
slope	S_o, S_f	$-$	$-$
specific gravity	$G = \gamma_s/\gamma$	$-$	$-$
Reynolds number	$\mathrm{Re} = Vh/v$	$-$	$-$
grain shear Reynolds number	$\mathrm{Re}_* = u_* d_s/v$	$-$	$-$
Froude number	$\mathrm{Fr} = V/\sqrt{gh}$	$-$	$-$
Shields parameter	$\tau_* = \tau/(\gamma_s - \gamma)d_s$	$-$	$-$
concentration	C_v, C_w, C	$-$	$-$

For example, one millimeter (mm) stands for 0.001 m and one mega watt (MW) equals one million watts (1,000,000 W).

In the English system of units, the time unit is a second, the fundamental units of length and mass are respectively the foot (ft), equal to 30.48 cm, and the slug, equal to 14.59 kg. The force required to accelerate a mass of one slug at one foot per second squared is a pound force (lb) used throughout this text. The temperature in degree Celsius T_C° is converted to the temperature in degree Fahrenheit T_F° using $T_F^\circ = 32^\circ\text{F} + 1.8 T_C^\circ$.

Table 2.2. *Conversion of units*

Unit	kg, m, s	N, Pa, Watt
1 acre	$= 4046.87$ m^2	
1 acre foot (acre-ft)	$= 1233.5$ m^3	
1 atmosphere	$= 101{,}325$ kg/ms^2	$= 101.3$ k Pa
1 Btu $= 778$ lb ft	$= 1055$ kg m^2/s^2	$= 1055$ Nm
1 bar	$= 100{,}000$ kg/ms^2	$= 100$ k Pa
1 °Celsius $= (T_F^\circ - 32^\circ)5/9$	$= 1^\circ$K	
1 °Fahrenheit $= 32 + 1.8T_C^\circ$	$= 0.555556^\circ$K	
1 day $= 1$ d	$= 86{,}400$ s	
1 drop	$= 61$ mm^3	
1 dyne	$= 0.00001$ kgm/s^2	$= 1 \times 10^{-5}$ N
1 dyne/cm^2	$= 0.1$ kg/ms^2	$= 0.1$ Pa
1 fathom	$= 1.8288$ m	
1 fluid ounce	$= 2.957 \times 10^{-5}$ m^3	
1 foot $= 1$ ft	$= 0.3048$ m	
1 ft^3/s	$= 0.0283$ m^3/s	
1 gallon (U.S., liquid) (gal)	$= 0.0037854$ m^3	
1 gallon per minute (gpm)	$= 6.31 \times 10^{-5}$ m^3/s	
1 mgd $= 1$ million gal/day $= 1.55$ ft^3/s	$= 0.04382$ m^3/s	
1 horse power $= 550$ lb ft/s	$= 745.70$ kg m^2/s^3	$= 745.7$ W
1 inch $= 1$ in	$= 0.0254$ m	
1 in of mercury	$= 3386.39$ kg/ms^2	$= 3386.39$ Pa
1 in of water	$= 248.84$ kg/ms^2	$= 248.84$ Pa
1 Joule	$= 1$ kg m^2/s^2	$= 1$ Nm $= 1$ J
1 kip $= 1000$ lb	$= 4448.22$ kg m/s^2	$= 4448.22$ N
1 knot	$= 0.5144$ m/s	
1 liter $= 1$l	$= 0.001$ m^3	
1 micron (μm)	$= 1 \times 10^{-6}$ m	
1 mile (nautical)	$= 1852$ m	
1 mile (statute)	$= 1609$ m	
1 Newton	$= 1$ kg m/s^2	1 N
1 ounce	$= 0.02835$ kg	
1 Pa	$= 1$ kg/ms^2	1 N/m^2
1 pint	$= 0.0004732$ m^3	
1 Poise $= 1$ P	$= 0.1$ kg/ms	0.1 Pa·s
1 pound-force (lb)	$= 4.448$ kg m/s^2	$= 4.448$ N
1 lb ft	$= 1.356$ kg m^2/s^2	$= 1.356$ Nm
1 psf (lb per ft^2)	$= 47.88$ kg/ms^2	$= 47.88$ Pa
1 psi (lb per in^2)	$= 6894.76$ kg/ms^2	$= 6894.76$ Pa
1 pound-force per ft^3	$= 157.09$ kg/m^{2}s^2	$= 157.09$ N/m^3
1 quart	$= 0.00094635$ m^3	
1 slug	$= 14.59$ kg	
1 slug/ft^3	$= 515.4$ kg/m^3	

Table 2.2 *(cont.)*

Unit	kg, m, s	N, Pa, Watt
1 Stoke $= 1\text{cm}^2/\text{s}$	$= 0.0001 \text{ m}^2/\text{s}$	
1 metric ton	$= 1{,}000$ kg	
1 short ton (2 kip mass)	$= 907.2$ kg	
1 short ton $= 2{,}000$ lb (weight)	$= 8900 \text{ kg m/s}^2$	8.9 kN
1 long ton (UK)	$= 1016.05$ kg	
1 Watt US (W)	$= 1 \text{ kg m}^2/\text{s}^3$	1 W
1 yard (yd)	$= 0.9144$ m	
1 year (yr)	31,536,000 s	

Most physical variables can be described in terms of three fundamental dimensions (M, L, T). Variables are classified as geometric, kinematic, dynamic, and dimensionless variables as shown in Table 2.1. Geometric variables involve length dimensions only and describe the geometry of a system through length, area, and volume. Kinematic variables describe the motion of fluid and solid particles and these variables can be depicted by only two fundamental dimensions, namely L and T. Dynamic variables involve mass terms in the fundamental dimensions. Force, pressure, shear stress, work, energy, power, mass density, specific weight, and dynamic viscosity are common examples of dynamic variables. Several conversion factors are listed in Table 2.2.

2.2 Physical properties of water

The principal properties of a nearly incompressible fluid like water are sketched on Figure 2.1.

Mass density of a fluid, ρ

The mass of fluid per unit volume is referred to as the mass density. The maximum mass density of water at 4°C is 1000 kg/m³ and varies slightly with temperature as shown in Table 2.3. In comparison, the mass density of sea water is 1,025 kg/m³, and at sea level, the mass density of air is $\rho_{\text{air}} = 1.2$ kg/m³ at 0°C. The conversion factor is 1 slug/ft³ $= 515.4$ kg/m³.

Specific weight of a fluid, γ

The weight of fluid per unit volume of fluid defines the specific weight, described by the symbol γ (gamma). At 4°C, water has a specific weight $\gamma = 9810$ N/m³ or

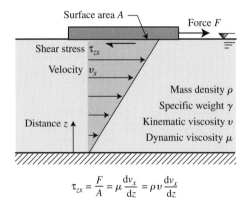

$$\tau_{zx} = \frac{F}{A} = \mu\frac{\mathrm{d}v_x}{\mathrm{d}z} = \rho v\frac{\mathrm{d}v_x}{\mathrm{d}z}$$

Figure 2.1. Newtonian fluid properties

62.4 lb/ft³ (1 lb/ft³ = 157.09 N/m³). Specific weight varies slightly with temperature as given in Table 2.3. The specific weight γ equals the product of the mass
density ρ times the gravitational acceleration $g = 32.2$ ft/s² $= 9.81$ m/s².

$$\gamma = \rho g \tag{2.1}$$

Dynamic viscosity, μ

As a fluid is brought into deformation, the velocity of the fluid at any boundary
equals the velocity of the boundary. The ensuing rate of fluid deformation causes a
shear stress τ_{zx} proportional to the dynamic viscosity μ and the rate of deformation
of the fluid, $\mathrm{d}v_x/\mathrm{d}z$.

$$\tau_{zx} = \mu\frac{\mathrm{d}v_x}{\mathrm{d}z} \tag{2.2}$$

The fundamental dimensions of the dynamic viscosity μ are M/LT which is
a dynamic variable. As indicated in Table 2.3, the dynamic viscosity of water
decreases with temperature. Fluids without yield stress for which the dynamic viscosity remains constant regardless of the rate of deformation are called Newtonian
fluids. The dynamic viscosity of clear water at 20°C is 1 centipoise: 1cP = 0.01 P
= 0.001 Ns/m² = 0.001 Pas (1 lb·s/ft² = 47.88 Ns/m² = 47.88 Pas).

Kinematic viscosity, v

When the dynamic viscosity of a fluid μ is divided by the mass density ρ of the
same fluid, the mass terms cancel out.

$$\mu = \rho v \tag{2.3a}$$

Table 2.3. *Physical properties of clear water at atmospheric pressure*

Temperature °C	Mass density ρ kg/m^3	Specific weight γ N/m^3 or kg/m^{2}s^2	Dynamic viscosity μ Ns/m^2 or kg/ms	Kinematic viscosity ν m^2/s
−30	921	9,035	Ice	Ice
−20	919	9,015	Ice	Ice
−10	918	9,005	Ice	Ice
0	999.9	9,809	1.79×10^{-3}	1.79×10^{-6}
4	1,000	9,810	1.56×10^{-3}	1.56×10^{-6}
5	999.9	9,809	1.51×10^{-3}	1.51×10^{-6}
10	999.7	9,809	1.31×10^{-3}	1.31×10^{-6}
15	999	9,800	1.14×10^{-3}	1.14×10^{-6}
20	998	9,790	1.00×10^{-3}	1.00×10^{-6}
25	997	9,781	8.91×10^{-4}	8.94×10^{-7}
30	996	9,771	7.97×10^{-4}	8.00×10^{-7}
35	994	9,751	7.20×10^{-4}	7.25×10^{-7}
40	992	9,732	6.53×10^{-4}	6.58×10^{-7}
50	988	9,693	5.47×10^{-4}	5.53×10^{-7}
60	983	9,643	4.66×10^{-4}	4.74×10^{-7}
70	978	9,594	4.04×10^{-4}	4.13×10^{-7}
80	972	9,535	3.54×10^{-4}	3.64×10^{-7}
90	965	9,467	3.15×10^{-4}	3.26×10^{-7}
100	958	9,398	2.82×10^{-4}	2.94×10^{-7}
°F	slug/ft^3	lb/ft^3	lb·s/ft^2	ft^2/s
0	1.78	57.40	Ice	Ice
10	1.78	57.34	Ice	Ice
20	1.78	57.31	Ice	Ice
30	1.77	57.25	Ice	Ice
32	1.931	62.40	3.75×10^{-5}	1.93×10^{-5}
40	1.938	62.43	3.23×10^{-5}	1.66×10^{-5}
50	1.938	62.40	2.73×10^{-5}	1.41×10^{-5}
60	1.936	62.37	2.36×10^{-5}	1.22×10^{-5}
70	1.935	62.30	2.05×10^{-5}	1.06×10^{-5}
80	1.93	62.22	1.80×10^{-5}	0.930×10^{-5}
100	1.93	62.00	1.42×10^{-5}	0.739×10^{-5}
120	1.92	61.72	1.17×10^{-5}	0.609×10^{-5}
140	1.91	61.38	0.981×10^{-5}	0.514×10^{-5}
160	1.90	61.00	0.838×10^{-5}	0.442×10^{-5}
180	1.88	60.58	0.726×10^{-5}	0.385×10^{-5}
200	1.87	60.12	0.637×10^{-5}	0.341×10^{-5}
212	1.86	59.83	0.593×10^{-5}	0.319×10^{-5}

This results in the kinematic viscosity ν with dimensions L^2/T, which is also shown in Table 2.3 to decrease with temperature. The viscosity of clear water at 20°C is 1 centistoke $= 0.01$ cm^2/s $= 1 \times 10^{-6}$ m^2/s (1 ft^2/s $= 0.0929$ m^2/s).

$$\nu = \frac{1.78 \times 10^{-6} \text{m}^2/S}{\left[1 + 0.0337\,T_C^o + 0.0002217\,T_C^{o2}\right]} \qquad (2.3b)$$

It is important to remember that both the density and viscosity of water decrease with temperature. Comparatively, the kinematic viscosity of air is approximately 1.6×10^{-4} ft^2/s or about 1.6×10^{-5} m^2/s at 20°C and increases slightly with temperature.

2.3 Physical properties of sediment

This section describes the physical properties of sediment as: single particle (Section 2.3.1), sediment mixture (Section 2.3.2), and sediment suspension (Section 2.3.3).

2.3.1 Single particle

The physical properties of a single solid particle of volume $\forall_s$ are sketched on Figure 2.2.

Mass density of solid particles, ρ_s

The mass density of a solid particle ρ_s describes the solid mass per unit volume. The mass density of quartz particles 2,650 kg/m^3 (1 slug/ft^3 $= 515.4$ kg/m^3) does not vary significantly with temperature and is assumed constant in most calculations. It must be kept in mind, however, that heavy minerals like iron, copper, etc. have much larger values of mass density.

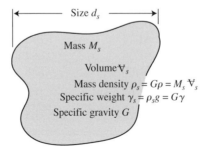

Figure 2.2. Physical properties of a single particle

Specific weight of solid particles, γ_s

The particle specific weight γ_s corresponds to the solid weight per unit volume of solid. Typical values of γ_s are 26.0 kN/m^3 or 165.4 lb/ft^3 (1 lb/ft^3 = 157.09 N/m^3). The specific weight of solid γ_s also equals the product of the mass density of a solid particle ρ_s times the gravitational acceleration g, thus

$$\gamma_s = \rho_s g \qquad (2.4a)$$

Submerged specific weight of a particle, γ_s'

Owing to Archimedes' principle, the specific weight of a solid particle γ_s submerged in a fluid of specific weight γ equals the difference between the two specific weights; thus,

$$\gamma_s' = \gamma_s - \gamma = \gamma(G-1) \qquad (2.4b)$$

Specific gravity, G

The ratio of the specific weight of a solid particle to the specific weight of fluid at a standard reference temperature defines the specific gravity G. With common reference to water at 4°C, the specific gravity of quartz particles is

$$G = \frac{\gamma_s}{\gamma} = \frac{\rho_s}{\rho} = 2.65 \qquad (2.5)\blacklozenge$$

Specific gravity is a dimensionless ratio of specific weights, and thus its value remains independent of the system of units. Table 2.4 lists values of specific gravity for various minerals.

Sediment size, d_s

The most important physical property of a sediment particle is its size. Table 2.5 shows the grade scale commonly used in sedimentation. Note that the size scales are arranged in geometric series with a ratio of 2. The conversion factor is 1 in = 25.4 mm.

The size of particles can be determined in a number of ways; the nominal diameter refers to the diameter of a sphere of the same volume as the particle usually measured by the displaced volume of a submerged particle, the sieve diameter is the minimum length of the square sieve opening, through which a particle will fall, the fall diameter is the diameter of an equivalent sphere of specific gravity $G = 2.65$ having the same terminal settling velocity in water at 24°C. Figure 2.3a shows the mass and weight of single spherical particles at $G = 2.65$. Figure 2.3b shows a

Table 2.4. *Specific gravity of different materials*

Material	Specific gravity (G)	Material	Specific gravity (G)
Air	0.0012	Humus	< 1.5
Alumina	3.6 – 3.9	Ice	0.9
Alumina (Kyanite)	3.58	Ilmenite	4.5 – 5
Alumina (Sandy)	3.95	Iron	7.2
Aluminum	2.69	Kaolinite	2.6
Anhydrite	2.9	Lead	11.38
Apatite	3.1 – 3.3	Limestone	2.7
Asbestos	2.4 – 2.5	Limonite	3.6 – 4
Basalt	3.3	Magnetite	5.2 – 6.5
Biotite	2.8 – 3.2	Mercury	13.57
Calcite	2.72	Mica	2.7 – 3.3
Chlorite	2.6 – 3	Muscovite	2.8 – 3.1
Chromite	4.3 – 4.6	Oil	0.9
Clay (Kaolin)	2.6	Olivine	3.2 – 3.6
Concrete	2.4	Porphyry	2.7
Copper	8.9	Pumice	<1.5
Copper Pellets	8.9	Pyrite	5 – 5.1
Diabas	3.3	Pyroxene	3.2 – 3.5
Diamond	3.5	Pyrrhotite	4.6 – 4.7
Dolomite	2.87	Quartz	2.66
Feldspar	2.5 – 2.8	Sandstone	2.1 – 2.2
Ferric Oxide	4.3	Silica Sand	2.6
Fluorite	3.2	Steel	7.83
Gabbro	> 3.2	Titanium	3.08
Garnet	3.5 – 4.3	Topaz	3.4 – 3.6
Glass	2.59	Tourmaline	3 – 3.3
Gold	15 – 19.3	Water (salt)	1.03
Granite	2.7	Wood (hard oak)	0.8
Hematite	5.3	Wood (soft pine)	0.48
Hornblende	3 – 3.3	Zircon	4.2 – 4.9

range of sampling mass required for the appropriate determination of grain size in gravel and cobble-bed streams.

Sphericity, Sp

The shape of sediment particles can be described by measurements of the longest axis ℓ_a, the intermediate ℓ_b, and the shortest axis ℓ_c. The sphericity $Sp = \left(\frac{\ell_b \ell_c}{\ell_a^2} \right)^{1/3}$ is used to define the equivalent side of a cube $\ell_a Sp$ that has the same volume as the particle. Also, the volume of the parallelepiped particle can be calculated as $\forall_p = Sp^3 \ell_a^3$. Figure 2.4 shows the type of particle shapes given ℓ_b/ℓ_a and ℓ_c/ℓ_b. The Corey shape factor $Co = \ell_c/\sqrt{\ell_a \ell_b}$ is always less than unity, and values of 0.7 are typical for natural particles.

Table 2.5. *Sediment grade scale*

Class name		millimeter (mm)	micron (μm)	inch (in)	US standard sieve
		\multicolumn{3}{c}{Size range}		
Boulder	Very large	4096–2048		160–80	
	Large	2048–1024		80–40	
	Medium	1024–512		40–20	
	Small	512–256		20–10	
Cobble	Large	256–128		10–5	
	Small	128–64		5–2.5	
Gravel	Very coarse	64–32		2.5–1.3	
	Coarse	32–16		1.3–0.6	
	Medium	16–8		0.6–0.3	5
	Fine	8–4		0.3–0.16	10
	Very fine	4–2		0.16–0.08	
Sand	Very coarse	2.000–1.000			18
	Coarse	1.000–0.500			35
	Medium	0.500–0.250			60
	Fine	0.250–0.125			120
	Very fine	0.125–0.062			230
Silt	Coarse		62–31		
	Medium		31–16		
	Fine		16–8		
	Very fine		8–4		
Clay	Coarse		4–2		
	Medium		2–1		
	Fine		1–0.5		
	Very fine		0.5–0.24		

2.3.2 Bed sediment mixture

The properties of a bed sediment mixture are sketched in Figure 2.5.

Particle size distribution

The particle size distribution in Figure 2.6 shows the percentage by weight of material finer than a given sediment size. The sediment size d_{50} for which 50% by weight of the material is finer is called the median grain size. Likewise d_{90} and d_{10} are values of grain size for which 90% and 10% of the material is finer, respectively. The examples shown in Figure 2.6 are for the Mississippi River near Tarbert Landing in suspension (A) and bed material (B). Also, the material of Little Granite Creek, Wyoming shows bedload (C), sub-surface (D), and surface material (E).

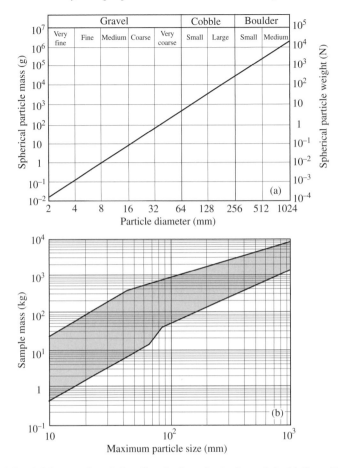

Figure 2.3. a) Mass and weight of a single spherical particle b) Sampling mass versus maximum particle size (modified after Bunte and Abt, 2001)

Sieve analysis

Sieving is considered a semidirect method of particle size measurement because it does not measure the particle size precisely for at least four reasons: (1) because of irregularities in shape, particles larger or smaller (larger for cylinders and smaller for disks) than the spherical equivalent diameter may pass through the sieve openings; (2) inaccuracies in size and shape of the sieve openings; (3) the sieving operation is for a finite duration and therefore some particles may not have an opportunity to pass a given sieve opening; and (4) small particles may cling to large ones, thereby changing the percentage of material reaching the sieves with smaller openings.

A wet-sieve method keeps the sieve screen and sand completely submerged. The equipment may consist of six or more 10-cm ceramic dishes, a set of 3-in. (7.5 cm) sieves, and a thin glass tube. All sieves are washed with a wetting solution

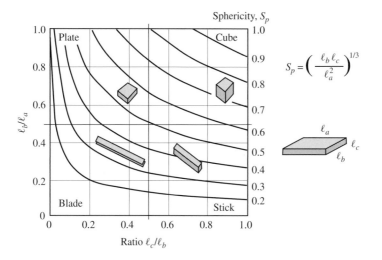

Figure 2.4. Particle shape factor

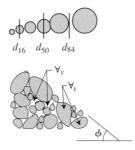

Figure 2.5. Properties of a sediment mixture

(detergent) and then raised gently with distilled water so that a membrane of water remains across all openings. The first or largest sieve is immersed in a ceramic dish with distilled water to a depth of about 1/4 in. (1/2 cm) above the screen. The sediment is washed onto the wet sieve and agitated somewhat vigorously in several directions until all particles smaller than the sieve openings have a chance to fall through the sieve. Material passing through the sieve with its wash water is then poured onto the next smaller size sieve. Particles retained on each sieve and those passing through the 0.062-mm sieve are transferred to containers that are suitable for drying the material and for obtaining the net weight of each fraction.

The dry-sieve method is less laborious than the wet-sieve method because a mechanical shaker can be used with a nest of sieves for simultaneous separation of all sizes of interest. It requires only that the dry sand be poured over the coarsest sieve and the nest of sieves shaken for 10 min on a shaker having both lateral and vertical movements. United States standard sieve numbers are listed in Table 2.5.

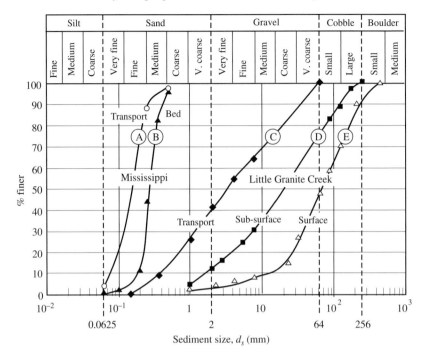

Figure 2.6. Particle size distribution examples

Gradation coefficients, σ_g and Gr

The gradation of the sediment mixture can be described by

$$\sigma_g = \left(\frac{d_{84}}{d_{16}}\right)^{1/2} \tag{2.6a}$$

or by the gradation coefficient

$$Gr = \frac{1}{2}\left(\frac{d_{84}}{d_{50}} + \frac{d_{50}}{d_{16}}\right) \tag{2.6b}$$

Angle of repose ϕ

The angle of repose of submerged loose material is the side slope, with respect to the horizontal, or a cone of material under incipient sliding condition. The angle of repose of granular material varies between 30° and 40°. Specific values are discussed in Chapter 7.

Porosity, p_o

The porosity p_o is a measure of the volume of void $\forall_v$ per total volume $\forall_t = \forall_v + \forall_s$. The volume of solid particles $\forall_s = C_v \forall_t = (1 - p_o) \forall_t$, thus

$$p_o = \frac{\forall_v}{\forall_t} = \frac{e}{1 + e} \qquad (2.7)$$

Void ratio, e

The void ratio e is a measure of the volume of void $\forall_v$ per volume of solid $\forall_s$ or

$$e = \frac{\forall_v}{\forall_s} = \frac{p_o}{1 - p_o} \qquad (2.8)$$

Dry specific weight of a mixture, γ_{md}

The dry specific weight of a bed sediment mixture is the weight of solid per unit total volume, including the volume of solids and voids. The dry specific weight of a mixture γ_{md} is a function of p_o as

$$\gamma_{md} = \frac{M_s g}{\forall_t} = \gamma_s (1 - p_o) = \gamma G (1 - p_o) \qquad (2.9)$$

Dry specific mass of a mixture, ρ_{md}

The dry specific mass of a mixture is the mass of solid per unit total volume. The dry specific mass of a mixture can be defined as a function of p_o as

$$\rho_{md} = \frac{M_s}{\forall_t} = \frac{\gamma_{md}}{g} = \rho_s (1 - p_o) = \rho G (1 - p_o) \qquad (2.10)$$

2.3.3 Sediment suspension

The properties of a sediment suspension are sketched in Figure 2.7, with the volume of void $\forall_v$ equal to the volume of water $\forall_w$, and the mass of the voids M_v equal to the mass of water M_w.

Volumetric sediment concentration, C_v

The volumetric sediment concentration C_v of a suspension is defined as the volume of solids $\forall_s$ over the total volume $\forall_t$, or

$$C_v = \frac{\forall_s}{\forall_t} = \frac{\forall_s}{\forall_s + \forall_v} \qquad (2.11)\blacklozenge$$

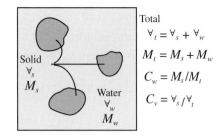

Figure 2.7. Properties of a suspension

Conversions to concentration by weight C_w, C_{ppm}, and $C_{mg/\ell}$ are presented in Section 10.1.

Specific weight of a mixture, γ_m

The specific weight of a submerged mixture is the total weight of solid and water in the voids per unit total volume. The specific weight of a mixture, γ_m is a function of the volumetric sediment concentration, C_v as

$$\gamma_m = \frac{M_t g}{\forall_t} = \frac{\gamma_s \forall_s + \gamma \forall_w}{\forall_s + \forall_w} = \gamma_s C_v + \gamma(1 - C_v) = \gamma(1 + (G-1)C_v) \qquad (2.12)$$

Specific mass of a mixture, ρ_m

The specific mass of a suspension is the total mass of solid and water in the voids per unit total volume. The specific mass of a mixture, ρ_m can be defined as a function of C_v as

$$\rho_m = \frac{M_t}{\forall_t} = \frac{\gamma_m}{g} = \rho_s C_v + \rho(1 - C_v) = \rho(1 + (G-1)C_v) \qquad (2.13)$$

Dynamic viscosity of a Newtonian mixture μ_m

The dynamic viscosity of a Newtonian mixture, μ_m increases with the concentration of sediment in suspension. Albert Einstein suggested the following function of volumetric sediment concentration C_v:

$$\mu_m = \mu(1 + 2.5C_v) \qquad (2.14)$$

This relationship is very approximate, even at low volumetric concentrations ($C_v < 0.05$). More details on the viscosity of hyperconcentrations is presented in Chapter 10 (Section 10.6).

Table 2.6. *Properties of water–sediment mixtures*

Variable	$M, \forall$	Relationships
$\rho = \dfrac{\gamma}{g}$	$\dfrac{M_w}{\forall_w}$	$C_v = 1 - p_o = 1/(1+e)$
$\rho_s = \dfrac{\gamma_s}{g}$	$\dfrac{M_s}{\forall_s}$	$G = \dfrac{\rho_s}{\rho} = \dfrac{\gamma_s}{\gamma}$
$\rho_m = \dfrac{\gamma_m}{g}$	$\dfrac{M_t}{\forall_t}$	$\gamma_m = \rho_m g = \gamma C_v(G-1) + \gamma$
$\rho_{md} = \dfrac{\gamma_{md}}{g}$	$\dfrac{M_s}{\forall_t}$	$\gamma_{md} = \rho_{md} g = \gamma_s C_v = \gamma_s(1 - p_o)$
$p_o = \dfrac{\forall_w}{\forall_T}$	$\dfrac{\forall_w}{\forall_t}$	$p_o = e/(1+e) = 1 - C_v$
$e = \dfrac{\forall_w}{\forall_s}$	$\dfrac{\forall_w}{\forall_s}$	$e = p_o/(1 - p_o) = (1 - C_v)/C_v$

Kinematic viscosity of a Newtonian mixture, $v_m = \mu_m/\rho_m$

The kinematic viscosity of a Newtonian mixture, v_m is obtained by dividing the dynamic viscosity of a Newtonian mixture μ_m by the mass density of the mixture ρ_m. A summary of the basic relationships for water–sediment mixtures is given in Table 2.6.

2.4 Dimensional analysis

Dimensional analysis is a method by which we deduce information about a phenomenon with the single premise that it can be described by a dimensionally correct group of variables. Dimensional analysis meets the double objective to: (1) reduce the number of variables for subsequent analysis of the problem; and (2) provide dimensionless parameters whose numerical values are independent of any system of units. Neither a complete solution to any physical investigation nor a clear understanding of any inner mechanism can be revealed by dimensional reasoning alone. Dimensional analysis is, however, a useful mathematical tool in the analysis of sedimentation problems. The following combines the contributions of Buckingham (1914), Rayleigh (1915), Hunsaker and Rightmire (1947), Langhaar (1951), and Sedov (1959).

Buckingham's Π theorem allows us to rearrange n variables in which there are j fundamental dimensions into $n - j$ dimensionless parameters designated by the Greek letter Π. Let the dependent variable Z_1 be related by any functional relationship $\mathscr{F}$ to the independent variables $Z_2, \ldots, Z_n$ such that:

$$Z_1 = \mathscr{F}(Z_2, \ldots, Z_n) \tag{2.15}$$

with j fundamental dimensions involved, for example $j = 3$ when their n parameters combine mass M, length L, and time T, the relation can be reduced to a function of $n - j$ dimensionless Π parameters in the form:

$$\Pi_1 = \mathscr{F}\left(\Pi_2, \Pi_3, \dots, \Pi_{n-j}\right) \tag{2.16}$$

The method of determining the Π parameters is to select j repeating variables among the Z variables such that: (1) all j fundamental dimensions can be found in the set of repeating variables; and (2) all repeating variables must have different fundamental dimensions. For instance, one cannot select both the width and the length as repeating variables because both parameters have the same fundamental dimension L.

The steps in a dimensional analysis can be summarized as follows:

(1) Select the dependent variable Z_1 as a function of the independent variables $Z_2, \dots, Z_n$ in the functional relationship $Z_1 = \mathscr{F}(Z_2, Z_3, \dots, Z_n)$.
(2) Write the variables in terms of fundamental dimensions and select the j repeating variables. These variables must contain the j fundamental dimensions of the problem and the dependent variable should not be selected as a repeating variable. Solve the fundamental dimensions in terms of the j repeating variables.
(3) Obtain the Π parameters by dividing the non-repeating variables by their fundamental dimensions written in terms of repeating variables.
(4) Write the functional relation $\mathscr{F}(\Pi_1, \Pi_2, \dots, \Pi_{n-j}) = 0$ or $\Pi_1 = \mathscr{F}(\Pi_2, \dots, \Pi_{n-j})$, and recombine if desired to alter the form of the dimensionless parameters Π, keeping the same number of independent parameters.

The dimensional analysis method is illustrated in the following examples: (1) drag force exerted on a sphere by relative fluid motion in Example 2.1; and (2) soil erosion by overland flow in Example 2.2.

Example 2.1 Drag force on a sphere

Consider the drag force F_D exerted on a sphere in motion through a homogeneous mixture (Figure E-2.1.1). The drag force is thought to vary with the relative velocity u_∞, the spherical particle diameter d_s, the mass density of the fluid mixture ρ_m, and the dynamic viscosity of the mixture μ_m. Use the method of dimensional analysis to identify the dimensionless parameter.

Step 1. The dependent variable F_D is a function of four independent variables, with a total of $n = 5$ variables:

$$F_D = \mathscr{F}(u_\infty, \, d_s, \, \rho_m, \, \mu_m) 1 \tag{E-2.1.1}$$

in which $\mathscr{F}$ represents an unspecified function.

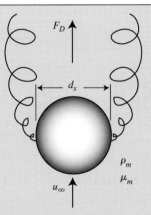

Figure E-2.1.1 Drag force on a moving sphere

Step 2. After selecting d_s, u_∞, and ρ_m as repeating variables, the three fundamental dimensions $(j = 3)$ are then rewritten in terms of repeating variables:

$$\left.\begin{array}{l} d_s = L \\ u_\infty = L/T \\ \rho_m = M/L^3 \end{array}\right\} \quad \left\{\begin{array}{l} L = d_s \\ T = d_s/u_\infty \\ M = \rho_m d_s^3 \end{array}\right.$$

Step 3. After substituting the relationships for M, L, T into the non-repeating variables divided by their fundamental dimensions, two Π terms $(n - j = 5 - 3 = 2)$ are obtained respectively from F_D and μ_m:

$$\Pi_1 = \frac{F_D T^2}{ML} = \frac{F_D d_s^2}{\rho_m d_s^3 d_s u_\infty^2} = \frac{F_D}{\rho_m u_\infty^2 d_s^2} \tag{E-2.1.2}$$

It turns out that Π_1 is defining the drag coefficient C_D, or $\Pi_1 = \pi C_D/8$. The dimensionless parameter for dynamic viscosity is

$$\Pi_2 = \frac{\mu_m LT}{M} = \frac{\mu_m d_s d_s}{\rho_m d_s^3 u_\infty} = \frac{\mu_m}{\rho_m u_\infty d_s} = \frac{1}{\mathrm{Re}_p}$$

The parameter Π_2 can be simply replaced by the Reynolds number of the particle Re_p.

Step 4. $\Pi_1 = \mathscr{F}(\Pi_2)$, or $C_D = \mathscr{F}(\mathrm{Re}_p)$, results in

$$F_D = \mathscr{F}(\mathrm{Re}_p) \frac{\pi d_s^2}{4} \frac{\rho_m u_\infty^2}{2} \tag{E-2.1.3}\blacklozenge$$

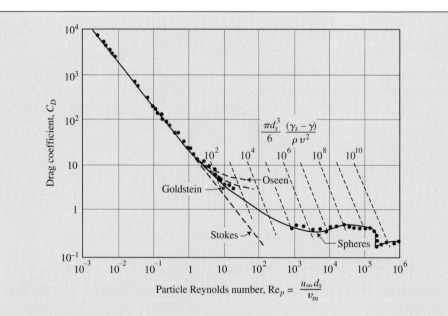

Figure E-2.1.2 Drag coefficient for spheres

The advantage of the method of dimensional analysis in this case has been to reduce the number of parameters from five in Equation (E-2.1.1) to two dimensionless parameters, $\Pi_1 = C_D$ and $\Pi_2 = \text{Re}_p$. Each of those two parameters is dimensionless and a unique graph can be made with data from different systems of units. The method, however, fails to provide any indication as to what kind of relationship may exist between C_D and Re_p. For further analysis, the scientist must carry out experiments and collect laboratory or field measurements of C_D versus Re_p. Measurements of drag coefficients around spheres versus the particle Reynolds number $\text{Re}_p = u_\infty d_s / v_m$ are shown in Figure E-2.1.2. Note that a similar plot for natural sand particles is also shown in Figure 5.2. Chapter 5 shows that the flow around a particle is laminar when $\text{Re}_p < 1$ and turbulent at large particle Reynolds numbers.

Example 2.2 Soil erosion by overland flow

Consider the problem of sheet erosion induced by rainfall on a bare soil surface (Fig. E-2.2.1). The method of dimensional analysis is first used to reduce the number of variables and define dimensionless parameters.

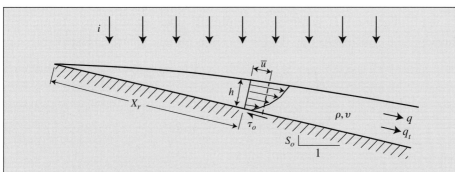

Figure E-2.2.1 Sheet erosion

Step 1. The rate of sediment transport by sheet erosion q_{sm} is written as a function of the geometric, fluid flow and soil variables:

$$q_{sm} = \mathscr{F}\left(S_o,\ q,\ i,\ X_r,\ \rho,\ v,\ \frac{\tau_c}{\tau_o}\right) \qquad \text{(E-2.2.1)}$$

in which q_{sm} is the rate of mass transport per unit width, q is the unit discharge, i is the rainfall intensity, X_r is the length of runoff, ρ is the mass density of the fluid, v is the kinematic viscosity of the fluid, τ_c and τ_o are respectively the critical and applied boundary shear stresses and S_o is the bed surface slope. The critical shear stress τ_c is the applied shear stress that is required to detach soil particles and bring them into motion.

Besides the two dimensionless variables in Equation (E-2.2.1) ($\Pi_5 = \tau_c/\tau_o$, $\Pi_2 = S_o$) the remaining variables ($n = 6$) are functions of three fundamental dimensions (M, L, T, thus $j = 3$), and can be transformed into three ($n - j = 3$) dimensionless parameters. Each variable is written in terms of the fundamental dimensions M, L, and T as follows:

$$q_{sm} = M/LT;\ q = L^2/T;\ i = L/T;$$
$$X_r = L;\ v = L^2/T;\ \rho = M/L^3$$

Step 2. The fundamental dimensions can be written in terms of the repeated variables X_r, v, and ρ

$$\left.\begin{array}{l} X_r = L \\ v = L^2/T \\ \rho = M/L^3 \end{array}\right\} \ thus \ \left\{\begin{array}{l} M = \rho X_r^3 \\ L = X_r \\ T = X_r^2/v \end{array}\right.$$

Step 3. The three Π parameters are directly obtained from substituting the fundamental dimensions into the relationships for q_{sm}, q, and i, respectively

$$\Pi_1 = \frac{q_{sm}}{M} LT = \frac{q_{sm} X_r X_r^2}{\rho X_r^3 v} = \frac{q_{sm}}{\rho v}$$

$$\Pi_3 = \frac{qT}{L^2} = \frac{q X_r^2}{X_r^2 v} = \frac{q}{v} = Re$$

$$\Pi_4 = \frac{iT}{L} = \frac{i X_r^2}{X_r v} = \frac{i X_r}{v}$$

Step 4. The five dimensionless parameters can thus be written

$$\frac{q_{sm}}{\rho v} = \mathscr{F}\left(S_o, \frac{q}{v}, \frac{i X_r}{v}, \frac{\tau_c}{\tau_o}\right) \tag{E-2.2.2}$$

The final result from this dimensional analysis is a dimensionless sediment transport parameter function of the soil surface slope, the Reynolds number, a dimensionless rainfall parameter, and the soil characteristics.

Further progress can only be achieved through physical understanding of the erosion processes and through laboratory or field experiments. For instance, the rate of sediment transport in sheet flow is assumed to be proportional to the product of the powers of the Π parameters

$$\left(\frac{q_{sm}}{\rho v}\right) = e1 S_o^{e2} \left(\frac{q}{v}\right)^{e3} \left(\frac{i X_r}{v}\right)^{e4} \left(1 - \frac{\tau_c}{\tau_o}\right)^{e5} \tag{E-2.2.3}$$

in which $e1, e2, e3, e4$, and $e5$ are coefficients to be determined from laboratory or field investigations.

The first three factors (S_o, q, i) of Equation (E-2.2.3) represent the potential erosion or sediment transport capacity of overland flow. It is interesting to note that for one-dimensional overland flow on impervious surfaces, $q = iX_r$. The sediment transport capacity is reduced by the last factor reflecting the soil resistance to erosion. When τ_c remains small compared to τ_o and with $q = i, X_r$, Equation (E-2.2.3) can be rearranged in the following form:

$$q_{sm} = \tilde{e}1 S_o^{e2} q^{e3} \tag{E-2.2.4}$$

The experiment on sandy soils by Kilinc (1972) at Colorado State University showed that q_{sm} (metric ton/ms) $= 2.55 \times 10^4\ S^{1.66} q^{2.035}$. This relationship will be used again in Chapter 11. At a given field site (constant slope S_o),

Equation E-2.2.4 further reduces to

$$q_{sm} \sim q^{e3} \tag{E-2.2.5}$$

This relationship defines the sediment-rating curve, and will also be used for alluvial channels in Chapter 11. From field observations in rivers, the value of the exponent $e3$ typically varies between 1.3 and 2.

Exercises

♦2.1 Erosion losses from pasture areas are considered excessive when they exceed 5 tons/acre-year. Determine the equivalent annual losses in metric tons per hectare per year and metric tons per square kilometer per year.

♦2.2 A sidecasting dredge operator tries to maintain the specific weight of the dredged material in the pipeline at 1.5 times that of water. Determine the volumetric concentration of sediment in this short pipeline.

2.3 Long pipelines tend to plug at volumetric sediment concentrations around 0.2. Determine the corresponding specific weight of the mixture in lb/ft^3 and kN/m^3.

Problems

Problem 2.1

Determine the mass density, specific weight, dynamic viscosity, and kinematic viscosity of clear water at 20°C: (a) in SI; and (b) in the English system of units.
Answer:

(a) $\rho = 998 \text{kg/m}^3$, $\gamma = 9790 \text{N/m}^3$, $\mu = 1.0 \times 10^{-3} \text{Ns/m}^2$,
$\nu = 1 \times 10^{-6} \text{m}^2/\text{s}$

(b) $\rho = 1.94 \text{slug/ft}^3$, $\gamma = 62.3 \text{lb/ft}^3$, $\mu = 2.1 \times 10^{-5} \text{lb·s/ft}^2$,
$\nu = 1.1 \times 10^{-5} \text{ft}^2/\text{s}$

♦Problem 2.2

Determine the sediment size, weight, mass density, specific weight, and submerged specific weight of small quartz cobbles: (a) in SI units; and (b) in the English system of units.

♦ Problem 2.3

The volumetric sediment concentration of a sample is $C_v = 0.05$. Determine the corresponding: (a) porosity p_o; (b) void ratio e; (c) specific weight γ_m; (d) specific mass ρ_m; (e) dry specific weight γ_{md}; and (f) dry specific mass ρ_{md}.
Answer: $p_o = 0.95$, $e = 19$, $\gamma_m = 10.6 \text{ kN/m}^3$, $\rho_m = 1082 \text{ kg/m}^3$
$\gamma_{md} = 1.29 \text{ kN/m}^3$, $\rho_{md} = 132 \text{ kg/m}^3$ (see also Table 10.3 p. 240).

♦♦Problem 2.4

A 50g bed-sediment sample from Big Sand Creek, Mississippi, is analyzed for particle size distribution.

Size fraction (mm)	mass (g)	Cumulative mass (g)	% finer
$d_s < 0.15$	0.9	0.9	1.8
$0.15 < d_s < 0.21$	2.9	3.8	7.6
$0.21 < d_s < 0.30$	16.0	19.8	39.6
$0.30 < d_s < 0.42$	20.1	39.9	79.6
$0.42 < d_s < 0.60$	8.9	48.8	97.6
$0.60 < d_s$	1.2	50.0	100

(a) Plot the sediment size distribution;
(b) determine d_{16}, d_{35}, d_{50}, d_{65}, and d_{84}; and
(c) calculate the gradation coefficients σ_g and Gr.

♦♦Problem 2.5

Consider energy losses ΔH_L in a straight open channel. The energy gradient $\Delta H_L/X_c$ in a smooth channel with turbulent flow depends upon the mean flow velocity V, the flow depth h, the gravitational acceleration g, the mass density ρ, and the dynamic viscosity μ. Determine the general form of the energy gradient equation from dimensional analysis.

Answer: $\dfrac{\Delta H_L}{X_c} = \mathscr{F}\left(\text{Reynolds number Re} = \dfrac{\rho V h}{\mu}; \text{ Froude number Fr} = \dfrac{V}{\sqrt{gh}} \right)$

♦Problem 2.6

Consider a near-bed turbulent velocity profile. The time-average velocity u at a distance z from the bed depends on the bed-material size d_s, the flow depth h, the dynamic viscosity of the fluid μ, the mass density ρ, and the boundary shear stress τ_o. Use the method of dimensional analysis to obtain a complete set of dimensionless parameters.

Hint: Select h, ρ, and τ_o as repeating variables. Also notice that the problem reduces to a kinematic problem after defining the shear velocity $u_* = \sqrt{\tau_o/\rho}$ and $\nu = \mu/\rho$.

Answer: $\mathscr{F}\left(u\sqrt{\dfrac{\rho}{\tau_o}}, \dfrac{\rho d_s}{\mu}\sqrt{\dfrac{\tau_o}{\rho}}, \dfrac{z}{h}, \dfrac{d_s}{h} \right) = 0$

♦♦**Problem 2.7**

A mass of 200 kg of sand is added to a cubic meter of water at 10°C in a container that is 0.5m × 0.5m at the base.

(a) If the sand is maintained in suspension through constant mixing, determine the following properties of the mixture in SI units: total volume, concentration by volume, concentration in mg/l, mass density, and specific weight of the mixture.

(b) Stop mixing and wait until all the sediment has settled at a dry specific weight of 93 pounds per cubic foot. Determine the following properties of the sediment deposit in SI units: height of the deposit, dry specific mass, void ratio, porosity, and volumetric concentration.

3

Mechanics of sediment-laden flows

This chapter summarizes some fundamental principles in fluid mechanics applied to sediment-laden flows. The major topics reviewed include: kinematics of flow (Section 3.1); continuity (Section 3.2); equations of motion (Section 3.3); Euler and Bernoulli equations (Sections 3.4 and 3.5); momentum equations (Section 3.6); and the power equation expressing the rate of work done (Section 3.7). Ten examples illustrate these theoretical concepts with applications.

3.1 Kinematics of flow

The kinematics of flow describes the motion in terms of velocity and type of deformation of fluid elements. The three most common orthogonal coordinate systems are: (1) Cartesian (x, y, z); (2) cylindrical (r, θ, z); and (3) spherical (r, θ, φ), as shown in Figure 3.1.

The rate of change in the position of the center point of a fluid element is a measure of its velocity. Velocity is defined as the ratio between the displacement ds and the corresponding increment of time dt. Velocity is a vector quantity which varies both in space (x, y, z) and time (t). Its scalar magnitude v at a given time equals the square root of the sum of the squares of its orthogonal components $v = \sqrt{v_x^2 + v_y^2 + v_z^2}$. The differential velocity components over an infinitesimal distance ds (dx, dy, dz) and time increment dt are:

$$dv_x = \frac{\partial v_x}{\partial t} dt + \frac{\partial v_x}{\partial x} dx + \frac{\partial v_x}{\partial y} dy + \frac{\partial v_x}{\partial z} dz \tag{3.1a}$$

$$dv_y = \frac{\partial v_y}{\partial t} dt + \frac{\partial v_y}{\partial x} dx + \frac{\partial v_y}{\partial y} dy + \frac{\partial v_y}{\partial z} dz \tag{3.1b}$$

$$dv_z = \underbrace{\frac{\partial v_z}{\partial t} dt}_{local} + \underbrace{\frac{\partial v_z}{\partial x} dx + \frac{\partial v_z}{\partial y} dy + \frac{\partial v_z}{\partial z} dz}_{convective} \tag{3.1c}$$

28

The flow is steady when the local terms are zero. Uniform flow is obtained when the convective terms are zero.

Now consider translation, linear deformation, angular deformation, and rotation of a fluid element as represented in Figure 3.2. The rate of linear deformation is indicated by the quantities $\ominus_x$, $\ominus_y$, and $\ominus_z$ defined from the velocity components

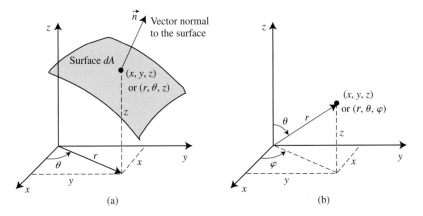

Figure 3.1. a) Cartesian and cylindrical coordinates b) Spherical coordinates

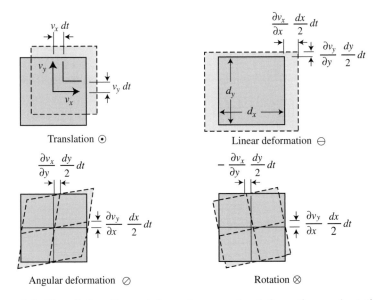

Figure 3.2. Translation, linear deformation, angular deformation, and rotation of a fluid element

(v_x, v_y, v_z) as:

$$\ominus_x = \frac{\partial v_x}{\partial x}; \; \ominus_y = \frac{\partial v_y}{\partial y}; \; \ominus_z = \frac{\partial v_z}{\partial z} \tag{3.2}$$

The velocity gradients in transverse directions (e.g. $\partial v_x/\partial y$) represent the rate of angular deformation of the element. The rates of angular deformation $\oslash$ in the respective planes are

$$\oslash_x = \frac{\partial v_z}{\partial y} + \frac{\partial v_y}{\partial z}; \; \oslash_y = \frac{\partial v_x}{\partial z} + \frac{\partial v_z}{\partial x}; \; \oslash_z = \frac{\partial v_y}{\partial x} + \frac{\partial v_x}{\partial y} \tag{3.3}$$

The rates of rotation $\otimes$ in their respective planes are defined as

$$\otimes_x = \left(\frac{\partial v_z}{\partial y} - \frac{\partial v_y}{\partial z} \right); \; \otimes_y = \left(\frac{\partial v_x}{\partial z} - \frac{\partial v_z}{\partial x} \right); \; \otimes_z = \left(\frac{\partial v_y}{\partial x} - \frac{\partial v_x}{\partial y} \right) \tag{3.4}$$

The components $\otimes_x$, $\otimes_y$ and $\otimes_z$ of the vorticity vector $\vec{\otimes}$ correspond to clockwise rotation rates about the Cartesian axes. The differential velocity components can be written as a function of local, linear, angular, and rotational acceleration terms

$$dv_x = \frac{\partial v_x}{\partial t} dt + \ominus_x dx + \frac{\oslash_z}{2} dy + \frac{\oslash_y}{2} dz + \frac{1}{2} (\otimes_y dz - \otimes_z dy) \tag{3.5a}$$

$$dv_y = \frac{\partial v_y}{\partial t} dt + \ominus_y dy + \frac{\oslash_z}{2} dx + \frac{\oslash_x}{2} dz + \frac{1}{2} (\otimes_z dx - \otimes_x dz) \tag{3.5b}$$

$$dv_z = \underbrace{\frac{\partial v_z}{\partial t} dt}_{local} + \underbrace{\ominus_z dz}_{linear} + \underbrace{\frac{\oslash_y}{2} dx + \frac{\oslash_x}{2} dy}_{angular} + \underbrace{\frac{1}{2} (\otimes_x dy - \otimes_y dx)}_{rotational} \tag{3.5c}$$

Accelerations are obtained from the time derivatives of velocity in Equations (3.1) and (3.5).

3.2 Equation of continuity

The equation of continuity is based on the law of conservation of mass, stating that mass cannot be created or destroyed. The continuity equation can be written in either differential or integral form.

3.2.1 Differential continuity equation

In differential form, consider an infinitesimal control volume $d\forall = dx\,dy\,dz$ on Figure 3.3 filled with fluid of mass density ρ_m. The difference between the mass fluxes leaving and entering the differential control volume equal the rate of increase

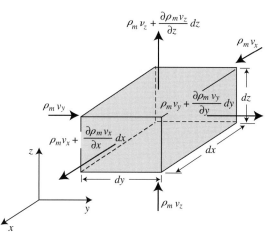

Figure 3.3. Infinitesimal element of fluid

of internal mass. For instance in the x direction, the mass flux (M/T) entering the control volume is $\rho_m v_x \, dy \, dz$. The mass flux leaving the control volume is $\frac{\partial \rho_m v_x}{\partial x} \, dx \, dy \, dz$ in excess of the entering mass flux. This process is repeated in the y and z directions, and the rate of increase of internal mass is $\frac{\partial \rho_m \, d\forall}{\partial t}$. The assumption of a continuous fluid medium yields the following differential relationships:
Cartesian coordinates (x, y, z)

$$\frac{\partial \rho_m}{\partial t} + \frac{\partial}{\partial x}(\rho_m v_x) + \frac{\partial}{\partial y}(\rho_m v_y) + \frac{\partial}{\partial z}(\rho_m v_z) = 0 \tag{3.6a}$$

Cylindrical coordinates (r, θ, z)

$$\frac{\partial \rho_m}{\partial t} + \frac{1}{r}\frac{\partial}{\partial r}(\rho_m r v_r) + \frac{1}{r}\frac{\partial}{\partial \theta}(\rho_m v_\theta) + \frac{\partial}{\partial z}(\rho_m v_z) = 0 \tag{3.6b}$$

Spherical coordinates (r, θ, φ)

$$\frac{\partial \rho_m}{\partial t} + \frac{1}{r^2}\frac{\partial}{\partial r}(\rho_m r^2 v_r) + \frac{1}{r\sin\theta}\frac{\partial}{\partial \theta}(\rho_m v_\theta \sin\theta) + \frac{1}{r\sin\theta}\frac{\partial}{\partial \varphi}(\rho_m v_\varphi) = 0 \tag{3.6c}$$

For incompressible fluids, the continuity equation reduces to

$$\frac{\partial v_x}{\partial x} + \frac{\partial v_y}{\partial y} + \frac{\partial v_z}{\partial z} = 0 \tag{3.6d}$$

The conservation of solid mass is defined in Example 3.1. The continuity equations for solids are identical with Equations (3.6a–c) after replacing ρ_m with C_v. For homogeneous incompressible suspensions without settling, the mass density is independent of space and time (ρ_s, ρ, ρ_m = constant), consequently, $\partial\rho_m/\partial t = 0$ and the continuity equation reduces to Equation (3.6d).

It is also interesting, though far less important, to consider that the properties of the vorticity vector $\vec{\otimes}$ and the velocity vector $\vec{v}$ for homogeneous incompressible fluids are strikingly similar:

$$\mathrm{div}\,\vec{v} = \frac{\partial v_x}{\partial x} + \frac{\partial v_y}{\partial y} + \frac{\partial v_z}{\partial z} = 0; \tag{3.7a}$$

$$\mathrm{div}\,\vec{\otimes} = \frac{\partial\otimes_x}{\partial x} + \frac{\partial\otimes_y}{\partial y} + \frac{\partial\otimes_z}{\partial z} = 0 \tag{3.7b}$$

Likewise, streamlines and vortex lines are respectively defined as

$$\frac{dx}{v_x} = \frac{dy}{v_y} = \frac{dz}{v_z} \tag{3.8a}$$

$$\frac{dx}{\otimes_x} = \frac{dy}{\otimes_y} = \frac{dz}{\otimes_z} \tag{3.8b}$$

A line tangent to the velocity vector at every point at a given instant is known as a streamline. The path line of a fluid element is the locus of the element through time. A streak line is defined as the line connecting all fluid elements that have passed successively at a given point in space.

Example 3.1 Differential sediment continuity equation

Derive the governing sediment continuity equation given the volumetric sediment concentration C_v. The net mass of sediment inside the control volume is $dm = \rho_s C_v\,dx\,dy\,dz$. Let's also consider a possible internal source of sediment within the control volume at a rate $\dot{m} = M/T$ added to the sediment suspension in a cubic element $dx\,dy\,dz$. This internal mass change can be due to a chemical reaction, flocculation of dissolved solids, or a phase change. This analysis can also apply to chemicals and contaminants, such that a conservative substance is one where $\dot{m} = 0$. The total mass change per unit time inside the control volume

$$\dot{m} = (d/dt)(\rho_s C_v)\,dx\,dy\,dz \tag{E-3.1.1}$$

Let's consider only advective fluxes as shown in Figure E-3.1.1. The mass flux entering the control volume by advection in the x direction is $\rho_s C_v v_x\,dy\,dz$ and the mass flux leaving the control volume in the x direction is $(\rho_s C_v v_x + \frac{\partial}{\partial x}(\rho_s C_v v_x)\,dx)\,dy\,dz$. With similar considerations in the y and z directions, similar to the x-direction fluxes shown in Figure E-3.1.1, the rate of mass change

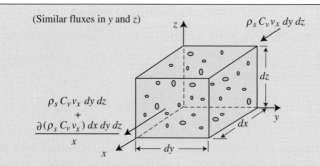

Figure E-3.1.1 Differential control volume for sediment continuity

inside the control volume equals the net mass flux entering from all three directions, thus

$$\dot{m} = \frac{\partial}{\partial t}(\rho_s C_v)\, dx\, dy\, dz$$

$$- \rho_s C_v v_x\, dy\, dz + \left[\rho_s C_v v_x + \frac{\partial}{\partial x}(\rho_s C_v v_x)\, dx \right] dy\, dz$$

$$- \rho_s C_v v_y\, dx\, dz + \left[\rho_s C_v v_y + \frac{\partial}{\partial y}(\rho_s C_v v_y)\, dy \right] dx\, dz$$

$$- \rho_s C_v v_z\, dx\, dy + \left[\rho_s C_v v_z + \frac{\partial}{\partial z}(\rho_s C_v v_z)\, dz \right] dx\, dy \qquad \text{(E-3.1.2)}$$

$$\frac{\partial}{\partial t}(\rho_s C_v) + \frac{\partial}{\partial x}(\rho_s C_v v_x) + \frac{\partial}{\partial y}\left(\rho_s C_v v_y\right) + \frac{\partial}{\partial z}(\rho_s C_v v_z) = \frac{\dot{m}}{dx\, dy\, dz}$$

$$\text{(E-3.1.3)}$$

which for a constant mass density of sediment ρ_s reduces to

$$\frac{\partial C_v}{\partial t} + \frac{\partial (C_v v_x)}{\partial x} + \frac{\partial (C_v v_y)}{\partial y} + \frac{\partial (C_v v_z)}{\partial z} = \dot{C}_v \qquad \text{(E-3.1.4)}$$

where $\dot{C}_v = \frac{\dot{\forall}_s}{\forall_t} = \frac{\dot{m}}{\rho_s \forall_t}$ is the volumetric source of sediment per unit time.

It is important to remember that this derivation only considers advective fluxes. In the case of sediment transport, diffusion and mixing can induce sediment fluxes even when all velocities are zero. Therefore, this term $\dot{C}_v$ can include the following processes: (1) diffusion, mixing, and dispersion (to be discussed later in Section 10.2); (2) phase change of the substance (e.g. change from dissolved solids to particle solids, like flocculation); (3) chemical reactions causing phase changes in the case of metal or contaminant transport; and (4) decay functions of substances or $\dot{m} \neq 0$ (e.g. radioactive material).

3.2.2 *Integral continuity equation*

The integral form of the continuity equation is simply the integral of the differential form (Eq. 3.6a) over a control volume $\forall$. For an incompressible fluid, the integral form of conservation of mass is

$$\int_\forall \frac{\partial \rho_m}{\partial t} d\forall + \int_\forall \left(\frac{\partial \rho_m v_x}{\partial x} + \frac{\partial \rho_m v_y}{\partial y} + \frac{\partial \rho_m v_z}{\partial z} \right) d\forall = 0 \qquad (3.9)$$

This volume integral of velocity gradients can be transformed into surface integrals owing to the divergence theorem applied to an argument F of a partial space derivative

$$\int_\forall \frac{\partial F}{\partial x} d\forall = \int_A F \frac{\partial x}{\partial n} dA \qquad (3.10) \blacklozenge\blacklozenge$$

in which $\partial x/\partial n$ is the cosine of the angle between the coordinate x and the normal vector pointing outside of the control volume as shown in Figure 3.1. Example 3.2 illustrates how the integral continuity equation can be directly applied to open channels. Example 3.3 shows an application of conservation of sediment mass in open channels.

Example 3.2 Integral continuity equation

Consider the impervious rectangular channel of length ΔX sketched in Figure E-3.2.1. The differential continuity equation of (Eq. 3.6a) is multiplied by $d\forall$ integrated over the control volume $\forall = Wh\Delta X$

$$\int_\forall \frac{\partial \rho_m}{\partial t} d\forall + \int_\forall \left(\frac{\partial \rho_m v_x}{\partial x} + \frac{\partial \rho_m v_y}{\partial y} + \frac{\partial \rho_m v_z}{\partial z} \right) d\forall = 0 \qquad (E\text{-}3.2.1)$$

Considering that the free surface can rise at a rate dh/dt, the first integral for incompressible fluids ($\rho_m \cong \rho$) corresponds to the mass change inside the control volume $\rho_m \partial (W \Delta Xh)/\partial t$. The divergence theorem (Equation 3.10) is applied to the second integral, which reduces to

$$\rho_m \frac{\partial (W \Delta Xh)}{\partial t} + \int_A \left(\rho_m v_x \frac{\partial x}{\partial n} + \rho_m v_y \frac{\partial y}{\partial n} + \rho_m v_z \frac{\partial z}{\partial n} \right) dA = 0.$$

The values of $\partial x/\partial n$, $\partial y/\partial n$, and $\partial z/\partial n$, are the cosines of the angle between the vector normal to the surface $\bar{n}$ pointing outside of the control volume and the Cartesian coordinates x, y, and z, respectively. For instance, Figure E-3.2.1 illustrates the direction cosines on the downstream face.

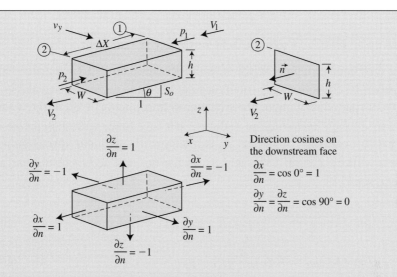

Figure E-3.2.1 Rectangular open-channel flow

Thus for a channel of length ΔX and flow depth h, the net flux ΔQ leaving the control volume in the x direction is $A_2 V_2 - A_1 V_1$. The net flux entering laterally in the y direction is $\Delta X q_\ell$, where the unit lateral discharge $q_\ell = h v_y$ and the change in control volume of fluid is $\frac{\Delta (W \Delta X h)}{\Delta t}$. The equation of conservation of mass can be expressed as

$$A_2 V_2 - A_1 V_1 - \Delta X q_\ell + \frac{\Delta (W \Delta X h)}{\Delta t} = 0 \qquad \text{(E-3.2.2)}$$

After dividing by ΔX given $Q = AV$ and the cross-sectional area $A_x = Wh$, this equation reduces to

$$\frac{\Delta Q}{\Delta X} + \frac{\Delta A_x}{\Delta t} = q_\ell \qquad \text{(E-3.2.3a)}\blacklozenge\blacklozenge$$

At a constant channel width without lateral inflow $(q_\ell = 0)$

$$\frac{\Delta h}{\Delta t} = \frac{-1 \Delta Q}{W \Delta x} \qquad \text{(E-3.2.3b)}$$

For steady flow without lateral inflow, $\Delta Q = 0$ and

$$Q = A_1 V_1 = A_2 V_2 \qquad \text{(E-3.2.4)}\blacklozenge$$

This integral continuity equation is only applicable to steady impervious open channels.

Example 3.3 Continuity of sediment

In a rectangular channel reach of width W and length ΔX, consider sediment transport by advection only (no molecular diffusion and no turbulent mixing) of sediment particles (without phase change or flocculation). Couple the volumetric settling flux Q_s = volume of sediment per unit time, with the change in bed elevation over time. Note that q_{sx} is a volumetric sediment discharge per unit width with dimension L^2/T and $q_{s\ell}$ is the net unit sediment discharge from lateral sources.

$$(a) \quad \frac{-\Delta q_{sx}\Delta XW}{\Delta X} + q_{s\ell}\Delta X - Q_{settling} = \frac{\Delta\,(C_v W\,\Delta Xh)}{\Delta t} \qquad \text{(E-3.3.1)}$$

$$(b) \quad Q_{settling} = W\,\Delta X \frac{\Delta z_b}{\Delta t}(1-p_o) \qquad \text{(E-3.3.2)}$$

Combining these two equations gives a formulation describing two-dimensional continuity of sediment where changes in unit sediment discharges by advection fluxes correspond to changes in bed elevation or change in sediment in suspension in the water column.

$$\underbrace{\Delta q_{sx}W - q_{s\ell}\Delta X}_{advection\ fluxes} + \underbrace{(1-p_o)\frac{\Delta\,(z_b W\,\Delta X)}{\Delta t}}_{change\ in\ bed\ elevation} + \underbrace{\frac{\Delta\,(C_v\forall_t)}{\Delta t}}_{\substack{internal\ mass \\ change\ in \\ suspension}} = 0 \qquad \text{(E-3.3.3)}$$

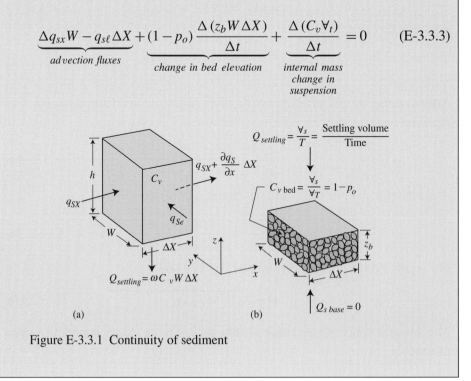

(a)   (b)

Figure E-3.3.1 Continuity of sediment

After dividing by ΔX, this gives an equivalent formulation as a function of the total volumetric sediment discharges $Q_{sx} = q_{sx}W$ and $\overline{A}_x = \mathbb{V}_t/\Delta X = Wh$ as the reach-average cross-sectional area.

$$\frac{\Delta Q_{sx}}{\Delta X} + (1 - p_o)\frac{\Delta Wz_b}{\Delta t} + \frac{\Delta\left(C_v\overline{A}_x\right)}{\Delta t} = q_{s\ell} \qquad \text{(E-3.3.4)}\blacklozenge\blacklozenge$$

In the case of steady non-uniform one-dimensional flow in the x direction, one can consider that $q_{s\ell} = 0$ and $dC_v/dt = 0$. The corresponding relationship reduces to

$$\frac{\Delta z_b}{\Delta t} = \frac{-1}{(1 - p_o)}\frac{\Delta q_{sx}}{\Delta X} \qquad \text{(E-3.3.5)}$$

This formulation is often referred to as the Exner equation. Notice that it assumes that all the sediment in the water column will deposit within the distance ΔX and q_{sx} is a volumetric unit sediment discharge. It also assumes that there is no lateral influx of sediment. This relationship clearly states that an increase in sediment flux in the downstream direction results in lowering the bed elevation, also called riverbed degradation. Conversely, a downstream decrease in sediment flux results in riverbed aggradation and bed deposition of sediment.

3.3 Equations of motion

The Cartesian acceleration components are obtained directly after dividing the terms of the velocity equations in Equation (3.1) by dt.

$$a_x = \frac{dv_x}{dt} = \frac{\partial v_x}{\partial t} + v_x\frac{\partial v_x}{\partial x} + v_y\frac{\partial v_x}{\partial y} + v_z\frac{\partial v_x}{\partial z} \qquad (3.11a)$$

$$a_y = \frac{dv_y}{dt} = \frac{\partial v_y}{\partial t} + v_x\frac{\partial v_y}{\partial x} + v_y\frac{\partial v_y}{\partial y} + v_z\frac{\partial v_y}{\partial z} \qquad (3.11b)$$

$$a_z = \frac{dv_z}{dt} = \underbrace{\frac{\partial v_z}{\partial t}}_{local} + \underbrace{v_x\frac{\partial v_z}{\partial x} + v_y\frac{\partial v_z}{\partial y} + v_z\frac{\partial v_z}{\partial z}}_{convective} \qquad (3.11c)$$

The convective terms of the acceleration Equation (3.11a) can also be separated into rotational and irrotational terms by adding and subtracting the terms $v_y\partial v_y/\partial x$ and $v_z\partial v_z/\partial x$, and by substituting $\otimes_y$ and $\otimes_z$ from Equation (3.4). Similar substitutions can also be done in the y and z directions to give

$$a_x = \frac{\partial v_x}{\partial t} + v_z \otimes_y - v_y \otimes_z + \frac{\partial(v^2/2)}{\partial x} \qquad (3.11\text{d})$$

$$a_y = \frac{\partial v_y}{\partial t} + v_x \otimes_z - v_z \otimes_x + \frac{\partial(v^2/2)}{\partial y} \qquad (3.11\text{e})$$

$$a_z = \underbrace{\frac{\partial v_z}{\partial t}}_{\text{local}} + \underbrace{v_y \otimes_x - v_x \otimes_y}_{\substack{\text{convective} \\ \text{rotational}}} + \underbrace{\frac{\partial(v^2/2)}{\partial z}}_{\substack{\text{convective} \\ \text{irrotational}}} \qquad (3.11\text{f})$$

Equations (3.11d and f) show that the total acceleration can be separated into local and convective acceleration terms, while the convective acceleration terms can be subdivided into rotational and irrotational components.

   As sketched in Figure 3.4, the forces acting on a Cartesian element of fluid and sediment (dx, dy, dz) are classified as either internal forces or external forces. The internal accelerations, or body forces per unit mass, acting at the center of mass of the element are denoted g_x, g_y, and g_z. The external forces per unit area applied on each face of the element are subdivided into normal and tangential stress components. The normal stresses σ_x, σ_y, and σ_z are positive for tension. Six shear stresses $\tau_{xy}, \tau_{yx}, \tau_{xz}, \tau_{zx}, \tau_{yz}, \tau_{zy}$, with two orthogonal components on each face are applied, as shown in Figure 3.4. The first subscript indicates

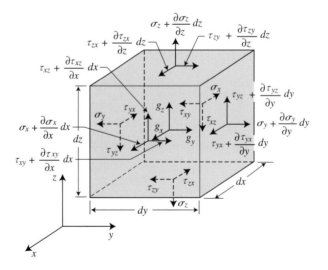

Figure 3.4. Surface stresses on a fluid element

the direction normal to the face and the second subscript designates the direction of the applied stress component. The following identities: $\tau_{xy} = \tau_{yx}$; $\tau_{xz} = \tau_{zx}$; $\tau_{yz} = \tau_{zy}$ result from the sum of moments of shear stresses around the centroid. If they were not equal, these stresses would spin an elementary fluid volume infinitely fast.

The element of fluid in Figure 3.4 is considered in equilibrium when the sum of the forces per unit mass in each direction x, y, and z equals the corresponding Cartesian acceleration component a_x, a_y, and a_z:

$$a_x = g_x + \frac{1}{\rho_m}\frac{\partial \sigma_x}{\partial x} + \frac{1}{\rho_m}\frac{\partial \tau_{yx}}{\partial y} + \frac{1}{\rho_m}\frac{\partial \tau_{zx}}{\partial z} \tag{3.12a}$$

$$a_y = g_y + \frac{1}{\rho_m}\frac{\partial \sigma_y}{\partial y} + \frac{1}{\rho_m}\frac{\partial \tau_{xy}}{\partial x} + \frac{1}{\rho_m}\frac{\partial \tau_{zy}}{\partial z} \tag{3.12b}$$

$$a_z = g_z + \frac{1}{\rho_m}\frac{\partial \sigma_z}{\partial z} + \frac{1}{\rho_m}\frac{\partial \tau_{xz}}{\partial x} + \frac{1}{\rho_m}\frac{\partial \tau_{yz}}{\partial y} \tag{3.12c}$$

These equations of motion are general without any restriction as to compressibility, viscous shear, turbulence, or other effects.

The normal stresses can be rewritten as a function of the pressure p and additional normal stresses τ_{xx}, τ_{yy}, and τ_{zz} accompanying deformation:

$$\sigma_x = -p + \tau_{xx} \tag{3.13a}$$

$$\sigma_y = -p + \tau_{yy} \tag{3.13b}$$

$$\sigma_z = -p + \tau_{zz} \tag{3.13c}$$

where all shear stress components will be defined in Chapter 5.

After expanding the acceleration components a_x, a_y, and a_z from Equation (3.11), the equations of motion in Cartesian, cylindrical, and spherical coordinates can be written as Equations (3.14)–(3.16) in Table 3.1.

After substituting the equations of motion from Equation (3.11d–f) and the normal stresses from Equation (3.13) into Equation (3.12), the following equations can be obtained

$$\frac{\partial v_x}{\partial t} + \frac{1}{\rho_m}\frac{\partial p}{\partial x} - g_x + \frac{\partial \left(v^2/2\right)}{\partial x} = \frac{1}{\rho_m}\left(\frac{\partial \tau_{xx}}{\partial x} + \frac{\partial \tau_{yx}}{\partial y} + \frac{\partial \tau_{zx}}{\partial z}\right) + v_y \otimes_z - v_z \otimes_y$$

$$\tag{3.17a}$$

$$\frac{\partial v_y}{\partial t} + \frac{1}{\rho_m}\frac{\partial p}{\partial y} - g_y + \frac{\partial \left(v^2/2\right)}{\partial y} = \frac{1}{\rho_m}\left(\frac{\partial \tau_{xy}}{\partial x} + \frac{\partial \tau_{yy}}{\partial y} + \frac{\partial \tau_{zy}}{\partial z}\right) + v_z \otimes_x - v_x \otimes_z$$

$$\tag{3.17b}$$

Table 3.1. *Equations of motion*

Cartesian coordinates (x,y,z)

x - component

$$a_x = \frac{\partial v_x}{\partial t} + v_x \frac{\partial v_x}{\partial x} + v_y \frac{\partial v_x}{\partial y} + v_z \frac{\partial v_x}{\partial z} = g_x - \frac{1}{\rho_m} \frac{\partial p}{\partial x} + \frac{1}{\rho_m} \left(\frac{\partial \tau_{xx}}{\partial x} + \frac{\partial \tau_{yx}}{\partial y} + \frac{\partial \tau_{zx}}{\partial z} \right) \quad (3.14a)$$

y - component

$$a_y = \frac{\partial v_y}{\partial t} + v_x \frac{\partial v_y}{\partial x} + v_y \frac{\partial v_y}{\partial y} + v_z \frac{\partial v_y}{\partial z} = g_y - \frac{1}{\rho_m} \frac{\partial p}{\partial y} + \frac{1}{\rho_m} \left(\frac{\partial \tau_{xy}}{\partial x} + \frac{\partial \tau_{yy}}{\partial y} + \frac{\partial \tau_{zy}}{\partial z} \right) \quad (3.14b)$$

z - component

$$a_z = \frac{\partial v_z}{\partial t} + v_x \frac{\partial v_z}{\partial x} + v_y \frac{\partial v_z}{\partial y} + v_z \frac{\partial v_z}{\partial z} = g_z - \frac{1}{\rho_m} \frac{\partial p}{\partial z} + \frac{1}{\rho_m} \left(\frac{\partial \tau_{xz}}{\partial x} + \frac{\partial \tau_{yz}}{\partial y} + \frac{\partial \tau_{zz}}{\partial z} \right) \quad (3.14c)$$

Cylindrical coordinates (r,θ,z)

r - component

$$\frac{\partial v_r}{\partial t} + v_r \frac{\partial v_r}{\partial r} + \frac{v_\theta}{r} \frac{\partial v_r}{\partial \theta} - \frac{v_\theta^2}{r} + v_z \frac{\partial v_r}{\partial z} =$$

$$g_r - \frac{1}{\rho_m} \frac{\partial p}{\partial r} + \frac{1}{\rho_m} \left(\frac{1}{r} \frac{\partial}{\partial r}(r\tau_{rr}) + \frac{1}{r} \frac{\partial \tau_{\theta r}}{\partial \theta} - \frac{\tau_{\theta\theta}}{r} + \frac{\partial \tau_{zr}}{\partial z} \right) \quad (3.15a)$$

θ - component

$$\frac{\partial v_\theta}{\partial t} + v_r \frac{\partial v_\theta}{\partial r} + \frac{v_\theta}{r} \frac{\partial v_\theta}{\partial \theta} + \frac{v_r v_\theta}{r} + v_z \frac{\partial v_\theta}{\partial z} =$$

$$g_\theta - \frac{1}{\rho_m r} \frac{\partial p}{\partial \theta} + \frac{1}{\rho_m} \left(\frac{1}{r^2} \frac{\partial}{\partial r}(r^2 \tau_{r\theta}) + \frac{1}{r} \frac{\partial \tau_{\theta\theta}}{\partial \theta} + \frac{\partial \tau_{z\theta}}{\partial z} \right) \quad (3.15b)$$

z - component

$$\frac{\partial v_z}{\partial t} + v_r \frac{\partial v_z}{\partial r} + \frac{v_\theta}{r} \frac{\partial v_z}{\partial \theta} + v_z \frac{\partial v_z}{\partial z} = g_z - \frac{1}{\rho_m} \frac{\partial p}{\partial z} + \frac{1}{\rho_m} \left(\frac{1}{r} \frac{\partial (r\tau_{rz})}{\partial r} + \frac{1}{r} \frac{\partial \tau_{\theta z}}{\partial \theta} + \frac{\partial \tau_{zz}}{\partial z} \right)$$

$$(3.15c)$$

Spherical coordinates (r,θ,φ)

r – component

$$\frac{\partial v_r}{\partial t} + v_r \frac{\partial v_r}{\partial r} + \frac{v_\theta}{r} \frac{\partial v_r}{\partial \theta} + \frac{v_\varphi}{r\sin\theta} \frac{\partial v_r}{\partial \varphi} - \frac{v_\theta^2 + v_\varphi^2}{r} = g_r - \frac{1}{\rho_m} \frac{\partial p}{\partial r}$$

$$+ \frac{1}{\rho_m} \left(\frac{1}{r^2} \frac{\partial}{\partial r}(r^2 \tau_{rr}) + \frac{1}{r\sin\theta} \frac{\partial}{\partial \theta}(\tau_{r\theta}\sin\theta) + \frac{1}{r\sin\theta} \frac{\partial \tau_{r\varphi}}{\partial \varphi} - \frac{\tau_{\theta\theta} + \tau_{\varphi\varphi}}{r} \right) \quad (3.16a)$$

Table 3.1. *(Cont.)*

θ - component

$$\frac{\partial v_\theta}{\partial t} + v_r \frac{\partial v_\theta}{\partial r} + \frac{v_\theta}{r}\frac{\partial v_\theta}{\partial \theta} + \frac{v_\varphi}{r\sin\theta}\frac{\partial v_\theta}{\partial \varphi} + \frac{v_r v_\theta}{r} - \frac{v_\varphi^2 \cot\theta}{r} = g_\theta - \frac{1}{\rho_m r}\frac{\partial p}{\partial \theta}$$

$$+ \frac{1}{\rho_m}\left(\frac{1}{r^2}\frac{\partial}{\partial r}(r^2 \tau_{r\theta}) + \frac{1}{r\sin\theta}\frac{\partial}{\partial \theta}(\tau_{\theta\theta}\sin\theta) + \frac{1}{r\sin\theta}\frac{\partial \tau_{\theta\varphi}}{\partial \varphi} + \frac{\tau_{r\theta}}{r} - \frac{\tau_{\varphi\varphi}\cot\theta}{r}\right)$$

$$(3.16b)$$

φ - component

$$\frac{\partial v_\varphi}{\partial t} + v_r \frac{\partial v_\varphi}{\partial r} + \frac{v_\theta}{r}\frac{\partial v_\varphi}{\partial \theta} + \frac{v_\varphi}{r\sin\theta}\frac{\partial v_\varphi}{\partial \varphi} + \frac{v_\varphi v_r}{r} + \frac{v_\theta v_\varphi}{r}\cot\theta = g_\varphi - \frac{1}{\rho_m r\sin\theta}\frac{\partial p}{\partial \varphi}$$

$$+ \frac{1}{\rho_m}\left(\frac{1}{r^2}\frac{\partial}{\partial r}(r^2 \tau_{r\varphi}) + \frac{1}{r}\frac{\partial \tau_{\theta\varphi}}{\partial \theta} + \frac{1}{r\sin\theta}\frac{\partial \tau_{\varphi\varphi}}{\partial \varphi} + \frac{\tau_{r\varphi}}{r} + \frac{2\tau_{\theta\varphi}\cot\theta}{r}\right) \qquad (3.16c)$$

$$\underbrace{\frac{\partial v_z}{\partial t}}_{\substack{\text{Euler}\\(\text{Chapter 3})} } + \underbrace{\frac{1}{\rho_m}\frac{\partial p}{\partial z} - g_z + \frac{\partial(v^2/2)}{\partial z}}_{\substack{\text{Bernoulli sum}\\(\text{Chapters 3\&4})} } = \underbrace{\frac{1}{\rho_m}\left(\frac{\partial \tau_{xz}}{\partial x} + \frac{\partial \tau_{yz}}{\partial y} + \frac{\partial \tau_{zz}}{\partial z}\right)}_{\substack{\text{viscosity}\\(\text{Chapter 5})} }$$

$$+ \underbrace{v_x \otimes_y - v_y \otimes_x}_{\substack{\text{vorticity/turbulence}\\(\text{Chapter 6})} } \qquad (3.17c)$$

The presentation of the upcoming chapters has been designed to explain the main terms of the equations of motion. Chapter 3 will focus on the terms on the left-hand side of Equation (3.17). The concepts of buoyancy force, momentum, and energy in Chapter 3 will be primarily based on the analysis of the terms on the left-hand side of Equation (3.17). The concept of lift force in Chapter 4 will involve applications of the Bernoulli sum. The concept of drag force in Chapter 5 will involve the shear stress terms on the right-hand side of Equation (3.17). Finally, the concept of vorticity will be expanded into turbulence in Chapter 6 involving the rotational terms on the right-hand side of Equation (3.17).

3.4 Euler equations

The Euler equations are simplified forms of the equations of motion (Eqs. 3.14–3.16) for frictionless fluids. Without friction, the shear stress components due to deformation are zero (all stress components in $\tau = 0$) and the normal stresses are equal and opposite to the pressure ($\sigma_x = \sigma_y = \sigma_z = -p$). Substitution into the equations of motion yields, for Cartesian coordinates,

$$a_x = g_x - \frac{1}{\rho_m}\frac{\partial p}{\partial x} \tag{3.18a}$$

$$a_y = g_y - \frac{1}{\rho_m}\frac{\partial p}{\partial y} \tag{3.18b}$$

$$a_z = g_z - \frac{1}{\rho_m}\frac{\partial p}{\partial z} \tag{3.18c}$$

These equations, valid for inviscid fluids, are known as Euler equations, or

$$\frac{-\partial p}{\partial x} = \rho_m\,(a_x - g_x) \tag{3.19a}$$

$$\frac{-\partial p}{\partial y} = \rho_m\,(a_y - g_y) \tag{3.19b}$$

$$\frac{-\partial p}{\partial z} = \rho_m\,(a_z - g_z) \tag{3.19c}$$

An example of application of the Euler equations is the buoyancy force resulting from the integration of the pressure distribution around a submerged sphere for inviscid fluids without convective acceleration. The buoyancy force has three components, F_{Bx}, F_{By}, F_{Bz} that can be determined as follows from the divergence theorem Equation (3.10).

$$F_{Bx} = \int_A -p\frac{\partial x}{\partial n}\,dA = \int_\forall -\frac{\partial p}{\partial x}\,d\forall = \int_\forall \rho_m\,(a_x - g_x)\,d\forall \tag{3.20a}$$

$$F_{By} = \int_A -p\frac{\partial y}{\partial n}\,dA = \int_\forall -\frac{\partial p}{\partial y}\,d\forall = \int_\forall \rho_m\,(a_y - g_y)\,d\forall \tag{3.20b}$$

$$F_{Bz} = \int_A -p\frac{\partial z}{\partial n}\,dA = \int_\forall -\frac{\partial p}{\partial z}\,d\forall = \int_\forall \rho_m\,(a_z - g_z)\,d\forall \tag{3.20c}$$

Example 3.4 shows an application of the Euler equations for hydrostatic conditions and Example 3.5 is applied to an accelerated control volume.

Example 3.4. Buoyancy force on a sphere

Consider the hydrostatic pressure distribution around a sphere of radius R submerged in a fluid of mass density ρ_m (Figure E-3.4.1). The pressure at the center of the sphere is p_o and the vertical elevation $\hat{z}$ at the surface of the sphere is $R\cos\theta$. The hydrostatic pressure thus varies as $p = p_o - \rho_m g R\cos\theta$. After considering

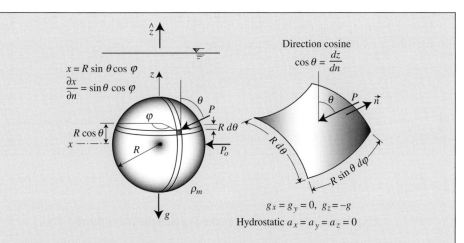

Figure E-3.4.1 Buoyancy force on a sphere

$z = \hat{z}$ as the vertical direction and $\partial z / \partial n = \cos\theta$, it is clear that $g_x = g_y = 0$ with $g_z = -g$ and hydrostatic conditions are described by $a_x = a_y = a_z = 0$. The buoyancy force F_B is calculated from the surface integral of the pressure along the surface of the sphere.

$$F_B = F_{Bz} = \int_A -p \frac{\partial z}{\partial n} dA = -\int_A p\cos\theta \, dA \qquad \text{(E-3.4.1)}$$

The elementary surface area is $dA = (R\sin\theta \, d\varphi)(R d\theta)$ and the integration is performed for $0 < \varphi < 2\pi$ and $0 < \theta < \pi$

$$F_B = -\int_0^\pi \int_0^{2\pi} p\cos\theta \, (R\sin\theta d\varphi)(R d\theta)$$

$$F_B = 0 + \rho_m g R^3 \int_0^\pi \cos^2\theta \sin\theta \, d\theta \int_0^{2\pi} d\varphi = 2\pi \gamma_m R^3 \times \frac{2}{3}$$

We learn that the integral of a constant on a closed surface is zero. The hydrostatic pressure distribution therefore gives the following buoyancy force.

$$F_B = \frac{4\pi}{3}\gamma_m R^3 = \gamma_m \forall_{sphere} \qquad \text{(E-3.4.2)}$$

It is interesting that the buoyancy force in Equation (E-3.4.1) can also be easily obtained from the divergence theorem as

$$F_B = \int_A -p\frac{\partial z}{\partial n}\,dA = \int_\forall -\frac{\partial p}{\partial z}\,d\forall = \int_\forall \rho_m (a_z - g_z)\,d\forall = \rho_m g\forall \qquad \text{(E-3.4.3)}$$

It is important to notice that the buoyancy force is different from Archimedes' principle when $a_z \neq 0$.

Example 3.5. Buoyancy force for accelerated fluids

Consider a neutrally buoyant sphere of radius R placed in a water container accelerated in the horizontal direction at $a_x = g/2$. Determine the buoyancy force on the sphere sketched in Figure E-3.5.1.

The convective acceleration terms for a neutrally buoyant sphere vanish because the fluid does not accelerate relative to the sphere. Acceleration components are $a_x = \partial v_x/\partial t = g/2$, $a_y = a_z = 0$. The gravitational acceleration components are $g_x = g_y = 0$, and $g_z = -g$. The buoyancy force component $F_{By} = 0$, the others are

$$F_{Bx} = \int_\forall \frac{-\partial p}{\partial x}\,d\forall = \int_\forall \rho\,(a_x - g_x)\,d\forall = \frac{\rho g}{2}\forall, \text{ and}$$

$$F_{Bz} = \int_\forall \rho\,(a_z - g_z)\,d\forall = \rho \int_\forall (0 + g)\,d\forall = \gamma\forall_{sphere}$$

The net buoyancy force from these two orthogonal components is $F_B = \frac{\sqrt{5}}{2}\gamma\forall_{sphere}$ acting at an angle of 26.56° from the vertical. The buoyancy force only equals the Archimedes' force in hydrostatic fluids.

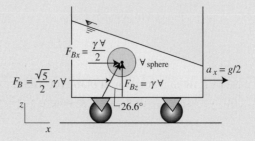

Figure E-3.5.1 Accelerated control volume

3.5 Bernoulli equation

The Bernoulli equation represents a particular form of the equations of motion for steady irrotational flow of frictionless fluids. A gravitation potential $\Omega_g = g\hat{z}$ can be defined with the axis $\hat{z}$ vertical upward such that the body acceleration components due to gravity are:

$$g_x = \frac{-\partial\Omega_g}{\partial x}; \; g_y = \frac{-\partial\Omega_g}{\partial y}; \; and \; g_z = \frac{-\partial\Omega_g}{\partial z}$$

Since g is a constant, the directional acceleration components are g times the cosine between $\hat{z}$ and the component direction. After considering the equations of motion (Eqs. 3.11 and 3.14), and the gravitation potential, the equations of motion for incompressible sediment-laden fluids of mass density ρ_m can be rewritten as follows, with the Bernoulli terms on the left-hand side:

$$\frac{\partial}{\partial x}\left(\frac{p}{\rho_m}+\Omega_g+\frac{v^2}{2}\right) = (v_y\otimes_z-v_z\otimes_y)-\frac{\partial v_x}{\partial t}+\frac{1}{\rho_m}\left(\frac{\partial\tau_{xx}}{\partial x}+\frac{\partial\tau_{yx}}{\partial y}+\frac{\partial\tau_{zx}}{\partial z}\right)$$
(3.21a)

$$\frac{\partial}{\partial y}\left(\frac{p}{\rho_m}+\Omega_g+\frac{v^2}{2}\right) = (v_z\otimes_x-v_x\otimes_z)-\frac{\partial v_y}{\partial t}+\frac{1}{\rho_m}\left(\frac{\partial\tau_{xy}}{\partial x}+\frac{\partial\tau_{yy}}{\partial y}+\frac{\partial\tau_{zy}}{\partial z}\right)$$
(3.21b)

$$\frac{\partial}{\partial z}\left(\frac{p}{\rho_m}+\Omega_g+\frac{v^2}{2}\right) = (v_x\otimes_y-v_y\otimes_x)-\frac{\partial v_z}{\partial t}+\frac{1}{\rho_m}\left(\frac{\partial\tau_{xz}}{\partial x}+\frac{\partial\tau_{yz}}{\partial y}+\frac{\partial\tau_{zz}}{\partial z}\right)$$
(3.21c)

For steady irrotational flow of frictionless fluids, the right-hand side of Equations (3.21a–c) vanishes and the Bernoulli sum H for homogeneous incompressible fluids is constant throughout the fluid

$$H = \frac{p}{\gamma_m}+\hat{z}+\frac{v^2}{2g} = ct \quad (3.22a)\blacklozenge$$

$$g\frac{\partial H}{\partial x} = 0 \quad (3.22b)$$

It is interesting to note that Equation (3.22a) describes a hydrodynamic formulation of pressure compared to the hydrostatic formulation obtained when $v=0$.

In the particular case of flow in a horizontal plane (constant $\hat{z}$) of a homogeneous fluid (constant ρ_m), the pressure at any point p where the velocity is v can be calculated from the pressure p_r at any reference point given the reference velocity v_r:

$$p = \frac{\rho_m}{2}\left(v_r^2-v^2\right)+p_r \quad (3.22c)$$

Accordingly, pressure is low when velocity is high and pressure is high at low velocity.

When the flow is steady, frictionless, but rotational, the right-hand side of Equation (3.21) equals zero only along a streamline (because of Eq. 3.8); hence for a homogeneous rotational incompressible fluid,

$$H = \frac{p}{\gamma_m} + \hat{z} + \frac{v^2}{2g} = ct; \; along \; a \; streamline \qquad (3.22\text{d})$$

For steady flow at a mean velocity V in a wide-rectangular channel at a bed slope θ, the first and third terms of the Bernoulli sum describe the specific energy function E defined as:

$$E = \frac{p}{\gamma_m} + \frac{V^2}{2g} = h\cos^2\theta + \frac{V^2}{2g} \qquad (3.23\text{a})$$

Considering the unit discharge $q = Vh$ and the Froude number Fr defined as $\mathrm{Fr}^2 = V^2/gh = q^2/gh^3$, the function E when θ is small is approximated by:

$$E = h + \frac{q^2}{2gh^3} = h\left(1 + \frac{\mathrm{Fr}^2}{2}\right) \qquad (3.23\text{b})$$

The properties of the function E are such that under constant unit discharge q, the critical depth h_c and the critical velocity V_c correspond to the minimum $\left(\frac{\partial E}{\partial h} = 0\right)$ of Equation (3.23b). The minimum value $E_{\min}$ is found when $\mathrm{Fr}_c^2 = \frac{V_c^2}{gh_c} = \frac{q^2}{gh_c^3} = 1$.

One can easily demonstrate that $q^2 = gh_c^3$ or $h_c = \sqrt[3]{\frac{q^2}{g}}$, and $E_{\min} = 1.5h_c$, resulting in the following identities for the Froude number

$$\mathrm{Fr} = \frac{V}{\sqrt{gh}} = \frac{q}{h\sqrt{gh}} = \left(\frac{h_c}{h}\right)^{3/2} \qquad (3.24)$$

An open-channel flow application of the Bernoulli equation with specific energy is presented in Example 3.6.

Example 3.6 Rapidly varied open channel flow

Consider steady flow in a wide-rectangular channel. Determine: (1) what is the maximum possible elevation of a sill Δz at section A that will not cause backwater?; and (2) what is the maximum lateral contraction of the channel at section B that will not cause backwater? The accelerating flow is shown with the specific energy diagram Figure E-3.6.1.

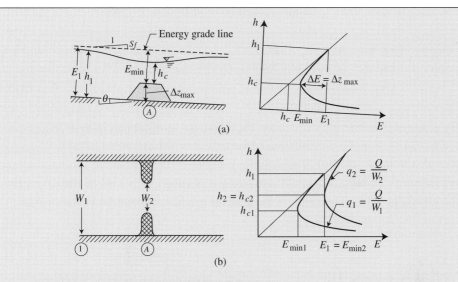

Figure E-3.6.1 a) Flow near sill b) and abutment

(1) The maximum elevation of the sill Δz_{max} at section A is such that the flow will be critical on top of the sill and $\Delta z_{max} + E_{min} = E_1$ or $\Delta z_{max} = \Delta E = E_1 - (3/2)h_c$;
(2) The minimum channel width W_2 at section A without causing backwater is such that the total discharge remains constant $Q = W_1 q_1 = W_2 q_2$ and the flow is critical in the contracted section A_2, or $h_{c2} = 0.67 E_1 = 0.67 E_{min2}$; and $Fr_{c2}^2 = 1 = q_2^2/gh_{c2}^3$, or $W_2 = Q/\sqrt{g(0.67E_1)^3}$.

3.6 Momentum equations

Momentum equations define the hydrodynamic forces exerted by sediment-laden flows. After multiplying the equations of motion (Eqs. 3.14 to 3.16) by the mass density of the mixture ρ_m, the volume integral of the terms on the left-hand side of the equations represent the rate of momentum change per unit volume, while the rate of impulse per unit volume is found on the right-hand side. Integration over the control volume $\forall$ shows that the rate of momentum change equals the impulse per unit time. For example, the x component in the Cartesian coordinates is:

$$\int_\forall \rho_m \left(\frac{\partial v_x}{\partial t} + v_x \frac{\partial v_x}{\partial x} + v_y \frac{\partial v_x}{\partial y} + v_z \frac{\partial v_x}{\partial z} \right) d\forall = \int_\forall \rho_m g_x \, d\forall - \int_\forall \frac{\partial p}{\partial x} d\forall$$

$$+ \int_\forall \left(\frac{\partial \tau_{xx}}{\partial x} + \frac{\partial \tau_{yx}}{\partial y} + \frac{\partial \tau_{zx}}{\partial z} \right) d\forall$$

$$(3.25)$$

The integrand on the left-hand side can be rewritten as follows:

$$\frac{\partial \rho_m v_x}{\partial t} + \frac{\partial \rho_m v_x^2}{\partial x} + \frac{\partial \rho_m v_x v_y}{\partial y} + \frac{\partial \rho_m v_x v_z}{\partial z} - v_x \left(\frac{\partial \rho_m}{\partial t} + \frac{\partial \rho_m v_x}{\partial x} + \frac{\partial \rho_m v_y}{\partial y} + \frac{\partial \rho_m v_z}{\partial z} \right)$$

$$(3.26)$$

By virtue of the continuity equation (Eq. 3.6a), the terms in parentheses in Equation (3.26) can be dropped. The integral of the time derivative is equal to the total derivative of a volume integral. The volume integral of the remaining momentum and stress terms can be transformed into surface integrals by means of the divergence theorem (Eq. 3.10). The result is the general impulse–momentum relationship.

x - component

$$\frac{d}{dt} \int_{\forall} \rho_m v_x \, d\forall + \int_A \rho_m v_x \left(v_x \frac{\partial x}{\partial n} + v_y \frac{\partial y}{\partial n} + v_z \frac{\partial z}{\partial n} \right) dA$$

$$= \int_{\forall} \rho_m g_x \, d\forall - \int_A p \frac{\partial x}{\partial n} \, dA + \int_A \left(\tau_{xx} \frac{\partial x}{\partial n} + \tau_{yx} \frac{\partial y}{\partial n} + \tau_{zx} \frac{\partial z}{\partial n} \right) dA \quad (3.27a)$$

y - component

$$\frac{d}{dt} \int_{\forall} \rho_m v_y \, d\forall + \int_A \rho_m v_y \left(v_x \frac{\partial x}{\partial n} + v_y \frac{\partial y}{\partial n} + v_z \frac{\partial z}{\partial n} \right) dA$$

$$= \int_{\forall} \rho_m g_y \, d\forall - \int_A p \frac{\partial y}{\partial n} \, dA + \int_A \left(\tau_{xy} \frac{\partial x}{\partial n} + \tau_{yy} \frac{\partial y}{\partial n} + \tau_{zy} \frac{\partial z}{\partial n} \right) dA \quad (3.27b)$$

z - component

$$\frac{d}{dt} \int_{\forall} \rho_m v_z \, d\forall + \int_A \rho_m v_z \left(v_x \frac{\partial x}{\partial n} + v_y \frac{\partial y}{\partial n} + v_z \frac{\partial z}{\partial n} \right) dA$$

$$= \int_{\forall} \rho_m g_z \, d\forall - \int_A p \frac{\partial z}{\partial n} \, dA + \int_A \left(\tau_{xz} \frac{\partial x}{\partial n} + \tau_{yz} \frac{\partial y}{\partial n} + \tau_{zz} \frac{\partial z}{\partial n} \right) dA \quad (3.27c)$$

It is observed that momentum is a vector quantity, the momentum change due to convection is embodied in the surface integral on the left-hand side of Equation (3.27), and all the stresses are expressed in terms of surface integrals. Example 3.7 provides a detailed application of the momentum equations to open-channel flows. To conclude this section, Example 3.8 introduces the concept of added mass.

Example 3.7 Momentum equations for open channels

With reference to the rectangular channel sketched on Figure E-3.7.1, the momentum relationship (Eq. 3.27a) in the downstream x direction is applied to an open channel, now subjected to rainfall at an angle θ_r and velocity V_r over the free surface area A_r, wind shear τ_w, upstream bank shear τ_s, and bed shear $\tau_b = \tau_o = \tau_{zx}$:

$$\frac{d}{dt}\int_\forall \rho_m v_x d\forall + \int_A \rho_m v_x \left(v_x \frac{\partial x}{\partial n} + v_y \frac{\partial y}{\partial n} + v_z \frac{\partial z}{\partial n} \right) dA$$

$$= \int_\forall \rho_m g_x d\forall - \int_A p \frac{\partial x}{\partial n} dA + \int_A \left(\tau_{xx} \frac{\partial x}{\partial n} + \tau_{yx} \frac{\partial y}{\partial n} + \tau_{zx} \frac{\partial z}{\partial n} \right) dA$$

Some integrals vanish for one-dimensional flow in impervious channels, $v_y = v_z = \tau_{xx} = 0$, except at the free surface where $v_z = -v_r \cos(\theta_r + \theta)$, leaving

$$\rho_m \Delta X \frac{dQ}{dt} + \int_A \rho_m v_x^2 \frac{\partial x}{\partial n} dA + \int_A \rho_m v_x v_z \frac{\partial z}{\partial n} dA$$

$$+ \int_A p \frac{\partial x}{\partial n} dA = \int_\forall \rho_m g_x d\forall + \int_A \tau_{zx} \frac{\partial z}{\partial n} dA + \int_A \tau_{yx} \frac{\partial y}{\partial n} dA$$

Consider an incompressible homogeneous fluid, constant ρ_m, and define the momentum correction factor β_m, also called the Boussinesq coefficient, given the cross-sectional averaged velocity V_x.

$$\beta_m = \frac{1}{A V_x^2} \int_A v_x^2 dA \qquad \text{(E-3.7.1)}\blacklozenge$$

For most practical applications, the value of β_m is generally close to unity, the reader can refer to Example 6.1 for a detailed calculation example. With average values of pressure p, velocity V, and area A at the upstream cross-section 1 and

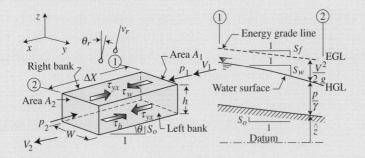

Figure E-3.7.1 Momentum equations in open channels

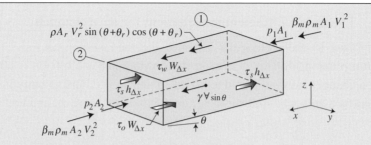

Figure E-3.7.2 Force balance in open channels

downstream cross-section 2. The integration of the momentum equation for this control volume $\forall$ of length ΔX, width W and height h yields:

$$\rho_m \Delta X \frac{\Delta Q}{\Delta t} + \beta_m \rho_m A_2 V_2^2 + p_2 A_2 - \beta_m \rho_m A_1 V_1^2 - p_1 A_1 - \rho A_r V_r^2 \sin(\theta + \theta_r)$$

$$\cos(\theta + \theta_r) = \gamma_m \forall \sin\theta - \tau_o W \Delta X - \tau_s 2h \Delta X + \tau_w W \Delta X \qquad \text{(E-3.7.2a)}$$

Notice here that on the right bank, $\tau_s = \tau_{yx}$ but $\partial y / \partial n = -1$, while on the left bank, $\tau_s = -\tau_{yx}$ but $\partial y / \partial n = +1$. The net result is that both shear forces are applied in the upstream (negative x) direction as shown in Figure E-3.7.2.

Assuming that the bed shear stress τ_0 equals the bank shear stress τ_s, the equation with negligible rainfall, $A_r \to 0$, without wind shear, $\tau_w \to 0$, can be rewritten when the channel inclination θ is small ($\sin\theta \cong$ the bed slope S_o) as

$$p_2 A_2 + \beta_m \rho_m A_2 V_2^2 + \tau_o (W + 2h) \Delta X$$

$$= p_1 A_1 + \beta_m \rho_m A_1 V_1^2 + \gamma_m \left(\frac{A_1 + A_2}{2} \right) \Delta X S_o \qquad \text{(E-3.7.2b)}$$

Further reduction of this equation is possible for uniform flow ($A = A_1 = A_2$), in which case, $p_1 A_1 = p_2 A_2$, $\beta_m \rho_m V_1^2 = \beta_m \rho_m V_2^2$ and the friction slope S_f equals both the water surface slope S_w and the bed slope S_o. The boundary shear stress τ_o is thus related to the friction slope S_f in the following manner

$$\tau_o = \gamma_m \frac{A}{(W + 2h)} S_f = \gamma_m \frac{A}{P} S_f = \gamma_m R_h S_f \qquad \text{(E-3.7.3)}\blacklozenge\blacklozenge$$

where the hydraulic radius $R_h = A/P$ is the ratio of the cross-sectional area $A = Wh$ to the wetted perimeter $P = W + 2h$, as shown in Figure E-3.7.3.

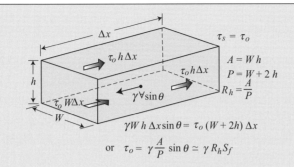

$$\gamma W h \Delta x \sin\theta = \tau_o (W + 2h) \Delta x$$

$$\text{or} \quad \tau_o = \gamma \frac{A}{P} \sin\theta \simeq \gamma R_h S_f$$

Figure E-3.7.3 Force balance for steady-uniform flow

Assuming that this shear stress relationship (Eq. E-3.7.3) is also applicable to gradually varied flow in wide channels, the Equation (E-3.7.2) can be rearranged for $V_r = 0$ and $\tau_w = 0$ as follows after considering $p_2 A_2 - p_1 A_1 = \rho_m g A \Delta h$ and $\beta_m \rho_m A_2 V_2^2 - \beta_m \rho_m A_1 V_1^2 = \rho_m \Delta \left(\beta_m Q^2 / A \right)$.

$$\rho_m \Delta X \frac{\Delta Q}{\Delta t} + \rho_m \Delta \left(\frac{\beta_m Q^2}{A} \right) + \rho_m g A \Delta h + \rho_m g A \Delta X S_f = \rho_m g A \Delta X S_o$$

(E-3.7.4a)

After dividing by $\rho_m \Delta X$, this reduces to

$$\frac{\Delta Q}{\Delta t} + \frac{\Delta}{\Delta X} \left(\frac{\beta_m Q^2}{A} \right) + gA \frac{\Delta h}{\Delta X} = gA \left(S_o - S_f \right) \qquad \text{(E-3.7.4b)}\blacklozenge$$

This equation is essentially describing force balance in gradually varied flow in a form equivalent to the Saint-Venant equations. For the particular case of steady-uniform flow, this reduces to $S_f = S_o$ and

$$\tau_{on} = \gamma_m R_h S_o \qquad \text{(E-3.7.5)}$$

where τ_{on} is the normal bed shear stress.

Resistance to flow

The Darcy–Weisbach friction factor is defined as $f = 8\tau_o / \rho_m V^2$, and from Equation (E-3.7.3) one obtains $\tau_o = (f/8)\rho_m V^2 = \rho_m g R_h S_f$, or

$$V = \sqrt{\frac{8g}{f}} R_h^{1/2} S_f^{1/2} \qquad \text{(E-3.7.6)}\blacklozenge\blacklozenge$$

For wide-rectangular channels, $W \gg h$ and $R_h = h$, this is equivalent to

$$S_f = \frac{f}{8}\frac{V^2}{gh} = \frac{f}{8}\mathrm{Fr}^2 = \frac{f}{8}\frac{q^2}{gh^3} \qquad \text{(E-3.7.7)}\blacklozenge\blacklozenge$$

where S_f is the friction slope and $\mathrm{Fr} = V/\sqrt{gh}$ is the Froude number. This Equation (E-3.7.7) can be combined with the unit discharge $q = Vh$ and solved for the normal flow depth h_n corresponding to $S_f = S_o = fq^2/8gh_n^3$, or

$$h_n = \left(fq^2/8gS_o\right)^{1/3} \qquad \text{(E-3.7.8a)}$$

In the case of gradually varied non-uniform flow, it is assumed that f, q, and g, remain constant, such that at any flow depth h, the corresponding friction slope S_f is given by $S_f = fq^2/8gh^3$, or

$$h_n = \left(fq^2/8gS_f\right)^{1/3} \qquad \text{(E-3.7.8b)}$$

From the ratio of h to h_n at constant unit discharge q and friction factor f, one obtains:

$$\frac{S_f}{S_o} = \left(\frac{h_n}{h}\right)^3 \qquad \text{(E-3.7.9)}\blacklozenge$$

The friction slope S_f in gradually varied flows with constant q and f can be approximated by Equation (E-3.7.9) which shows that $S_f < S_o$ when the flow depth exceeds the normal depth and increases very rapidly as $h < h_n$ as sketched in Figure E-3.7.4.

Likewise, a reasonable first approximation for bed shear stress τ_0 in gradually varied flows, with constant q and f, is compared with the bed shear stress at

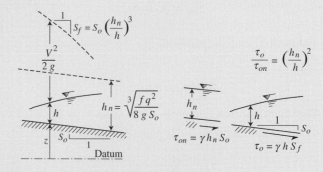

Figure E-3.7.4 Friction slope and shear stress relationships

normal flow depth τ_{0n}

$$\frac{\tau_0}{\tau_{0n}} \cong \frac{\gamma_m h S_f}{\gamma_m h_n S_o} = \frac{h}{h_n}\left(\frac{h_n}{h}\right)^3 = \left(\frac{h_n}{h}\right)^2 \qquad \text{(E-3.7.10)}$$

This shows that the bed shear stress increases ($\tau > \tau_n$) at flow depths less than normal depth ($h < h_n$).

The other important result from this analysis of gradually varied flow is that as long as the unit discharge q is constant, Equation (E-3.7.10) shows that the shear stress increases for converging flow ($\partial h/\partial x < 0$), and shear stress decreases for diverging flow ($\partial h/\partial x > 0$). One can thus expect sediment transport to increase in the downstream direction and cause degradation in converging flow, similarly sediment transport decreases and causes aggradation in diverging flows.

Example 3.8 Concept of added mass

One important property of the solid–fluid interaction is that the solid is not entirely free to move within the fluid. As sketched in Figure E-3.8.1, as the solid particle moves from position (a) to (b), an equal volume of fluid must move in the opposite direction. This implies that the motion of the particle is constrained by the motion of the same volume of fluid in the opposite direction. Therefore if the mass of the solid particle m_s is accelerated from (a) to (b), the mass of an equal volume of fluid m_f must also be accelerated. This concept is referred to as the added mass because the force F required to move the solid particle at an acceleration a is equal to $F = (m_s + m_f)a$. For instance, consider a submerged object of weight $F_w = \gamma_s \forall$ and $F_B = \gamma \forall$. The acceleration from rest is $a = g(G-1)/(G+1)$.

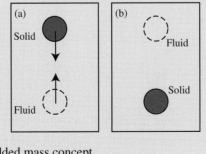

Figure E-3.8.1 Added mass concept

3.7 Power equation

Power is the rate of work done by fluid motion within a control volume $\forall$. It is obtained by integrating the product of the equations of force per unit volume $\rho_m a_x, \rho_m a_y, \rho_m a_z$ (from Eq. 3.12) and the velocity component v_x, v_y, v_z in the same direction. The Cartesian components in the three orthogonal directions are then added to give a single scalar equation of rate of work done, or power:

$$
\int_\forall \rho_m(a_x v_x + a_y v_y + a_z v_z)\, d\forall = \int_\forall \rho_m(v_x g_x + v_y g_y + v_z g_z)\, d\forall
$$

$$
+ \int_\forall \left[v_x \left(\frac{\partial \sigma_x}{\partial x} + \frac{\partial \tau_{yx}}{\partial y} + \frac{\partial \tau_{zx}}{\partial z} \right) + v_y \left(\frac{\partial \tau_{xy}}{\partial x} + \frac{\partial \sigma_y}{\partial y} + \frac{\partial \tau_{zy}}{\partial z} \right) \right.
$$

$$
\left. + v_z \left(\frac{\partial \tau_{xz}}{\partial x} + \frac{\partial \tau_{yz}}{\partial y} + \frac{\partial \sigma_z}{\partial z} \right) \right] d\forall \qquad (3.28)
$$

This is a lengthy derivation, but the vorticity components in the equations of motion (Eq. 3.11d–f) cancel out and the left-hand side of Equation 3.28 can be written solely as a function of the square of the velocity magnitude v^2. The elastic energy of the fluid per unit mass Ω_e is defined as follows:

$$
\rho_m \frac{d\Omega_e}{dt} = -p\left(\Theta_x + \Theta_y + \Theta_z\right) = -p\left(\frac{\partial v_x}{\partial x} + \frac{\partial v_y}{\partial y} + \frac{\partial v_z}{\partial z} \right) \qquad (3.29)
$$

This elastic energy Ω_e vanishes for incompressible fluids.

Using the divergence theorem, the equation of power (Eq. 3.28) is transformed to

$$
\frac{d}{dt} \int_\forall \rho_m \left(\frac{v^2}{2} + \Omega_g + \Omega_e \right) d\forall
$$

$$
+ \int_A \rho_m \left(\frac{v^2}{2} + \Omega_g + \Omega_e \right) \left(v_x \frac{\partial x}{\partial n} + v_y \frac{\partial y}{\partial n} + v_z \frac{\partial z}{\partial n} \right) dA
$$

$$
= \int_A \left[(v_x \sigma_x + v_y \tau_{xy} + v_z \tau_{xz}) \frac{\partial x}{\partial n} + (v_x \tau_{yx} + v_y \sigma_y + v_z \tau_{yz}) \frac{\partial y}{\partial n} \right.
$$

$$
\left. + (v_x \tau_{zx} + v_y \tau_{zy} + v_z \sigma_z) \frac{\partial z}{\partial n} \right] dA
$$

$$
- \int_\forall \left[\tau_{xx} \frac{\partial v_x}{\partial x} + \tau_{yy} \frac{\partial v_y}{\partial y} + \tau_{zz} \frac{\partial v_z}{\partial z} + \tau_{xy}(\oslash_z) + \tau_{xz}(\oslash_y) + \tau_{yz}(\oslash_x) \right] d\forall
$$

$$
\tag{3.30}\blacklozenge
$$

The rate of work done is a scalar quantity. Convective terms reduce to the net flux across the surfaces. The surface integral on the right-hand side represents the total rate of work done by the external stresses, which is conservative. The volume integral then indicates the rate at which mechanical energy gradually transforms into heat.

Example 3.9 applies the power equation to the rectangular open channel considered in Example 3.7. The analysis of backwater curves in Example 3.10 highlights the application of the energy equation to gradually varied flow.

Example 3.9 Rate of work done in open channels

Consider the application of the power equation (Eq. 3.30) to the open-channel flow illustrated in Example 3.7 (see Figure E-3.7.1). The first integral can be dropped for steady flow. Further simplification arises for incompressible fluids ($\Omega_e = 0$ from the continuity relationship) and one-dimensional flow ($v_y = v_z = 0$). If we assume that bank and transversal shear stresses are negligible, $\tau_{xx} = \tau_{yy} = \tau_{zz} = \tau_{zy} = \tau_{yz} = \tau_{yx} = \tau_{xy} = 0$, only the bed shear stress in the downstream direction is non-zero $\tau_{zx} = \tau_{xz} = \tau_o = \tau_b \neq 0$, and the energy equation reduces to

$$\int_A \rho_m \left(\frac{v^2}{2} + \Omega_g \right) v_x \frac{\partial x}{\partial n} dA = \int_A -v_x p \frac{\partial x}{\partial n} dA + \int_A v_x \tau_{zx} \frac{\partial z}{\partial n} dA - \int_\forall \tau_{xz} \frac{\partial v_x}{\partial z} d\forall$$

The last surface integral vanishes because v_x equals zero at the bed and τ_{zx} is zero at the free surface. After combining the first two surface integrals with $\Omega_g = g\hat{z}$, one obtains the integral form of the energy equation

$$\int_A \rho_m g \left(\frac{v^2}{2g} + \hat{z} + \frac{p}{\gamma_m} \right) v_x \frac{\partial x}{\partial n} dA = - \int_\forall \tau_{xz} \frac{\partial v_x}{\partial z} d\forall \qquad \text{(E-3.9.1)}$$

For one-dimensional flow, the energy correction factor α_e, also called the Coriolis coefficient, is defined as

$$\alpha_e = \frac{1}{V^3 A} \int_A v_x^3 \, dA \qquad \text{(E-3.9.2)} \blacklozenge$$

where V is the cross-sectional averaged velocity. The calculation example for α_e in Example 6.1 illustrates that numerical values of α_e remain close to unity. The volume integral of Equation (E-3.9.1) defines the head loss ΔH_L

$$\Delta H_L = \frac{1}{\gamma_m Q} \int_A \tau_{xz} \frac{\partial v_x}{\partial z} d\forall$$

Assuming a hydrostatic pressure distribution, the integral form of the energy equation (Eq. E-3.9.1) is rewritten as

$$\gamma_m Q \underbrace{\left(\frac{p_2}{\gamma_m} + \hat{z}_2 + \alpha_{e2} \frac{V_2^2}{2g} \right)}_{\hat{H}_2} - \gamma_m Q \underbrace{\left(\frac{p_1}{\gamma_m} + \hat{z}_1 + \alpha_{e1} \frac{V_1^2}{2g} \right)}_{\hat{H}_1} = -\gamma_m Q \Delta H_L$$

or

$$\Delta H_L = \tilde{H}_1 - \tilde{H}_2$$

in which the integral form of the Bernoulli sum $\tilde{H}$ resembles the differential formulation (Eq. 3.22d), except for the energy correction factor α_e. This integral form of the Bernoulli equation which includes head losses is appropriate for open-channel flows. The integral form of the specific energy $\tilde{E}$ follows

$$\frac{\Delta \tilde{H}}{\Delta x} = \frac{\Delta}{\Delta x}\left(\frac{p}{\gamma_m} + \hat{z} + \alpha_e \frac{V^2}{2g}\right) = -S_f$$

$$\frac{\Delta \tilde{E}}{\Delta x} = \frac{\Delta}{\Delta x}\left(\frac{p}{\gamma_m} + \alpha_e \frac{V^2}{2g}\right) = -\frac{\Delta \hat{z}}{\Delta x} - S_f = S_o - S_f$$

$$(E\text{-}3.9.3) \blacklozenge$$

in which $\tilde{E} = \frac{p}{\gamma_m} + \alpha_e \frac{V^2}{2g}$ is the integral form of the specific energy.

Notice the similarity between $\tilde{E}$ and E from Equation (3.23a), considering that α_e remains close to unity in most turbulent flows. As sketched in Figure E-3.9.1, the flow depth is added to the bed elevation to define the free-surface elevation, or hydraulic grade line HGL. The velocity head $\alpha_e V^2/2g$ is added to the HGL to define the energy grade line EGL. The slope of the EGL defines the friction slope S_f.

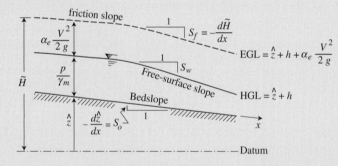

Figure E-3.9.1 Hydraulic and energy grade lines

Example 3.10 Backwater curves

Water surface elevation profiles commonly called backwater curves result from a direct application of the integral form of the energy equation. In the simplified case of steady one-dimensional flow, Equation (E-3.9.3) can be rewritten as

$$\frac{d\tilde{E}}{dx} = \frac{d\tilde{E}}{dh}\frac{dh}{dx} = S_o - S_f$$

From Equation (E-3.9.3), the derivative of $\tilde{E}$ with respect to h given that $p = \gamma_m h$ and $q = Vh$ is

$$\left(1 - \alpha_e \frac{q^2}{gh^3}\right) \frac{dh}{dx} = S_o - S_f$$

After substituting the Froude number (Eq. 3.24) and $\alpha_e \cong 1$, the relationship describing water surface elevation for steady one-dimensional flow of an incompressible sediment-laden fluid is

$$\frac{dh}{dx} = \frac{S_o - S_f}{1 - \mathrm{Fr}^2} \qquad \text{(E-3.10.1)} \blacklozenge\blacklozenge$$

Using the properties of critical flow depth h_c from Equation (3.24) and normal depth h_n from Equation (E-3.7.9), in wide-rectangular channels, $R_h = h$, the governing equation for steady flow, with constant q and f, becomes

$$\frac{dh}{dx} = \frac{S_o\left[1 - \left(\dfrac{h_n}{h}\right)^3\right]}{\left[1 - \left(\dfrac{h_c}{h}\right)^3\right]} \qquad \text{(E-3.10.2)}$$

$$\text{where } h_n = \left(\frac{fq^2}{8gS_o}\right)^{1/3} \text{ and } h_c = \left(\frac{q^2}{g}\right)^{1/3}$$

Notice that $dh/dx \to 0$ as the flow depth h approaches the normal depth h_n, as shown in Figure E-3.10.1. Also, $dh/dx \to \infty$ near critical depth as $h \to h_c$. The sign of dh/dx depends on the relative magnitude of $h, h_n,$ and h_c. Five types of backwater profiles are possible:

(1) H profiles for horizontal surfaces with $h_n \to \infty$
(2) M profiles for mild slopes when $h_n > h_c$
(3) C profiles for critical slopes when $h_n = h_c$
(4) S profiles for steep slopes when $h_n < h_c$
(5) A profiles for adverse slopes when $S_o < 0$

Typical water surface profiles for mild and steep slopes are shown in Figure E-3.10.1.

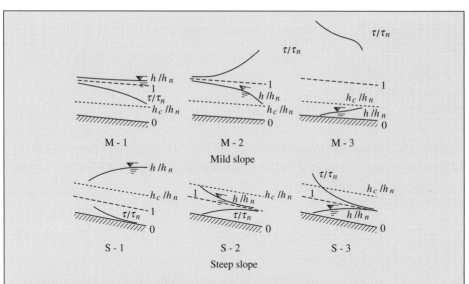

Figure E-3.10.1 Mild and steep slope backwater profiles

Numerical calculations using the direct step method can be initiated from a given flow depth h_1, and the distance increment Δx at which $h_2 = h_1 \pm \Delta h$ is approximated by

$$\Delta x \cong \frac{\Delta h \left[1 - \left(\dfrac{h_c}{h} \right)^3 \right]}{S_o \left[1 - \left(\dfrac{h_n}{h} \right)^3 \right]}$$

(E-3.10.3)

Alternatively, the standard step approach can be used whereby the flow depth increment Δh can be calculated at a fixed downstream distance increment Δx

$$\Delta h \cong \Delta x S_0 \left[1 - \left(\frac{h_n}{h} \right)^3 \right] \bigg/ \left[1 - \left(\frac{h_c}{h} \right)^3 \right]$$

(E-3.10.4)

Bed shear stress distributions for one-dimensional mild M and steep S back-water curves are also sketched in Figure E-3.10.1. The analysis based on Equation (E-3.7.10) shows that the shear stress increases in the downstream direction for converging flows (M-2 and S-2 backwater curves), and decreases for diverging flows (M-1, M-3, S-1, and S-3 backwater curves).

Exercises

3.1 With reference to Figure 3.2, determine which type of deformation is obtained when $v_x = 2y$ and $v_y = v_z = 0$.

Answer: Translation along x only; no linear deformation; angular deformation $\oslash_z = 2$; rotation, $\otimes_z = -2$.

♦♦3.2 Demonstrate that the equations of motion (Eqs. 3.11d and 3.11a) are identical. Also reduce Equation (3.5a) to Equation (3.1a).

♦♦3.3 Derive the x component of the equation of motion in Cartesian coordinates (Eq. 3.14a) from the Equations (3.11–3.13).

3.4 Derive the x component of the Bernoulli equation (Eq. 3.21a) from Equations (3.11d), (3.14a), and the gravitation potential.

3.5 Derive the x component of the momentum equation (Eq. 3.27a) from Equation (3.25).

♦3.6 Demonstrate, from the specific energy function E, that $q^2 = gh_c^3$ and $E_{min} = 3h_c/2$ for steady one-dimensional open-channel flow.

Problems

♦Problem 3.1

With reference to Example 3.1, a container 1 m × 1 m at the base and 10 m high is 90% filled with water. Sediment is added at 1 kg/minute: (a) what is the rate of change in volumetric concentration; and (b) when the container is filled and the sediment is well mixed, what is the sediment concentration of the overflow?

♦Problem 3.2

Assuming constant depth, what is the cross-sectional average flow velocity $\bar{V}$ of the main channel as a function of V on the labyrinth weir shown.

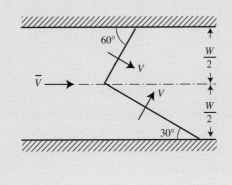

Figure P-3.2

♦Problem 3.3

Calculate the magnitude and direction of the buoyancy force applied on a neutrally buoyant sphere $(G = 1)$ submerged under steady one-dimensional flow $(v_y = v_z = 0)$ on a steep slope. Assume that the particle moves with the surrounding fluid of density ρ_m such that no shear stress is exerted at the edge of the sphere. Compare the results with Example 3.4. (*Hint*: integrate the Euler equations around the sphere from Equation (3.18) with $a_z = a_y = 0$, and $g_x = g\sin 30°$, $g_y = 0$ and $g_z = -g\cos 30°$).

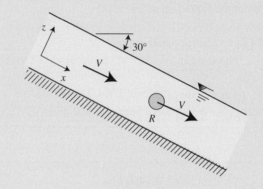

Figure P-3.3

Answer: $F_B = \gamma_m \forall \cos 30°$ in direction z, which is less than F_B in Example 3.4. Notice that $\partial p/\partial x = 0$, and the sphere will accelerate in the x direction at $a_x = g_x = g\sin 30° = 4.9$ m/s². On a spillway, the particle would accelerate until friction forces become important and $(1/\rho_m)(\partial\tau_{zx}/\partial z) = -g_x$. In all cases, the x component of the buoyancy force vanishes, and in contrast with Example 3.5, acceleration is due to the body force per unit mass.

♦♦Problem 3.4

Repeat the calculations of Example 3.5 when the sphere and the surrounding fluid are accelerated at $a_z = 3g$ upward. Finally, what is the buoyancy force when the fluid is in free fall?

♦Problem 3.5

With reference to Example 3.3 determine the rate of bed aggradation when the volumetric flux of settling sediment is $Q_{\text{settling}} = \omega W \Delta X C_{v \text{ susp}}$, where ω is the settling velocity and $C_{v \text{ susp}}$ is the volumetric concentration of sediment in suspension.

♦♦Problem 3.6

With reference to Example 3.3, link the two components (a) and (b) and define a sediment continuity relationship when the fraction T_E of sediment is in the deposit and the fraction $(1 - T_E)$ remains in suspension.

Problem 3.7

Redo the analysis of momentum equations in Equation (E-3.7.2) for steady uniform flow in a 1V:2H trapezoidal channel. Determine the relationship for shear stress as a function of the other variables.

♦♦Problem 3.8

Consider a half-cylinder of radius R and length L under hydrostatic pressure condition. If the pressure on the flat base is p_o, determine and plot the pressure distribution around the surface of the object. Integrate the pressure distribution and find the vertical buoyancy force. Also determine the horizontal buoyancy force F_{bx}.

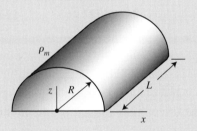

Figure P-3.8 Half-cylinder

♦Problem 3.9

Consider a quarter-sphere of radius R under hydrostatic pressure. Determine the pressure distribution around the surface if the pressure on the flat base is p_o. Integrate the pressure distribution to determine the buoyancy force components in the x and z directions.

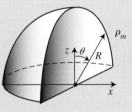

Figure P-3.9 Quarter-sphere

♦Problem 3.10

A plastic barrel contains 1,000 kg of water. It is neutrally buoyant, sealed, and submerged below the free surface. If you tie the container to a boat, starting from rest ($V = 0$) what force is required in the cable to accelerate the container at $a = g/4$? (*Hint*: neglect friction on the cable but consider added mass.)

♦♦Problem 3.11

A 30 lb fish is pulled vertically from rest ($v = 0$). Calculate the tension in the line when the fish is in the water. Consider the upward accelerations 0, $g/2$, g, $3g/2$, and $2g$. Repeat the calculations when you start from rest in the air. (*Hint*: consider the weight and added mass.)

	$a = 0$	$A = g/2$	$a = g$	$a = 3g/2$	$a = 2g$
air		45 lb	60 lb		
water			60 lb	90 lb	

♦Problem 3.12

Consider an 8 m³ cubic container filled with water and placed on a scale. A 500 kg solid copper sphere is held with a thin cable without mass and lowered in the fluid. The sphere is submerged at the center of the full container and the excess water spills over and off the scale. Determine for each of the following three conditions: (a) the force in the thread; (b) the hydrodynamic force on the sphere; and (c) the weight measured on the scale. Three conditions are examined: (1) the sphere is held stationary by the thread; (2) at the instant when the thread is cut and the sphere has no velocity; and (3) assume that the sphere settles at a constant fall velocity. Write the results in a table using rows for 1,2,3 and columns for a,b,c. Neglect the mass of the thread and the container. (*Hints:* consider three free-body diagrams: the sphere, the fluid only, and the container on the scale; and also consider in (2) that if the sphere moves relative to the fluid, an equivalent volume of fluid must move in the opposite direction.)

♦♦Problem 3.13

A spherical ball containing 0.1m³ of air at a density of 2 kg/m³ is held submerged at the bottom of a pool, 5 m below the water surface. What is the force required to hold it in place? Also, determine its acceleration when it is released from rest. (*Hint*: notice that without added mass, this ball would reach an acceleration of about $500g$!)

♦♦*Computer problem 3.1*

Consider steady flow ($q = 3.72$ m²/s) in the impervious rigid boundary channel. Assume a very wide channel and $f = 0.03$. Determine the distribution of the following parameters along the 25 km channel reach when the water surface elevation at the dam is 10 m above the bed elevation: (a) flow depth in m; (b) mean flow velocity in m/s; and (c) bed shear stress in N/m².

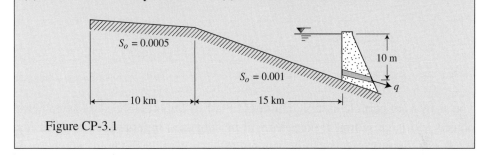

Figure CP-3.1

4

Particle motion in inviscid fluids

The analysis of particle motion in inviscid fluids is important because asymmetric objects and large sediment particles will be subjected to large lift forces. In real fluids, the viscous effects can be ignored at large particle Reynolds numbers. At every point on the surface of a submerged particle, the fluid exerts a force per unit area or stress. In Chapter 3, the stress vector was subdivided into a pressure component acting in the direction normal to the surface and two orthogonal shear stress components acting in the plane tangent to the surface. In this chapter, the stress vector for inviscid fluids is always normal to the surface, which means that there is only a pressure component and no shear stress.

The flow of inviscid fluids around submerged particles may be due either to the movement of the fluid, the movement of the particle, or a combination of both. The following discussion considers flow conditions made steady by application of the principle of relative motion of the fluid around a stationary particle. For steady flow of incompressible and inviscid fluids, the Bernoulli equation is applicable as long as the flow is irrotational. The approach in this chapter is first to define the flow field for irrotational flow. The Bernoulli equation applies throughout the flow field and then defines the pressure from the velocity. The lift force can then be calculated after integrating the pressure distribution around the solid surface. The analysis focuses on simple particle shapes, such as cylinders (Section 4.1) and spheres (Section 4.2), to provide basic understanding of the fundamental concept of lift force. Five examples provide applications on half-cylinders and half-spheres.

Irrotational flow properties

Irrotational flow is obtained when the rotational terms defined in Equation (3.4) equal zero. Steady two-dimensional flow is described mathematically by potential functions Φ and stream functions Ψ. A flow potential exists such that streamlines Ψ are always perpendicular to potential lines Φ and form orthogonal flow nets,

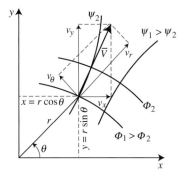

Figure 4.1. Flow net

as shown in Figure 4.1. The velocity components v_x and v_y are defined from the gradient of the velocity potential Φ and stream function Ψ as:

$$v_x = -\frac{\partial \Phi}{\partial x} = -\frac{\partial \Psi}{\partial xy} \tag{4.1a}$$

$$v_y = -\frac{\partial \Phi}{\partial y} = \frac{\partial \Psi}{\partial x} \tag{4.1b}$$

The continuity relationship for incompressible fluids $(\partial v_x/\partial x + \partial v_y/\partial y = 0)$ satisfies the condition that Ψ gives an exact differential expression and Φ obeys the Laplace equation $(\nabla^2 \Phi = \partial^2 \Phi/\partial x^2 + \partial^2 \Phi/\partial y^2 = 0)$. The stream function is therefore a direct consequence of the continuity equation and is applicable to both incompressible rotational and irrotational flows. The potential function Φ gives an exact differential expression only when the vorticity is zero $(\partial v_x/\partial y - \partial v_y/\partial x = 0)$. This potential function is thus a direct consequence of irrotational flow and it also defines the Laplace equation $(\otimes_z \equiv \nabla^2 \Psi = 0)$.

In cylindrical coordinates, the velocity components v_r in the radial direction and v_θ in the tangential direction (v_θ is positive in the direction of increasing θ) of a system are:

$$v_r = -\frac{\partial \Phi}{\partial r} = -\frac{1}{r}\frac{\partial \Psi}{\partial \theta} \tag{4.2a}$$

$$v_\theta = -\frac{1}{r}\frac{\partial \Phi}{\partial \theta} = \frac{\partial \Psi}{\partial r} \tag{4.2b}$$

Equations satisfying the Laplace equation are called harmonic. Equipotential lines (constant Φ) are orthogonal to streamlines (constant Ψ). The sum of several harmonic functions also satisfies the Laplace equation. This is a convenient property because harmonic functions are linear and can be superposed. The velocity components can thus be defined from the gradient of the sum of the harmonic functions.

4.1 Irrotational flow around a circular cylinder

This section describes the combination of fundamental two-dimensional flow nets which describe flow configuration around a circular cylinder. The flow field around a cylinder is defined in Section 4.1.1 and Section 4.1.2 focuses on the calculation of lift and drag forces.

Rectilinear flow (Fig. 4.2a)

Uniform flow velocity $v_x = -\partial\Phi/\partial x = u_\infty$ along the x direction, $v_y = -\partial\Phi/\partial y = 0$ and allows the definition of the flow potential from

$$\Phi = \int d\Phi = \int \frac{\partial\Phi}{\partial x} dx + \int \frac{\partial\Phi}{\partial y} dy = -\int v_x dx - \int v_y dy$$

or:

$$\Phi_\ell = -u_\infty x = -u_\infty r\cos\theta \tag{4.3a}$$

Similarly, the stream function is obtained from Equation (4.1) and

$$\Psi = \int d\Psi = \int \frac{\partial\Psi}{\partial x} dx + \int \frac{\partial\Psi}{\partial y} dy = +\int v_y dx - \int v_x dy$$

or:

$$\Psi_\ell = -u_\infty y = -u_\infty r\sin\theta \tag{4.3b}$$

Source (Fig. 4.2b)

The strength of a line source is equal to the volumetric flow rate $q = 2\pi r v_r$, with $v_\theta = 0$. From Equation (4.2), one obtains:

$$\Phi_{so} = -\frac{q}{4\pi}\ln(x^2 + y^2) = -\frac{q}{2\pi}\ln r \tag{4.4a}$$

$$\Psi_{so} = -\frac{q}{2\pi}\tan^{-1}\frac{y}{x} = -\frac{q\theta}{2\pi} \tag{4.4b}$$

Sink (Fig. 4.2c)

A sink is a negative source, or:

$$\Phi_{si} = \frac{q}{4\pi}\ln(x^2 + y^2) = \frac{q}{2\pi}\ln r \tag{4.5a}$$

$$\Psi_{si} = \frac{q}{2\pi}\tan^{-1}\frac{y}{x} = \frac{q\theta}{2\pi} \tag{4.5b}$$

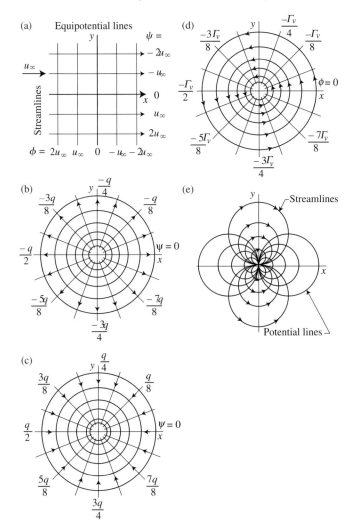

Figure 4.2. Two-dimensional flow net: a) rectilinear b) source c) sink d) vortex
e) doublet

Free vortex (Fig. 4.2d)

The positive counterclockwise free vortex of strength $\Gamma_v = 2\pi r v_\theta$, with $v_r = 0$,
and center at the origin is

$$\Phi_v = -\frac{\Gamma_v}{2\pi}\tan^{-1}\frac{y}{x} = -\frac{\Gamma_v \theta}{2\pi} \qquad (4.6a)$$

$$\Psi_v = \frac{\Gamma_v}{4\pi}\ln(x^2 + y^2) = \frac{\Gamma_v}{2\pi}\ln r \qquad (4.6b)$$

Dipole (Fig. 4.2e)

A dipole, or doublet, is obtained when a source and a sink of equal strength are brought together in such a way that the product of their strength and the distance separating them remains constant. The flow net given here without derivation is:

$$\Phi_d = \frac{\Gamma_d}{2\pi}\left(\frac{x}{x^2+y^2}\right) = \frac{\Gamma_d\cos\theta}{2\pi r} \tag{4.7a}$$

$$\Psi_d = -\frac{\Gamma_d}{2\pi}\left(\frac{y}{x^2+y^2}\right) = -\frac{\Gamma_d\sin\theta}{2\pi r} \tag{4.7b}$$

These fundamental flow nets are combined in Section 4.1.1 to define the flow field around a two-dimensional circular cylinder, like the near surface flow field around a vertical bridge pier.

4.1.1 Flow field around a circular cylinder

The flow net with circulation past a circular cylinder of radius R is obtained by combining a rectilinear flow with a doublet of constant strength $\Gamma_d = -2\pi u_\infty R^2$ and a counterclockwise free vortex of variable strength Γ_v. The streamlines for flow around a cylinder without circulation ($\Gamma_v = 0$) are shown on Figure 4.3a, the

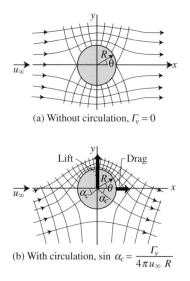

(a) Without circulation, $\Gamma_v = 0$

(b) With circulation, $\sin\alpha_c = \dfrac{\Gamma_v}{4\pi u_\infty R}$

Figure 4.3. Flow net around a cylinder a) without circulation, $\Gamma_v = 0$ b) with circulation, $\sin\dfrac{\Gamma_v}{4\pi u_\infty R}$

effect of clockwise circulation ($\Gamma_v < 0$) is shown on Figure 4.3b.

$$\Phi_{cyl} = -u_\infty \left(x + \frac{R^2 x}{x^2 + y^2} \right) - \frac{\Gamma_v}{2\pi} \tan^{-1} \frac{y}{x} = -u_\infty \left(r + \frac{R^2}{r} \right) \cos\theta - \frac{\Gamma_v \theta}{2\pi} \quad (4.8a)$$

$$\Psi_{cyl} = -u_\infty \left(y - \frac{R^2 y}{x^2 + y^2} \right) + \frac{\Gamma_v}{4\pi} \ln(x^2 + y^2) = -u_\infty \left(r - \frac{R^2}{r} \right) \sin\theta + \frac{\Gamma_v}{2\pi} \ln r$$

$$(4.8b)$$

The velocity components around the cylinder are obtained after applying the operator in Equation (4.2) to the flow net from Equation (4.8).

$$v_r = u_\infty \left(1 - \frac{R^2}{r^2} \right) \cos\theta \quad (4.9a)$$

$$v_\theta = -u_\infty \left(1 + \frac{R^2}{r^2} \right) \sin\theta + \frac{\Gamma_v}{2\pi r} \quad (4.9b)$$

where $x = r\cos\theta$, $y = r\sin\theta$, and v_θ is positive counterclockwise. At the cylinder surface, the radial velocity components $v_r = 0$, and v_θ vanish only at the stagnation points.

4.1.2 Lift and drag on a circular cylinder

The velocity at any point along the surface of a vertical cylinder results from Equation (4.9b) at $r = R$. The pressure at any point in a horizontal plane is then calculated from the velocity and the Bernoulli equation (Eq. 3.22a) given the reference pressure $p = p_\infty$ where $v = u_\infty$, the relative pressure $p_r = p - p_\infty$ is

$$p_r = p - p_\infty = \frac{\rho_m}{2} \left[u_\infty^2 - \left(-2u_\infty \sin\theta + \frac{\Gamma_v}{2\pi R} \right)^2 \right] \quad (4.10)$$

With reference to Figure 4.4, the drag force F_D per unit length is defined as the net fluid force in the direction parallel to the horizontal approach velocity u_∞ (toward the viewer), while the lift force F_L per unit length L is the net force in the normal direction (positive when $\theta = \pi/2$).

$$F_L = \int_0^{2\pi} -p_r \sin\theta L R \, d\theta$$

$$= \frac{\rho_m L R}{2} \int_0^{2\pi} -\left[u_\infty^2 - \left(-2u_\infty \sin\theta + \frac{\Gamma_v}{2\pi R} \right)^2 \right] \sin\theta \, d\theta = -\rho_m L u_\infty \Gamma_v$$

$$(4.11a)$$

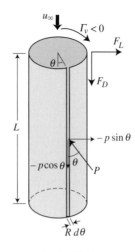

Figure 4.4. Lift and drag on a vertical cylinder

$$F_D = \int_0^{2\pi} -p_r \cos\theta LR\,d\theta$$

$$= \frac{\rho_m LR}{2} \int_0^{2\pi} -\left[u_\infty^2 - \left(-2u_\infty \sin\theta + \frac{\Gamma_v}{2\pi R}\right)^2\right]\cos\theta\,d\theta = 0 \qquad (4.11b)$$

It is concluded from this potential flow analysis for an inviscid fluid around a circular cylinder that circulation is necessary for the occurrence of lift forces. Drag forces cannot be generated in irrotational flows. Example 4.1 provides detailed calculations of lift and drag forces on a half-cylindrical surface. Without circulation, lift is caused by the symmetry of the object. Example 4.2 combines lift and buoyancy forces on a half-cylinder.

Example 4.1 Lift and drag forces on a vertical half-cylinder

Neglect end forces and calculate the lift force per unit mass on the outer surface of a long vertical half-cylinder of radius R, assuming potential flow without circulation ($\Gamma_v = 0$) (Figure E-4.1.1).

Step 1. The lift force $F_{L\frac{1}{2}}$ on the half-cylinder is $F_{L\frac{1}{2}} = \int_0^{\pi}(-p_r \sin\theta)LR\,d\theta$.

Substituting the pressure from Equation (4.10) without circulation ($\Gamma_v = 0$)

$$F_{L\frac{1}{2}} = \int_0^{\pi}\left(-\frac{\rho_m LR}{2}\right)(u_\infty^2 - 4u_\infty^2 \sin^2\theta)\sin\theta\,d\theta \qquad (\text{E-4.1.1})$$

$$F_{L\frac{1}{2}} = \frac{5}{3}\rho_m LR u_\infty^2$$

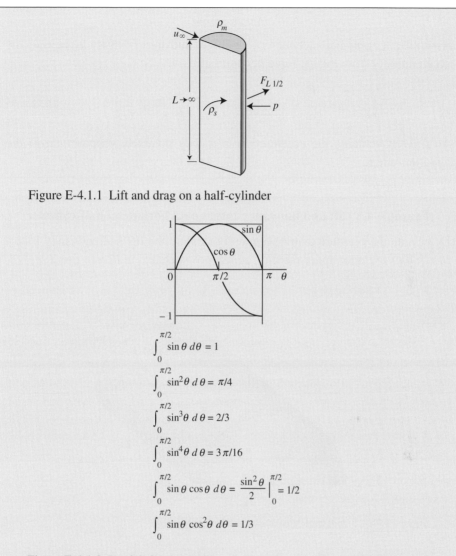

Figure E-4.1.1 Lift and drag on a half-cylinder

$$\int_0^{\pi/2} \sin\theta \, d\theta = 1$$

$$\int_0^{\pi/2} \sin^2\theta \, d\theta = \pi/4$$

$$\int_0^{\pi/2} \sin^3\theta \, d\theta = 2/3$$

$$\int_0^{\pi/2} \sin^4\theta \, d\theta = 3\pi/16$$

$$\int_0^{\pi/2} \sin\theta \cos\theta \, d\theta = \left.\frac{\sin^2\theta}{2}\right|_0^{\pi/2} = 1/2$$

$$\int_0^{\pi/2} \sin\theta \cos^2\theta \, d\theta = 1/3$$

Figure E-4.1.2 Useful sine-integrals

Some useful integrals are shown in Figure E-4.1.2 and Appendix B. Note that the velocity along the flat face is $v_b = u_\infty$. The relative pressure on the flat face is zero and does not yield any additional force in the lateral direction.

Step 2. The drag force $F_{D\frac{1}{2}}$ on the half-cylinder is

$$F_{D\frac{1}{2}} = \int_0^\pi (-p_r \cos\theta) LR \, d\theta = 0 \qquad \text{(E-4.1.2)}$$

Acceleration

Assuming a solid half-cylinder of radius R and mass density $\rho_s = G\rho$, the acceleration a from the lift force per solid unit mass is:

$$a = \frac{F_{L\frac{1}{2}}}{\text{mass}} = \frac{F_{L\frac{1}{2}}}{\rho_s \forall} = \frac{5}{3}\frac{2}{\pi}\frac{\rho_m L R u_\infty^2}{\rho_s \pi R^2 L} \simeq \frac{u_\infty^2}{GR} \qquad \text{(E-4.1.3)}$$

At a given velocity, the acceleration a is thus inversely proportional to the cylinder radius.

Example 4.2 Lift and buoyancy forces on a horizontal half-cylinder

Determine the pressure distribution and calculate the lift and buoyancy forces on the half-cylinder under the flow conditions sketched in Figure E-4.2.1.

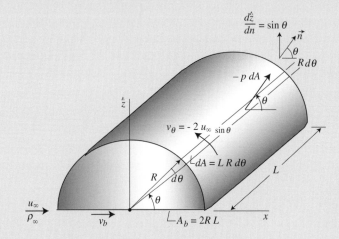

Figure E-4.2.1 Lift and buoyancy on a half-cylinder

Step 1. The flow velocity field is v_b at the base and $v = v_\theta = -2u_\infty \sin\theta$ around the curved surface.

Step 2. The Bernoulli equation can be applied between the distant point and any point along the curved surface.

$$\frac{p}{\gamma_m} + \hat{z} + \frac{v^2}{2g} = \frac{p_\infty}{\gamma_m} + \frac{u_\infty^2}{2g}$$

or

$$p = p_\infty - \gamma_m R \sin\theta + \rho_m \frac{u_\infty^2}{2} - \rho_m \frac{v^2}{2}$$

At the base, the pressure p_b is also obtained from the Bernoulli equation

$$p_b = p_\infty + \rho_m \frac{u_\infty^2}{2} - \rho_m \frac{v_b^2}{2}$$

Step 3. The vertical force on the half-cylinder is obtained from the sum of the components on the curved surface and at the base

$$F_z = \int_{A \text{ curved}} -p \sin\theta \, dA + p_b A_b$$

$$= -\int_0^\pi \left[p_\infty - \gamma_m R \sin\theta + \rho_m \frac{u_\infty^2}{2} - \rho_m \frac{v^2}{2} \right] RL \sin\theta \, d\theta + p_b A_b$$

$$= -RL \left[\int_o^\pi p_\infty \sin\theta \, d\theta - \gamma_m R \int_o^\pi \sin^2\theta \, d\theta + \rho_m \frac{u_\infty^2}{2} \int_o^\pi \sin\theta \, d\theta \right.$$

$$\left. - \rho_m \frac{4u_\infty^2}{2} \int_o^\pi \sin^3\theta \, d\theta \right] + p_b A_b \qquad \text{(E-4.2.1)}$$

For convenience some integrals in Figure E-4.1.2 can be used.

$$F_z = -RL p_\infty 2 + \gamma_m R^2 L \frac{2\pi}{4} - \frac{\rho_m u_\infty^2}{2} RL \, 2 + 2\rho_m u_\infty^2 \frac{4}{3} RL$$

$$+ 2RL \left(p_\infty + \frac{\rho_m u_\infty^2}{2} - \frac{\rho_m v_b^2}{2} \right)$$

$$= \underbrace{\gamma_m \frac{\pi R^2}{2} L}_{\substack{\text{Buoyancy} \\ \text{force}}} + \underbrace{\frac{5}{3}\rho_m u_\infty^2 RL}_{\substack{\text{Lift force on} \\ \text{curved surface}}} + \underbrace{\rho_m RL \left(u_\infty^2 - v_b^2 \right)}_{\substack{\text{Force at} \\ \text{the base}}} \qquad \text{(E-4.2.2)}$$

when $v_b = 0$, the net lift force $F_L = \frac{8}{3}\rho_m u_\infty^2 RL$, and when $v_b = u_\infty$, the net lift force $F_L = \frac{5}{3}\rho_m u_\infty^2 RL$.

This example shows that the integral of a constant, like p_∞, around a closed surface is zero. This explains why the use of the reference pressure in Section 4.1.2 and Example 4.1 are sufficient to calculate the lift and drag forces. The elevation term $\hat{z}$ of the Bernoulli equation in Step 2 resulted in the buoyancy force as determined in Chapter 3. It is the velocity head term of the Bernoulli equation that causes the lift force.

4.2 Irrotational flow around a sphere

This section describes fundamental three-dimensional flow patterns which can be combined to define the flow configuration around three-dimensional objects like a sphere. The flow field around a sphere is discussed in Section 4.2.1 and the lift and drag forces are presented in Section 4.2.2. The discussion pertains to axisymmetrical patterns for which $v_v = 0$. Flow patterns in three dimensions are more complex and most flow potential functions of this section are given without demonstration.

Three-dimensional rectilinear flow

Uniform flow velocity u_∞ along $x = r \cos\theta$ in spherical coordinates (r, θ, φ)

$$\Phi_{\ell 3} = -r u_\infty \cos\theta \tag{4.12a}$$

$$\Psi_{\ell 3} = -\frac{r^2 u_\infty}{2}\sin^2\theta \tag{4.12b}$$

Three-dimensional source and sink

The strength of a point source is equal to the volumetric flow rate $Q = 4\pi r^2 v_r$. The flow net is given by

$$\Phi_{s3} = \frac{Q}{4\pi r} \tag{4.13a}$$

$$\Psi_{s3} = \frac{Q}{4\pi}\cos\theta \tag{4.13b}$$

A three-dimensional sink is a negative source for which the flow net is obtained after considering a negative discharge $-Q$ in Equations (4.13a) and (4.13b).

Three-dimensional dipole

As in the previous two-dimensional case, the dipole is obtained when a source and sink of equal strength are brought together in such a way that the product of their strength and the distance separating them remains constant. The resulting flow net is described by

$$\Phi_{d3} = -\frac{\Gamma_{d3}\cos\theta}{r^2} \tag{4.14a}$$

$$\Psi_{d3} = \frac{\Gamma_{d3}\sin^2\theta}{r} \tag{4.14b}$$

4.2.1 Flow field around a sphere

For steady potential flow around a sphere of radius R in relative motion u_∞, the potential function Φ and the stream function Ψ in spherical coordinates, at a point (r,θ), are given from inserting a dipole of strength $\Gamma_{d3} = u_\infty R^3/2$ in three-dimensional rectilinear flow. The resulting flow net from Equations (4.12a,b) and (4.14a,b) is:

$$\Phi = \frac{-u_\infty R^3}{2r^2}\cos\theta - u_\infty r\cos\theta \qquad (4.15a)$$

$$\Psi = \frac{u_\infty R^3}{2r}\sin^2\theta - u_\infty\frac{r^2}{2}\sin^2\theta \qquad (4.15b)$$

In three-dimensions, the velocity components v_r and $v_\theta (v_\nu = 0)$, in spherical coordinates are given by the stream function Ψ and the potential function Φ from

$$v_r = \frac{-1}{r^2\sin\theta}\frac{\partial\Psi}{\partial\theta} = -\frac{\partial\Phi}{\partial r} \qquad (4.16a)$$

$$v_\theta = \frac{1}{r\sin\theta}\frac{\partial\Psi}{\partial r} = -\frac{1}{r}\frac{\partial\Phi}{\partial\theta} \qquad (4.16b)$$

The velocity field is described by

$$v_r = u_\infty\cos\theta - u_\infty\frac{R^3}{r^3}\cos\theta \qquad (4.17a)$$

$$v_\theta = -u_\infty\sin\theta - \frac{u_\infty}{2}\frac{R^3}{r^3}\sin\theta \qquad (4.17b)$$

The velocity v at any point on the surface of the sphere $(r = R)$ is obtained from $v = v_\theta$ because $v_r = 0$.

$$v = -\frac{3}{2}u_\infty\sin\theta \qquad (4.18)\blacklozenge$$

4.2.2 Lift and drag forces on a sphere

The pressure distribution over the surface of the sphere is calculated from this velocity distribution, knowing that the Bernoulli equation is applicable. Integrating the hydrostatic pressure distribution results in the buoyancy force obtained in Example 3.4. The integral of a constant pressure p_∞ around a closed surface is zero, which is equivalent to assuming that the reference pressure $p_\infty = 0$. The hydrodynamic pressure distribution associated with the velocity distribution around the

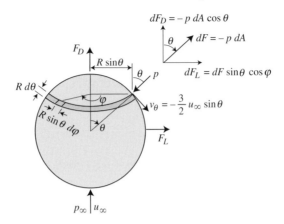

Figure 4.5. Lift and drag on a sphere

sphere is given by:

$$p_r = p - p_\infty = \frac{\rho_m u_\infty^2}{2}\left[1 - \frac{9}{4}\sin^2\theta\right] \tag{4.19}$$

With reference to Figure 4.5, the lift force F_L is calculated from the integration of relative pressure in the direction perpendicular to the main flow direction, $-p_r \sin\theta\cos\varphi$, multiplied by the elementary surface area $Rd\theta R\sin\theta\,d\varphi$ and integrated over the entire sphere, $0 < \varphi < 2\pi$, and $0 < \theta < \pi$.

$$F_L = \int_0^\pi \int_0^{2\pi} -p_r \sin\theta\cos\varphi R^2\sin\theta\,d\varphi\,d\theta = 0 \tag{4.20a}$$

The drag force F_D is calculated from the integral of the relative pressure component in the main flow direction $-p\cos\theta$,

$$F_D = \int_0^\pi \int_0^{2\pi} -p_r \cos\theta R^2\sin\theta\,d\varphi\,d\theta = 0 \tag{4.20b}$$

Irrotational flow around a sphere generates neither lift nor drag forces. Irrotational flow around asymmetrical surfaces, however, generates lift and drag forces. An instructive application for a half-sphere is presented in Example 4.3. The analysis of lift forces leads to particle equilibrium in Example 4.4, and a comparison of lift coefficients in Example 4.5.

Example 4.3 Lift and drag forces on a half-sphere

Calculate the drag force and the lift force from the pressure distribution of irrotational flow without circulation around the curved surface of a half-sphere (Figure E-4.3.1). Consider that the flow velocity at the base of the half-sphere is $v_b = u_\infty$.

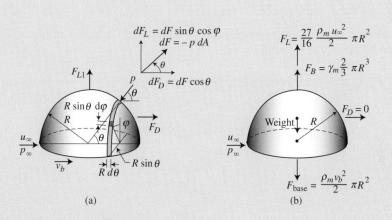

Figure E-4.3.1 Lift and drag on a half-sphere

Drag force

The relative pressure component $-p_r \cos\theta$ in the flow direction is multiplied by the surface area $R d\theta R \sin\theta d\varphi$ and integrated over the entire surface of the half-sphere $-\pi/2 < \varphi < \pi/2$ and $0 < \theta < \pi$.

$$F_D = \int_A -p_r \cos\theta\, dA$$

$$F_D = -\int_{-\pi/2}^{\pi/2} \int_0^\pi \frac{\rho_m u_\infty^2}{2}\left(1 - \frac{9}{4}\sin^2\theta\right)\cos\theta R d\theta R \sin\theta\, d\varphi$$

$$F_D = \pi R^2 \rho_m u_\infty^2 [0] = 0 \qquad\qquad\text{(E-4.3.1)}$$

Lift force

The lift force on a whole sphere is zero because of the symmetry. It is however instructive to calculate the lift force on a half sphere (Figure E-4.3.1). Using Bernoulli, the pressure at the base is $p_b = p_\infty + \frac{\rho_m}{2}\left(u_\infty^2 - v_b^2\right)$.

The pressure on the curved surface of elementary area $dA = R^2 \sin\theta \, d\theta \, d\varphi$ is $p = p_\infty - \gamma_m R \sin\theta \cos\varphi + \frac{\rho_m u_\infty^2}{2}\left(1 - \frac{9}{4}\sin^2\theta\right)$. The upward vertical force at the base is

$$F_b = p_b A_b = \left(p_\infty + \frac{\rho_m}{2}\left[u_\infty^2 - v_b^2\right]\right)\pi R^2.$$

The upward vertical force component on the curved surface is obtained from

$$F_{sz} = \int_A -p\,dA \sin\theta \cos\varphi$$

$$= -p_\infty R^2 \int_o^\pi \sin\theta \sin\theta \left[\int_{-\pi/2}^{\pi/2} \cos\varphi\,d\varphi\right]d\theta$$

$$+ \gamma_m R R^2 \int_o^\pi \sin^3\theta \left[\int_{-\pi/2}^{\pi/2} \cos^2\varphi\,d\varphi\right]d\theta$$

$$- \frac{\rho_m u_\infty^2}{2} R^2 \int_o^\pi \sin^2\theta \left[\int_{-\pi/2}^{\pi/2} \cos\varphi\,d\varphi\right]d\theta$$

$$+ \frac{\rho_m u_\infty^2}{2} R^2 \frac{9}{4} \int_o^\pi \sin^2\theta \sin^2\theta \left[\int_{-\pi/2}^{\pi/2} \cos\varphi\,d\varphi\right]d\theta$$

$$= \left(-p_\infty R^2 \frac{\pi}{2} \times 2\right) + \left(\gamma_m R^3 2 \times \frac{2}{3} \times \frac{\pi}{2}\right) - \left(\frac{\rho_m u_\infty^2}{2} R^2 \frac{\pi}{2} \times 2\right)$$

$$+ \left(\frac{\rho_m u_\infty^2}{2} R^2 \frac{9}{4} \times 2 \times \frac{3\pi}{16} \times 2\right)$$

The net vertical force $F_V = F_b + F_{sz}$ shows that the integral of a constant vanishes, and the remaining terms are:

$$F_V = \underbrace{\gamma_m \frac{2}{3}\pi R^3}_{\substack{Buoyancy \\ force}} + \underbrace{\rho_m \frac{u_\infty^2}{2}\pi R^2 \frac{27}{16}}_{\substack{Lift\ force\ on \\ curved\ surface}} - \underbrace{\rho_m \frac{v_b^2}{2}\pi R^2}_{\substack{Force\ at\ the \\ base}} \qquad \text{(E-4.3.2)}\blacklozenge$$

The results are quite similar to Equation (E-4.2.2). The first term relates to buoyancy, the net lift force where $v_b = 0$ is $F_L = \frac{27}{32}\pi\rho_m u_\infty^2 R^2$, and when $v_b =$

u_∞, the lift force reduces to $F_L = \frac{11}{32}\pi\rho_m u_\infty^2 R^2$. It is thus becoming clear that the maximum lift force of a particle is when $v_b = 0$.The second conclusion is that the magnitude of the lift force depends on particle shape.

Acceleration

When considering a solid upper half-sphere, of mass density $\rho_s = \rho G$ and volume $\forall = 2\pi R^3/3$, the acceleration a can be determined from the lift force per unit mass when $v_b = u_\infty$:

$$a = \frac{F_{L1}}{\rho_s \forall} = \frac{11\pi\rho_m}{32\rho_s}\frac{u_\infty^2 R^2}{2\pi R^3}3 = \frac{33\rho_m}{64\rho_s}\frac{u_\infty^2}{R} \simeq 0.5\frac{u_\infty^2}{GR} \qquad \text{(E-4.3.3)}\blacklozenge$$

This expression for the acceleration, or lift force per unit mass, resembles that for flow around a cylindrical section (Eq. E-4.1.3). The lift coefficient and vertical acceleration depend on particle shape. Another conclusion from this analysis is that for a given relative velocity u_∞, the lift force per unit volume (or mass) is inversely proportional to the size of the particle. It is thus concluded that, at a given flow velocity, fine particles are subjected to larger hydrodynamic accelerations than coarse particles.

Example 4.4 Lift, weight, and buoyancy forces on a sphere

Consider the hydrodynamic forces exerted on the upper half of the sphere illustrated in Figure E-4.4.1 and determine the critical approach velocity u_c at which the sphere will be moved out of the pocket. The submerged weight F_S is given by subtracting the buoyancy force from the particle weight $F_S = F_W - F_B = (\gamma s - \gamma_m)\frac{4}{3}\pi R^3$. The lift force F_{L1} on a half-sphere from Equation (E-4.3.2) with $v_b = 0$ is $F_{L1} = \frac{27\pi}{32}\rho_m u_\infty^2 R^2$. Equilibrium is obtained when $F_W = F_s$, which corresponds to $u_\infty = u_c$,

$$(\gamma_s - \gamma_m)\frac{4\pi}{3}R^3 = \frac{27\pi}{32}\rho_m u_c^2 R^2$$

$$u_c = \sqrt{(\gamma_s - \gamma_m)\frac{4}{3}R\frac{32}{27\rho_m}}$$

$$u_c \cong 1.4\sqrt{(G-1)gd_s} \qquad \text{(E-4.4.1)}$$

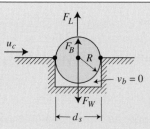

Figure E-4.4.1 Equilibrium of a sphere

This approximation is quite informative, and we will obtain very similar results
for the incipient motion of coarse particles in Chapter 7.

Example 4.5. Lift force and lift coefficient

The concept of lift coefficient can be examined from dimensional analysis (see
Example 2.1). The relationship for the lift coefficient can be defined as a function
of the base area of the object A_p as $C_L = \frac{F_L}{\rho_m u_\infty^2 A_p}$.

The base area of a half-cylinder, $A_p = 2LR$ and a half-sphere, $A_p = \pi R^2$,
the corresponding forces, drag coefficients, and forces/volume are shown in the
table below.

	Half-cylinder	Half-sphere
$v_b = u_\infty$	$F_L = \dfrac{5}{3}\rho_m u_\infty^2 LR$	$F_L = \dfrac{11\pi R^2}{32}\rho_m u_\infty^2$
	$C_L = \dfrac{5}{6} = 0.83$	$C_L = \dfrac{11}{32} = 0.34$
	$\dfrac{F_L}{\rho_s \forall} = \dfrac{10}{3\pi}\dfrac{u_\infty^2}{GR}$	$\dfrac{F_L}{\rho_s \forall} = \dfrac{33}{64}\dfrac{u_\infty^2}{GR}$
$v_b = 0$	$F_L = \dfrac{8}{3}\rho_m u_\infty^2 LR$	$F_L = \dfrac{27\pi R^2}{32}\rho_m u_\infty^2$
	$C_L = \dfrac{8}{6} = 1.33$	$C_L = \dfrac{27}{32} = 0.84$
	$\dfrac{F_L}{\rho_s \forall} = \dfrac{16}{3\pi}\dfrac{u_\infty^2}{GR}$	$\dfrac{F_L}{\rho_s \forall} = \dfrac{81}{64}\dfrac{u_\infty^2}{GR}$

It can be concluded that for a given flow velocity, the lift coefficients vary largely
with particle shape and with the position relative to other particles and surfaces.
For instance, objects in suspension will have a lower lift coefficient than objects

placed against a flat surface. As a consequence, coarse sand particles can be ejected from the surface and then fall back from the suspension resulting in saltation. Because of the high Reynolds requirement for this approximation, this is in practice valid for particle sizes coarser than about 1 mm. Viscous effects are discussed in Chapter 5.

Exercises

4.1 Calculate the angle α_c from $v_\theta = 0$ in Equation (4.9b) with clockwise circulation ($\Gamma_v < 0$), and compare with Figure 4.3.

4.2 Derive the potential and stream functions Equation (4.4) for a source from Equation (4.2) when $v_r = q/2\pi r$ and $v_\theta = 0$.

♦4.3 Derive the velocity components around a sphere (Eq. 4.17a and b) from the flow potential Φ in Equations (4.15) and (4.16).

♦♦4.4 Repeat the calculations of Example 4.1 with $F_{L\frac{1}{2}} = \int_o^\pi -p\sin\theta\, dA$, and explain why the use of the reference pressure p_r yields the same result as pressure p.

Problems
Problem 4.1

In Example 4.5 consider added mass and estimate the vertical acceleration of a 4 mm semi-spherical gravel particle when $u_4 = 2u_c$. Once the particle is ejected, consider that the base velocity is u_∞ and determine the acceleration. (*Hint*: notice the change in lift force depending on base flow conditions.)

♦*Problem 4.2*

Calculate the lift force in SI and English units on a 4 m-diameter semi-spherical tent under a 100 km/h wind. Compare with the lift force of a 4 m-long semi-cylindrical tent that has the same volume. (*Hint*: find the mass density of air and assume that there is no velocity at the base of the tent.)

♦♦*Problem 4.3*

Plot and compare the distribution of surface velocity, relative pressure, and boundary shear stress for irrotational flow without circulation around a cylinder and a sphere of radius R.

$$pr_{\max} = \frac{\rho u_\infty^2}{2}; \; p_{r_{\max}} \quad \text{and } v = 0 \text{ at } \theta = 0° \text{ and } 180°$$

(*Answer:* $p_{r_{\min}}$ and $v_{\max}$ at $\theta = 90°$ and $270°$; $\tau = 0$ everywhere;

$v_{\max}$ cylinder $= 2u_\infty$; and $v_{\max}$ sphere $= 1.5u_\infty$)

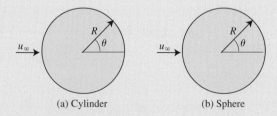

(a) Cylinder   (b) Sphere

Figure P-4.3 a) cylinder, b) sphere

◆Problem 4.4

Integrate the pressure distribution and calculate the net forces applied on a quarter-sphere with cut planes aligned with the flow.

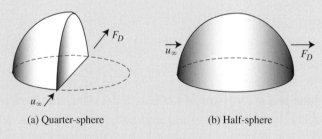

(a) Quarter-sphere   (b) Half-sphere

Figure P-4.4 a) quarter-sphere b) half-sphere

◆ Problem 4.5

A farmer covers a 30 ft-diameter pile of hay with tarp. If the long semi-cylindrical pile is oriented N–S, what weight would anchor the top under 50 mph westerly winds? Also, would the weight be less if the hay were to be piled in a half-spherical shape? Finally, which option would require less tarp?

◆◆ Problem 4.6

A semi-cylindrical tunnel is built across a large river to allow traffic of vehicles from one side of the river to the other. If the tunnel diameter is 30 ft and the tunnel weighs 15 tons per linear foot of length, what is the safe operational range in river flow velocities for this tunnel? How much weight would have to be added when the flow velocity reaches 15 ft/s during large floods?

♦ Problem 4.7

The CSU-CHILL radar can be seen at the website http://chill.colostate.edu. The radar is protected by a near-spherical vinyl-coated Dacron radome with a diameter of 73 ft. The radome is inflated with air pumps and pressurized at 9.75 inches of water pressure differential when the wind speed exceeds 40 miles per hour. Wind speeds up to 80 miles per hour have been measured at the site in the past 10 years. Neglect the drag force and determine the lift force in the upper half-sphere at an 80 mph wind speed. Also determine the maximum wind speed that would tear the vinyl at the "equator" if the tear-strength of the vinyl is 650 lb/in. (*Hint*: only consider the upper half of the spherical radome, also consider $\rho_{air} = 1.3$ kg/m^3.)

5

Particle motion in Newtonian fluids

In contrast with Chapter 4, this chapter examines cases where viscous forces are dominant compared to inertial forces. In sedimentation terms, this will describe flow conditions around small particles like silts and clays. At low rates of deformation and low concentrations, sediment-laden flows obey Newton's law of deformation. The governing equations of motion are called the Navier–Stokes equations (Section 5.1) which are applied around a sphere to determine: the flow field (Section 5.2); the drag force (Section 5.3); the fall velocity (Section 5.4); the rate of energy dissipation (Section 5.5); and laboratory measurements (Section 5.6).

Viscosity is a fluid property that differentiates real fluids from ideal fluids. The viscosity of a fluid is a measure of resistance to flow. The dynamic viscosity of a Newtonian mixture μ_m is defined as the ratio of shear stress to the rate of deformation, thus its dimensions are mass per unit length and time. The kinematic viscosity of a mixture v_m is defined as the ratio of dynamic viscosity to mass density of the mixture ($v_m = \mu_m/\rho_m$) and the dimensions, L^2/T do not involve mass.

In a Newtonian fluid the shear stress τ_{yx} acting in the x direction is proportional to the gradient in the y direction of the velocity component v_x in the x direction:

$$\tau_{yx} = \mu_m \frac{dv_x}{dy} \tag{5.1}$$

This basic description is true for one-dimensional flow, but incomplete for two-dimensional flow because τ_{yx} is also a function of $\partial v_y/\partial x$.

5.1 Navier–Stokes equations

Shear stresses in a Newtonian fluid equal the product of the dynamic viscosity of the mixture μ_m and the rate of angular deformation defined in Section 3.1,

Table 5.1. *Navier–Stokes equations in Cartesian coordinates* (x,y,z)

x - component

$$\frac{\partial v_x}{\partial t} + v_x \frac{\partial v_x}{\partial x} + v_y \frac{\partial v_x}{\partial y} + v_z \frac{\partial v_x}{\partial z} = g_x - \frac{1}{\rho_m}\frac{\partial p}{\partial x} + \frac{\mu_m}{\rho_m}\left(\frac{\partial^2 v_x}{\partial x^2} + \frac{\partial^2 v_x}{\partial y^2} + \frac{\partial^2 v_x}{\partial z^2}\right)$$

y - component

$$\frac{\partial v_y}{\partial t} + v_x \frac{\partial v_y}{\partial x} + v_y \frac{\partial v_y}{\partial y} + v_z \frac{\partial v_y}{\partial z} = g_y - \frac{1}{\rho_m}\frac{\partial p}{\partial y} + \frac{\mu_m}{\rho_m}\left(\frac{\partial^2 v_y}{\partial x^2} + \frac{\partial^2 v_y}{\partial y^2} + \frac{\partial^2 v_y}{\partial z^2}\right)$$

z - component

$$\frac{\partial v_z}{\partial t} + v_x \frac{\partial v_z}{\partial x} + v_y \frac{\partial v_z}{\partial y} + v_z \frac{\partial v_z}{\partial z} = g_z - \frac{1}{\rho_m}\frac{\partial p}{\partial z} + \frac{\mu_m}{\rho_m}\left(\frac{\partial^2 v_z}{\partial x^2} + \frac{\partial^2 v_z}{\partial y^2} + \frac{\partial^2 v_z}{\partial z^2}\right)$$

thus

$$\tau_{yx} = \tau_{xy} = \mu_m \left(\frac{\partial v_x}{\partial y} + \frac{\partial v_y}{\partial x}\right) = \mu_m \varnothing_z \tag{5.2a}$$

$$\tau_{zy} = \tau_{yz} = \mu_m \left(\frac{\partial v_y}{\partial z} + \frac{\partial v_z}{\partial y}\right) = \mu_m \varnothing_x \tag{5.2b}$$

and

$$\tau_{xz} = \tau_{zx} = \mu_m \left(\frac{\partial v_z}{\partial x} + \frac{\partial v_x}{\partial z}\right) = \mu_m \varnothing_y \tag{5.2c}$$

The normal stresses of isotropic Newtonian fluids also relate to pressure p, viscosity μ_m, and velocity gradients. The relationships for the normal stresses σ_x, σ_y and σ_z are

$$\sigma_x = -p + 2\mu_m \frac{\partial v_x}{\partial x} - \frac{2\mu_m}{3}\left(\frac{\partial v_x}{\partial x} + \frac{\partial v_y}{\partial y} + \frac{\partial v_z}{\partial z}\right) \tag{5.3a}$$

$$\sigma_y = -p + 2\mu_m \frac{\partial v_y}{\partial y} - \frac{2\mu_m}{3}\left(\frac{\partial v_x}{\partial x} + \frac{\partial v_y}{\partial y} + \frac{\partial v_z}{\partial z}\right) \tag{5.3b}$$

$$\sigma_z = -p + 2\mu_m \frac{\partial v_z}{\partial z} - \frac{2\mu_m}{3}\left(\frac{\partial v_x}{\partial x} + \frac{\partial v_y}{\partial y} + \frac{\partial v_z}{\partial z}\right) \tag{5.3c}$$

In incompressible fluids, the terms in parentheses of Equation 5.3 can be dropped because they correspond to the continuity equation (Eq. (3.6d)).

Substitution of these stress tensor relationships (Eqs. (5.2) and (5.3)) into the equations of motion (Eq. (3.12)) gives the complete set of equations of motion shown in Tables 5.1–5.3. These equations are valid for Newtonian fluids in the

Table 5.2. *Stress tensor and Navier–Stokes equations in cylindrical coordinates* (r,θ,z)

$$\sigma_{rr} = -p + \mu_m\left[2\frac{\partial v_r}{\partial r} - \frac{2}{3}(\nabla\bullet v)\right]$$

$$\sigma_{\theta\theta} = -p + \mu_m\left[2\left(\frac{1}{r}\frac{\partial v_\theta}{\partial \theta} + \frac{v_r}{r}\right) - \frac{2}{3}(\nabla\bullet v)\right]$$

$$\sigma_{zz} = -p + \mu_m\left[2\frac{\partial v_z}{\partial z} - \frac{2}{3}(\nabla\bullet v)\right]$$

$$\tau_{r\theta} = \tau_{\theta r} = \mu_m\left[r\frac{\partial}{\partial r}\left(\frac{v_\theta}{r}\right) + \frac{1}{r}\frac{\partial v_r}{\partial \theta}\right]$$

$$\tau_{\theta z} = \tau_{z\theta} = \mu_m\left[\frac{\partial v_\theta}{\partial z} + \frac{1}{r}\frac{\partial v_z}{\partial \theta}\right]$$

$$\tau_{zr} = \tau_{rz} = \mu_m\left[\frac{\partial v_z}{\partial r} + \frac{\partial v_r}{\partial z}\right]$$

where $(\nabla\bullet v) = \frac{1}{r}\frac{\partial}{\partial r}(rv_r) + \frac{1}{r}\frac{\partial v_\theta}{\partial \theta} + \frac{\partial v_z}{\partial z}$

r-component

$$\frac{\partial v_r}{\partial t} + v_r\frac{\partial v_r}{\partial r} + \frac{v_\theta}{r}\frac{\partial v_r}{\partial \theta} - \frac{v_\theta^2}{r} + v_z\frac{\partial v_r}{\partial z} = g_r - \frac{1}{\rho_m}\frac{\partial p}{\partial r}$$

$$+\frac{\mu_m}{\rho_m}\left[\frac{\partial}{\partial r}\left(\frac{1}{r}\frac{\partial}{\partial r}(rv_r)\right) + \frac{1}{r^2}\frac{\partial^2 v_r}{\partial \theta^2} - \frac{2}{r^2}\frac{\partial v_\theta}{\partial \theta} + \frac{\partial^2 v_r}{\partial z^2}\right]$$

θ - component

$$\frac{\partial v_\theta}{\partial t} + v_r\frac{\partial v_\theta}{\partial r} + \frac{v_\theta}{r}\frac{\partial v_\theta}{\partial \theta} + \frac{v_r v_\theta}{r} + v_z\frac{\partial v_\theta}{\partial z} = g_\theta - \frac{1}{r\rho_m}\frac{\partial p}{\partial \theta}$$

$$+\frac{\mu_m}{\rho_m}\left[\frac{\partial}{\partial r}\left(\frac{1}{r}\frac{\partial}{\partial r}(rv_\theta)\right) + \frac{1}{r^2}\frac{\partial^2 v_\theta}{\partial \theta^2} + \frac{2}{r^2}\frac{\partial v_r}{\partial \theta} + \frac{\partial^2 v_\theta}{\partial z^2}\right]$$

z - component

$$\frac{\partial v_z}{\partial t} + v_r\frac{\partial v_z}{\partial r} + \frac{v_\theta}{r}\frac{\partial v_z}{\partial \theta} + v_z\frac{\partial v_z}{\partial z} = g_z - \frac{1}{\rho_m}\frac{\partial p}{\partial z}$$

$$+\frac{\mu_m}{\rho_m}\left[\frac{1}{r}\frac{\partial}{\partial r}\left(\frac{r\partial v_z}{\partial r}\right) + \frac{1}{r^2}\frac{\partial^2 v_z}{\partial \theta^2} + \frac{\partial^2 v_z}{\partial z^2}\right]$$

Table 5.3. *Stress tensor and Navier–Stokes equations in spherical coordinates (r,θ,φ)*

$$\sigma_{rr} = -p + \mu_m \left[2\frac{\partial v_r}{\partial r} - \frac{2}{3}(\nabla\bullet v) \right]$$

$$\sigma_{\theta\theta} = -p + \mu_m \left[2\left(\frac{1}{r}\frac{\partial v_\theta}{\partial \theta} + \frac{v_r}{r} \right) - \frac{2}{3}(\nabla\bullet v) \right]$$

$$\sigma_{\varphi\varphi} = -p + \mu_m \left[2\left(\frac{1}{r\sin\theta}\frac{\partial v_\varphi}{\partial \varphi} + \frac{v_r}{r} + \frac{v_\theta \cot\theta}{r} \right) - \frac{2}{3}(\nabla\bullet v) \right]$$

$$\tau_{r\theta} = \tau_{\theta r} = \mu_m \left[r\frac{\partial}{\partial r}\left(\frac{v_\theta}{r} \right) + \frac{1}{r}\frac{\partial v_r}{\partial \theta} \right]$$

$$\tau_{\theta\varphi} = \tau_{\varphi\theta} = \mu_m \left[\frac{\sin\theta}{r}\frac{\partial}{\partial \theta}\left(\frac{v_\varphi}{\sin\theta} \right) + \frac{1}{r\sin\theta}\frac{\partial v_\theta}{\partial \varphi} \right]$$

$$\tau_{\varphi r} = \tau_{r\varphi} = \mu_m \left[\frac{1}{r\sin\theta}\frac{\partial v_r}{\partial \varphi} + r\frac{\partial}{\partial r}\left(\frac{v_\varphi}{r} \right) \right]$$

$$(\nabla\bullet v) = \frac{1}{r^2}\frac{\partial}{\partial r}\left(r^2 v_r \right) + \frac{1}{r\sin\theta}\frac{\partial}{\partial \theta}(v_\theta \sin\theta) + \frac{1}{r\sin\theta}\frac{\partial v_\varphi}{\partial \varphi}$$

$$\nabla^2 = \frac{1}{r^2}\frac{\partial}{\partial r}\left(r^2\frac{\partial}{\partial r} \right) + \frac{1}{r^2\sin\theta}\frac{\partial}{\partial \theta}\left(\sin\theta\frac{\partial}{\partial \theta} \right) + \frac{1}{r^2\sin^2\theta}\left(\frac{\partial^2}{\partial \varphi^2} \right)$$

r-component

$$\frac{\partial v_r}{\partial t} + v_r\frac{\partial v_r}{\partial r} + \frac{v_\theta}{r}\frac{\partial v_r}{\partial \theta} + \frac{v_\varphi}{r\sin\theta}\frac{\partial v_r}{\partial \varphi} - \frac{v_\theta^2 + v_\varphi^2}{r}$$

$$= g_r - \frac{1}{\rho_m}\frac{\partial p}{\partial r} + \frac{\mu_m}{\rho_m}\left[\nabla^2 v_r - \frac{2}{r^2}v_r - \frac{2}{r^2}\frac{\partial v_\theta}{\partial \theta} - \frac{2}{r^2}v_\theta \cot\theta - \frac{2}{r^2\sin\theta}\frac{\partial v_\varphi}{\partial \varphi} \right]$$

θ - component

$$\frac{\partial v_\theta}{\partial t} + v_r\frac{\partial v_\theta}{\partial r} + \frac{v_\theta}{r}\frac{\partial v_\theta}{\partial \theta} + \frac{v_\varphi}{r\sin\theta}\frac{\partial v_\theta}{\partial \varphi} + \frac{v_r v_\theta}{r} - \frac{v_\varphi^2 \cot\theta}{r}$$

$$= g_\theta - \frac{1}{\rho_m r}\frac{\partial p}{\partial \theta} + \frac{\mu_m}{\rho_m}\left[\nabla^2 v_\theta + \frac{2}{r^2}\frac{\partial v_r}{\partial \theta} - \frac{v_\theta}{r^2\sin^2\theta} - \frac{2\cos\theta}{r^2\sin^2\theta}\frac{\partial v_\varphi}{\partial \varphi} \right]$$

φ - component

$$\frac{\partial v_\varphi}{\partial t} + v_r\frac{\partial v_\varphi}{\partial r} + \frac{v_\theta}{r}\frac{\partial v_\varphi}{\partial \theta} + \frac{v_\varphi}{r\sin\theta}\frac{\partial v_\varphi}{\partial \varphi} + \frac{v_\varphi v_r}{r} + \frac{v_\theta v_\varphi}{r}\cot\theta$$

$$= g_\varphi - \frac{1}{\rho_m r\sin\theta}\frac{\partial p}{\partial \varphi} + \frac{\mu_m}{\rho_m}\left(\nabla^2 v_\varphi - \frac{v_\varphi}{r^2\sin^2\theta} + \frac{2}{r^2\sin\theta}\frac{\partial v_r}{\partial \varphi} + \frac{2\cos\theta}{r^2\sin^2\theta}\frac{\partial v_\theta}{\partial \varphi} \right)$$

laminar regime (low Reynolds numbers) and were developed by Navier, Cauchy, Poisson, Saint-Venant, and Stokes. They are commonly referred to as the Navier–Stokes equations.

5.2 Newtonian flow around a sphere

The vorticity components $\otimes_x$, $\otimes_y$, and $\otimes_z$ defined previously (Eq. (3.4)) have been shown to satisfy Equation (3.7) which can be used to rewrite the Navier–Stokes equations from Table 5.1 for an incompressible fluid in a form similar to Equations (3.21a–c):

$$\frac{\partial v_x}{\partial t} - v_y \otimes_z + v_z \otimes_y = -\frac{g \partial H}{\partial x} + v_m \nabla^2 v_x \tag{5.4a}$$

$$\frac{\partial v_y}{\partial t} - v_z \otimes_x + v_x \otimes_z = -\frac{g \partial H}{\partial y} + v_m \nabla^2 v_y \tag{5.4b}$$

$$\frac{\partial v_z}{\partial t} - v_x \otimes_y + v_y \otimes_x = -\frac{g \partial H}{\partial z} + v_m \nabla^2 v_z \tag{5.4c}$$

in which H represents the Bernoulli sum from Equation (3.22a).

After elimination of the Bernoulli sum H through cross-differentiation (see Exercise 5.3), the foregoing equations become

$$\frac{d \otimes_x}{dt} = \otimes_x \frac{\partial v_x}{\partial x} + \otimes_y \frac{\partial v_x}{\partial y} + \otimes_z \frac{\partial v_x}{\partial z} + v_m \nabla^2 \otimes_x \tag{5.5a}$$

$$\frac{d \otimes_y}{dt} = \otimes_x \frac{\partial v_y}{\partial x} + \otimes_y \frac{\partial v_y}{\partial y} + \otimes_z \frac{\partial v_y}{\partial z} + v_m \nabla^2 \otimes_y \tag{5.5b}$$

$$\frac{d \otimes_z}{dt} = \otimes_x \frac{\partial v_z}{\partial x} + \otimes_y \frac{\partial v_z}{\partial y} + \otimes_z \frac{\partial v_z}{\partial z} + v_m \nabla^2 \otimes_z \tag{5.5c}$$

These are the equations governing the diffusion of vorticity.

Irrotational flow of a viscous fluid is possible because the conditions $\otimes_x = \otimes_y = \otimes_z = 0$ are also solutions to the Navier–Stokes equations from which the vorticity equations were derived. However, such flows are only possible when the solid boundary moves at the same velocity as the fluid at the boundary.

The equations of diffusion of vorticity are analogous to the law of conduction of heat. It is evident from this analogy that vorticity cannot originate from the interior of a viscous fluid, but must diffuse inward from the boundary when the fluid moves relative to the boundary.

The vorticity transport equations can also be written in terms of the stream function Ψ. From the definition of the two-dimensional stream function ($v_x = -\partial \Psi / \partial y$ and $v_y = \partial \Psi / \partial x$) the continuity equation is satisfied and $\otimes_z = \nabla^2 \Psi$. The vorticity transport equation becomes

$$\frac{\partial \nabla^2 \Psi}{\partial t} - \frac{\partial \Psi}{\partial y}\frac{\partial \nabla^2 \Psi}{\partial x} + \frac{\partial \Psi}{\partial x}\frac{\partial \nabla^2 \Psi}{\partial y} = \nu_m \nabla^4 \Psi \tag{5.6}$$

This fourth-order nonlinear partial differential equation contains only one unknown in Ψ. The inertia terms on the left-hand side are balanced by the frictional terms on the right-hand side of the equation. The physical significance of the stream function is that lines of constant value of Ψ are streamlines.

When the viscous forces are considerably larger than the inertial forces, the left-hand side of Equation (5.6) vanishes and the vorticity transport equation reduces to the following linear equation:

$$\nabla^4 \Psi = 0 \tag{5.7}$$

Creeping motion of a small sphere of radius R in a Newtonian fluid can be obtained by applying Equation (5.7) to the steady motion of a sphere with small relative velocity u_∞ (see Fig. 5.1). This is solved by setting two boundary conditions for $v_r = v_\theta = 0$ at the edge of the sphere $(r = R)$, and two additional boundary conditions for $v_r = u_\infty \cos\theta$ and $v_\theta = -u_\infty \sin\theta$ at $r = \infty$.

The velocity distribution around a sphere in a creeping motion of radius R is given by:

$$\frac{v_r}{u_\infty} = \left[1 - \frac{3}{2}\left(\frac{R}{r}\right) + \frac{1}{2}\left(\frac{R}{r}\right)^3\right]\cos\theta \tag{5.8a}$$

and

$$\frac{v_\theta}{u_\infty} = -\left[1 - \frac{3}{4}\left(\frac{R}{r}\right) - \frac{1}{4}\left(\frac{R}{r}\right)^3\right]\sin\theta \tag{5.8b}$$

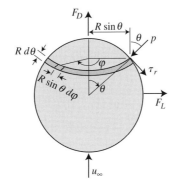

Figure 5.1. Creeping flow past a sphere

The corresponding pressure and shear stress distributions are found analytically from the stress tensor and the Navier–Stokes equations in spherical coordinates (Table 5.3). The pressure can be subdivided into three components: the ambient pressure p_r, the hydrostatic pressure p_h, and the dynamic pressure p_d, such that $p = p_h + p_d$.

$$p_h = p_r - \rho_m gR\cos\theta \qquad (5.9a)\blacklozenge$$

$$p_d = -\frac{3}{2}\frac{\mu_m u_\infty}{R}\left(\frac{R}{r}\right)^2\cos\theta \qquad (5.9b)\blacklozenge$$

$$\tau = \tau_{r\theta} = -\frac{3\mu_m u_\infty}{2R}\left(\frac{R}{r}\right)^4\sin\theta \qquad (5.9c)\blacklozenge$$

The quantity p_r is the ambient pressure far away from the sphere ($\theta = 90°$ or $270°$ and $R \to \infty$) and $-\rho_m gR\cos\theta$ is the hydrostatic pressure contribution due to the fluid mixture weight. Shear stress is positive in the direction of increasing angle θ.

5.3 Drag force on a sphere

Besides the buoyancy force resulting from integrating the hydrostatic pressure distribution on a sphere (Example 3.4), the drag force exerted by the moving fluid around the sphere is computed by integrating the stress tensor over the sphere surface. The drag force is subdivided into two components: (1) the surface drag results from the integration of the shear stress distribution; and (2) the form drag arises from the integration of the hydrodynamic pressure distribution.

5.3.1 Surface drag

With reference to Figure 5.1, the shear stress $\tau = \tau_{r\theta}$ at each point on the spherical surface $r = R$ is the tangential force in the increasing θ-direction per unit area of spherical surface. The shear stress component in the flow direction, $-\tau\sin\theta$, is multiplied by the elementary area, $R^2\sin\theta\, d\theta d\varphi$ and integrated over the spherical surface to give the surface drag F'_D:

$$F'_D = \int_0^{2\pi}\int_0^{\pi} -\tau\sin\theta R^2\sin\theta\, d\theta\, d\varphi \qquad (5.10)$$

The shear stress distribution τ at the surface of the sphere ($r = R$) from Equation (5.9c) is substituted into the integral in Equation (5.10) to give the surface drag force:

$$F'_D = 4\pi \mu_m R u_\infty \qquad (5.11)\blacklozenge$$

5.3.2 Form drag

At each point on the surface of the sphere, the dynamic pressure p_d acts perpendicularly to the surface, of which the upward vertical component is $-p_d \cos\theta$. We now multiply this local force per unit area by the surface area on which it acts, $R^2 \sin\theta\, d\theta\, d\varphi$, and integrate over the surface of the sphere to get the resultant vertical force called form drag force F''_D:

$$F''_D = \int_0^{2\pi} \int_0^\pi -p_d \cos\theta R^2 \sin\theta\, d\theta\, d\varphi \qquad (5.12)$$

Substituting the dynamic pressure p_d from Equation (5.9b) into the integral gives the form dragF''_D.

$$F''_D = 2\pi \mu_m R u_\infty \qquad (5.13)\blacklozenge$$

Hence the total drag force F_D exerted by the motion of a Newtonian viscous fluid around the sphere is given by the sum of Equations (5.11) and (5.13):

$$F_D = F'_D + F''_D = \underbrace{4\pi \mu_m R u_\infty}_{\text{surface drag}} + \underbrace{2\pi \mu_m R u_\infty}_{\text{form drag}} = \underbrace{6\pi \mu_m R u_\infty}_{\text{total drag}} \qquad (5.14)$$

The buoyancy force F_B resulting from the hydrostatic pressure distribution from Example 3.4 is added to the drag force F_D from Equation (5.14) to give the total force F:

$$F = F_B + F_D = \underbrace{\frac{4}{3}\pi \gamma_m R^3}_{\text{buoyancy}} + \underbrace{6\pi \mu_m R u_\infty}_{\text{total drag}} \qquad (5.15)\blacklozenge\blacklozenge$$

The total upward force F exerted on a sphere falling at a velocity u_∞ is thus the sum of the hydrostatic (buoyancy force) and hydrodynamic (total drag) components.

Example 5.1 Particle equilibrium in viscous flow

Consider the approach velocity u_∞ against the top half-sphere of radius R (Fig. E-5.1.1). Determine from equilibrium conditions the critical velocity u_c that will move the sphere out of the pocket. Equilibrium is governed by the sum of moments about point 0 because the rotation about 0 is induced by horizontal and vertical forces. Include both the distribution of pressure and shear stress. Notice that the lift force vanishes in viscous flow due to the symmetrical distribution of shear stress and pressure about the vertical axis z.

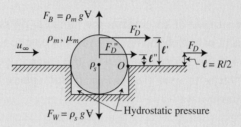

Figure E-5.1.1 Particle equilibrium

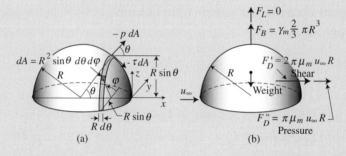

Figure E-5.1.2 Half-sphere

Step 1. With reference to Figure E-5.1.2, the element of force dF' due to the shear stress distribution is given by its unit vector components $\vec{i}$, $\vec{j}$, and $\vec{k}$ in Cartesian coordinates:

$$dF' = -(\tau\, dA \sin\theta)\,\vec{i} + (\tau\, dA \cos\theta \cos\varphi)\,\vec{j} + (\tau\, dA \cos\theta \sin\varphi)\,\vec{k}$$

The moment arm about point 0 is

$$\vec{r} = (-R + R\cos\theta)\,\vec{i} + (R\sin\theta \cos\varphi)\,\vec{j} + (R\sin\theta \sin\varphi)\,\vec{k}$$

The moment about point 0 is obtained as the determinant of the following matrix

$$dM'_D = \begin{vmatrix} \vec{i} & \vec{j} & \vec{k} \\ r_x & r_y & r_z \\ dF'_x & dF'_y & dF'_z \end{vmatrix}$$

The moment about the y-axis is obtained by the $\vec{j}$ component, or

$$dM'_{Dj} = -\vec{j}\left(r_x dF'_z - r_z dF'_x\right)$$

$$dM'_{Dj} = R(1 - \cos\theta)\,\tau\,dA\cos\theta\sin\varphi - R\sin\theta\sin\varphi\tau\,dA\sin\theta$$

The net moment about the y-axis is then obtained from the integral of dM'_D over the surface area of the half-sphere. After substituting $\tau = \frac{-3}{2}\mu_m\frac{u_\infty}{R}\sin\theta$ and $dA = R^2\sin\theta\,d\varphi\,d\theta$, one obtains:

$$M'_{Dj} = -\frac{3}{2}\mu_m u_\infty R^2 \int_o^\pi \sin^2\theta\,(1 - \cos\theta)\cos\theta \left[\int_o^\pi \sin\varphi\,d\varphi\right]d\theta$$

$$+ \frac{3}{2}\mu_m u_\infty R^2 \int_o^\pi \sin^4\theta \left[\int_o^\pi \sin\varphi\,d\varphi\right]d\theta$$

Which reduces to

$$M'_{Dj} = \frac{3}{2}\pi\,\mu_m u_\infty R^2$$

Given that $F'_D = 2\pi\,\mu_m u_\infty R$ on a half-sphere from Equation (5.11), the moment arm of the surface drag ℓ' is obtained from

$$\ell' = M'_{Dj}/F'_D = 3R/4$$

Step 2. Similarly, one can demonstrate that the moment component around the y-axis from the pressure distribution is zero, $M''_D = 0$. This can simply be understood from the fact that the pressure always acts through the center of the sphere, thus the form drag force $F''_D = \pi\mu_m u_\infty R^2$ on the upper half-sphere must pass through point 0, and $\ell'' = M''_{Dj}/F''_D = 0$. Similarly, the moment arm $\ell = M_D/F_D = R/2$.

Step 3. The buoyancy force $F_B = \rho_m g \forall$ and the particle weight $F_w = \rho_s g \forall$ act vertically in opposite directions through the center of the sphere.

Step 4. Equilibrium is defined when $u_\infty = u_c$ brings the sum of moments of all forces about point 0 to vanish:

$$\sum M_D = F_B R + M'_D + M''_D - F_W R = 0$$

$$(\gamma_s - \gamma_m)\frac{4}{3}\pi R^4 = \frac{3}{2}\pi\mu_m u_c R^2$$

And solving for u_c

$$u_c = \frac{2}{9}\frac{(\gamma_s - \gamma_m)\,d_s^2}{\mu_m}$$

This critical velocity is proportional to the settling velocity of fine particles in Equation (5.21).

5.4 Drag coefficient and fall velocity

5.4.1 Drag coefficient

With reference to the dimensional analysis in Example 2.1 describing flow around a particle of diameter d_s, it was inferred in Equation (E-2.1.3) that the drag coefficient C_D could be written as a function of the particle Reynolds number $\text{Re}_p = u_\infty d_s / v_m$. For the laminar flow of a mixture (ρ_m, μ_m) around a sphere, this relationship is obtained after substituting F_D from Equation (5.14) into Equation (E-2.1.2):

$$C_D = \frac{8F_D}{\rho_m \pi u_\infty^2 d_s^2} = \frac{24 v_m}{u_\infty d_s} = \frac{24}{\text{Re}_p} \qquad (5.16)\blacklozenge$$

This equation is valid when $\text{Re}_p < 0.1$. At high Re_p, from Figure E-2.1.2, the value of C_D for spheres becomes approximately 0.5,

$$C_D \simeq \frac{24}{\text{Re}_p} + 0.5 \qquad (5.17)$$

Natural particles are not spherical and several modifications have been proposed. After Oseen (1927) and Goldstein (1929), Rubey (1933) followed with a simple approximation of the drag coefficient $C_D = 2 + 24/\text{Re}_p$ encompassing a wide range of particle Reynolds numbers. When compared with the drag coefficient of large particles in Figure 5.2a, Rubey's equation is on the high side.

The drag coefficient is a function of the Corey shape factor $C_o = l_c / \sqrt{l_a l_b}$ where l_a, l_b, and l_c are the longest, intermediate, and smallest size of the particle, respectively. The drag coefficient can be approximated by

$$C_D = \frac{24}{\text{Re}_p} + \frac{0.5}{C_o^2} \qquad (5.18)$$

This relationship can be useful for gravels $(\text{Re}_p > 200)$ when l_a, l_b, and l_c are known. For natural sands and gravels, the experimental values of the drag coefficient of Engelund and Hansen (1967), shown in Figure 5.2b, can also be used as:

$$C_D = \frac{24}{\text{Re}_p} + 1.5 \qquad (5.19)$$

This corresponds to a Corey shape factor $C_o \cong 0.58$.

5.4.2 Fall velocity

Starting from rest, a particle of density exceeding that of the surrounding fluid $(G > 1)$ will accelerate in the downward direction until it reaches an equilibrium

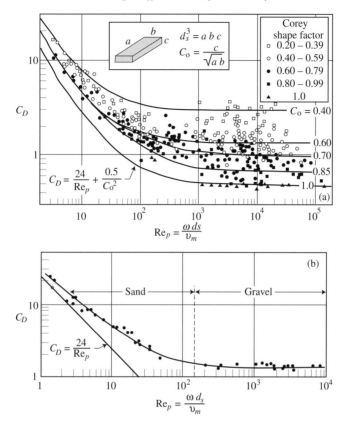

Figure 5.2. a) Drag coefficient of coarse particles (modified after Schultz *et al.* 1954) b) Drag coefficient for natural sands and gravels (modified after Engelund and Hansen, 1967).

fall velocity ω. The equilibrium fall velocity ($\omega = u_\infty$) of a small solid sphere falling in creeping motion under its own weight ($F_W = \gamma_s \forall$) in a viscous fluid is calculated by substituting the particle weight F_W to the force F in Equation (5.15).

$$F_W = \frac{\pi}{6} d_s^3 \rho_s \, g = \frac{\pi}{6} d_s^3 \rho_m g + 3\pi \mu_m d_s \omega \tag{5.20}$$

Solving Equation (5.20) for the fall velocity ω as a function of the particle diameter d_s gives Equation (5.21a) in a mixture of mass density ρ_m and Equation (5.21b) for the settling velocity ω_0 in clear water:

$$\omega = \frac{1}{18} \frac{(\gamma_s - \gamma_m)}{\mu_m} d_s^2; \text{ in a mixture where } Re_p < 0.1; \tag{5.21a}$$

$$\omega_0 = \frac{1}{18} \frac{(G-1)g}{v} d_s^2; \text{ in clear water where } Re_p < 0.1. \tag{5.21b}$$

This equation is valid for small particles ($d_s < 0.1$ mm in water) falling in viscous fluids ($Re_p < 0.1$). In a more general form, the fall velocity is expressed as a function of the drag coefficient C_D from Equations (5.16) and (5.20) after replacing F_D by $F_W - F_B$ with $G = \gamma_s/\gamma$.

$$\omega = \left[\frac{4}{3} \frac{(\gamma_s - \gamma_m)}{\gamma_m} \frac{gd_s}{C_D} \right]^{1/2} ; \text{ in a mixture.} \tag{5.22a}\blacklozenge$$

$$\omega_0 = \left[\frac{4}{3}(G-1)\frac{gd_s}{C_D} \right]^{1/2} ; \text{ in clear water.} \tag{5.22b}\blacklozenge$$

Notice that the fall velocity of natural coarse particles ($C_D = 1.33$, for $d_s \geq 1$ mm in water) is roughly equal to $\omega_0 \cong \sqrt{(G-1)gd_s}$. It is also instructive to compare this result with Equation (E-4.4.1).

Rubey's approximate formulation (Rubey, 1933) of the fall velocity in clear water based on $C_D = 2 + 24v/\omega d_s$ is given by

$$\omega_0 = \frac{1}{d_s}\left(\sqrt{\frac{2g}{3}(G-1)d_s^3 + 36v^2} - 6v \right) \tag{5.23a}\blacklozenge$$

or

$$\omega_0 = \left(\sqrt{\frac{2}{3} + \frac{36v^2}{(G-1)gd_s^3}} - \sqrt{\frac{36v^2}{(G-1)gd_s^3}} \right) \sqrt{(G-1)gd_s} \tag{5.23b}$$

Except for its use in Einstein's bedload equation (Section 9.1.3) this formulation is rarely seen in practice.

A similar formula based on the drag coefficient of sand particles on Figure 5.2a with $C_D = \tilde{C}_D + 24v/\omega d_s$ gives:

$$\omega = \frac{v_m}{d_s}\frac{12}{\tilde{C}_D}\left(\left(1 + \frac{\tilde{C}_D d_*^3}{108} \right)^{0.5} - 1 \right) \tag{5.23c}\blacklozenge\blacklozenge$$

where $\tilde{C}_D$ can be approximated by $\tilde{C}_D = 0.5/C_o^2$. This formulation is useful for particles of different shapes. Finally, when $\tilde{C}_D = 1.5$ for natural sands and gravels as shown in Figure 5.2b, it reduces to

$$\omega = \frac{8v_m}{d_s}\left(\left(1 + 0.0139d_*^3 \right)^{0.5} - 1 \right) \tag{5.23d}\blacklozenge\blacklozenge$$

where the dimensionless particle diameter d_* is defined as

$$d_* = d_s \left[\frac{(G-1)g}{v_m^2} \right]^{1/3} \tag{5.23e}$$

Equation (5.23d) estimates the fall velocity of particles under a wide range of Reynolds numbers. Approximate values of fall velocity in clear water are given in

Table 5.4. *Clear water fall velocity ω_0 and dimensionless particle diameter d_**

Class name	Particle diameter (mm)	ω_0 at 10°C (mm/s)	ω_0 at 20°C (mm/s)	d_* at 10°C	d_* at 20°C	$\dfrac{\omega}{\sqrt{(G-1)gd_s}}$
Boulder						
Very Large	> 2,048	5,430	5,430	43,271	51,807	0.94
Large	> 1,024	3,839	3,839	21,635	25,903	0.94
Medium	> 512	2,715	2,715	10,817	12,951	0.94
Small	> 256	1,919	1,919	5,409	6,475	0.94
Cobble						
Large	> 128	1,357	1,357	2,704	3,237	0.94
Small	> 64	959	959	1,352	1,618	0.94
Gravel						
Very Coarse	> 32	678	678	676	809	0.94
Coarse	> 16	479	479	338	404	0.94
Medium	> 8	338	338	169	202	0.94
Fine	> 4	237	238	84	101	0.93
Very Fine	> 2	164	167	42	50	0.91
Sand						
Very Coarse	> 1	109	112	21	25	0.86
Coarse	> 0.5	66.4	70.3	10.5	12.6	0.73
Medium	> 0.25	31.3	36	5.3	6.32	0.49
Fine	> 0.125	10.1	12.8	2.6	3.16	0.22
Very Fine	> 0.0625	2.66	3.47	1.3	1.58	0.08
Silt						
Coarse	> 0.031	0.67^a	0.88^a	0.66	0.79	0.03
Medium	> 0.016	0.167^a	0.22^a	0.33	0.395	0.01
Fine	> 0.008	0.042^a	0.055^a	0.165	0.197	0.003
Very Fine	> 0.004	0.010^a	0.014^a	0.082	0.099	0.001
Clay						
Coarse	> 0.002	$2.6 \times 10^{-3\ a}$	$3.4 \times 10^{-3\ a}$	0.041	0.049	4.6×10^{-4}
Medium	> 0.001	$6.5 \times 10^{-4\ a}$	$8.6 \times 10^{-4\ a}$	0.021	0.025	1.7×10^{-4}
Fine	> 0.0005	$1.63 \times 10^{-4\ a}$	$2.1 \times 10^{-4\ a}$	0.010	0.012	5.5×10^{-5}
Very Fine	> 0.00024	$4.1 \times 10^{-5\ a}$	$5.3 \times 10^{-5\ a}$	0.005	0.006	2×10^{-5}

[a] Possible flocculation (see Section 5.4.3).

Table 5.4. The effect of particle shape on the settling velocity of coarse particles can be estimated from $\tilde{C}_D = 0.5/C_o^2 = 0.5ab/c^2$ in Equation (5.23c).

It is interesting to define the ratio of the settling velocity ω to $\sqrt{(G-1)gd_s}$, which for natural particles $\left(\tilde{C}_D = 1.5\right)$ is approximately

$$\frac{\omega}{\sqrt{(G-1)gd_s}} \cong \frac{8v_m}{d_s\sqrt{(G-1)gd_s}} \left(\left(1+\frac{d_*^3}{72}\right)^{0.5} - 1\right) = \frac{8}{d_*^{3/2}} \left(\left(1+\frac{d_*^3}{72}\right)^{0.5} - 1\right)$$

$$(5.24)$$

This ratio thus becomes solely a function of d_*. Values in Table 5.4 show that this ratio becomes close to unity for particles coarser than 1 mm. It can be concluded that $\omega \cong \sqrt{(G-1)gd_s}$ for gravels and cobbles. Also, as a second practical approximation from Table 5.4, the settling velocity of sands in mm/s is about 100 times the particle diameter in mm. Table 5.4 clearly shows that: (1) the settling velocity of gravels and cobbles increases with the square root of d_s; (2) the settling velocity of silts and clays decreases until the second power of d_s; and (3) temperature does affect the settling velocity of silts and clay, but does not affect gravels and cobbles.

5.4.3 Flocculation

Flocculation is the property of very fine sediments to aggregate and settle as a flocculated mass. In general, flocculation is enhanced at high sediment concentration, high salt content, and higher fluid temperature. According to Migniot (1989), the settling velocity of flocculated particles ω_f can be calculated given the settling velocity of dispersed particles ω from

$$\omega_f = \frac{250}{d_s^2}\omega; \text{ when } d_s < 40\mu\text{m} \tag{5.25a}$$

where d_s is the particle diameter in microns (1 μm $= 10^{-6}$m). Combined with Stokes law, the flocculated settling velocity is approximately 0.15 to 0.6 mm/s and does not vary much with the particle size. This flocculated settling velocity is

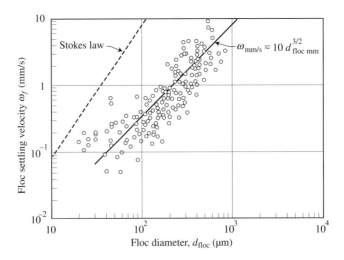

Figure 5.3. Flocculated settling velocity versus floc size (modified after Winterwerp, 1999)

comparable to the settling of medium to coarse silt particles. Flocculation is not important on individual particles larger than coarse silts ($d_s > 0.04$ mm).

Winterwerp (1999) compiled measurements of settling velocity as a function of floc size, as shown in Figure 5.3. As a first approximation

$$\omega_{f \text{ mm/s}} \cong 10 d_{\text{floc mm}}^{3/2} \qquad (5.25b)$$

This requires floc size measurements but nevertheless shows that settling velocities up to 10 mm/s are possible. Floc sizes larger than 1 mm are not common.

Deflocculation can be secured in laboratory settling experiments by adding a 1% dilution of a dispersing agent comprised of 35.7 g/l of sodium hexametaphosphate and 7.9 g/l of sodium carbonate. The comparison of settling with and without deflocculant is a useful laboratory procedure to determine whether flocculation is present.

5.4.4 Oden curve

When several size fractions are settling at the same time, the Oden curve method is used to separate the settling proportions of different size fractions. The following example is used to explain the process. Consider 50 g of particle size A settling in 10 seconds, and 50 g of particle size B settling in 20 seconds. When the two are mixed, 75 g will settle in 10 seconds and 100 g will settle in 20 seconds, as sketched in Figure 5.4a. To separate the relative masses of A and B from the total, the rate of settling after 10 seconds needs to be subtracted from the total settling at 10 seconds. Projecting the tangent at 10 seconds to the origin thus gives the correct proportions of A and B in the mixture. For instance, subtracting 25 g from the 75g (tangent line to the origin) gives the correct mass of 50 g for size A.

The Oden curve method thus requires drawing the tangent line to the settling curve on a linear scale. The reading at the origin determines the percentage of material for each size fraction. Care is needed in drawing the tangents to the Oden curve on a linear scale, as the curvature greatly affects the position of the intercept on the percentage scale. For samples containing clays, the slope of the curve does not approach zero at the time of the last scheduled withdrawal because clay particles are still settling. Obviously, the Oden curve should have a gradually decreasing slope. The Oden curve sketched in Figure 5.4b shows the percentage by weight of sediment in suspension as a function of time. If tangents are drawn to the Oden curve at any two consecutive times of withdrawal t_i and t_{i+1} (from Figure 5.4b), the tangents then intersect the ordinate axis at $\%W_i$ and $\%W_{i+1}$, respectively. The difference between the percentage, $\%W_{i+1}$ and $\%W_i$ represents the percentage by weight of material in the size range corresponding to the settling times t_i and t_{i+1}. For instance, Figure 5.4b shows that the sediment sample contains

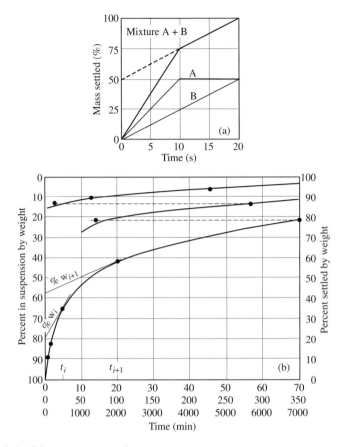

Figure 5.4. Oden curve example

78%–58% = 20% of coarse silt (0.0312 mm < d_s < 0.0625 mm), see Problem 5.4 for details.

5.5 Rate of energy dissipation

With reference to Section 3.7, the rate of energy dissipation can be expressed in terms of the stresses applied on a fluid element (Fortier, 1967). The rate of work done per unit mass χ by external forces on a fluid element is the product of the gradient of stress by the respective velocity component.

$$\chi = \left(\frac{\partial \sigma_x}{\partial x} + \frac{\partial \tau_{yx}}{\partial y} + \frac{\partial \tau_{zx}}{\partial z} \right) v_x + \left(\frac{\partial \tau_{xy}}{\partial x} + \frac{\partial \sigma_y}{\partial y} + \frac{\partial \tau_{zy}}{\partial z} \right) v_y$$

$$+ \left(\frac{\partial \tau_{xz}}{\partial x} + \frac{\partial \tau_{yz}}{\partial y} + \frac{\partial \sigma_z}{\partial z} \right) v_z \tag{5.26a}$$

Expanding these terms gives:

$$
\chi = \left[\frac{\partial}{\partial x}(\sigma_x v_x + \tau_{yx} v_y + \tau_{zx} v_z) + \frac{\partial}{\partial y}(\tau_{xy} v_x + \sigma_y v_y + \tau_{zy} v_z) \right.
$$

$$
\left. + \frac{\partial}{\partial z}(\tau_{xz} v_x + \tau_{yz} v_y + \sigma_z v_z) \right] - \left(\sigma_x \frac{\partial v_x}{\partial x} + \sigma_y \frac{\partial v_y}{\partial y} + \sigma_z \frac{\partial v_z}{\partial z} \right) - \tau_{zx}
$$

$$
\left(\frac{\partial v_z}{\partial x} + \frac{\partial v_x}{\partial z} \right) - \tau_{zy} \left(\frac{\partial v_z}{\partial y} + \frac{\partial v_y}{\partial z} \right) - \tau_{xy} \left(\frac{\partial v_x}{\partial y} + \frac{\partial v_y}{\partial x} \right) \quad (5.26b)
$$

The terms in brackets on the first line of Equation (5.26b) describe the rate of increase in mechanical energy of the fluid and are not dissipative. For Newtonian mixtures, the terms outside the brackets describe the rate of energy dissipation per unit mass χ_D, which from Equations (5.2) and (5.3) can be rewritten as

$$
\chi_D = -p \left(\frac{\partial v_x}{\partial x} + \frac{\partial v_y}{\partial y} + \frac{\partial v_z}{\partial z} \right) + 2\mu_m \left[\left(\frac{\partial v_x}{\partial x} \right)^2 + \left(\frac{\partial v_y}{\partial y} \right)^2 + \left(\frac{\partial v_z}{\partial z} \right)^2 \right]
$$

$$
- \frac{2}{3} \mu_m \left[\frac{\partial v_x}{\partial x} + \frac{\partial v_y}{\partial y} + \frac{\partial v_z}{\partial z} \right]^2
$$

$$
+ \mu_m \left[\left(\frac{\partial v_y}{\partial x} + \frac{\partial v_x}{\partial y} \right)^2 + \left(\frac{\partial v_z}{\partial y} + \frac{\partial v_y}{\partial z} \right)^2 + \left(\frac{\partial v_x}{\partial z} + \frac{\partial v_z}{\partial x} \right)^2 \right] \quad (5.27)
$$

The first term of Equation (5.27) expresses the rate at which the fluid is compressed and vanishes for incompressible fluids. The last three terms in brackets involve the fluid viscosity μ_m and their sum determines the rate at which energy is dissipated through viscous action per unit volume of fluid. The dissipation χ_D for incompressible Newtonian fluids in Cartesian, cylindrical, and spherical coordinates are given in Table 5.5. In Cartesian coordinates, the dissipation function then reduces to

$$
\chi_D = 2\mu_m \left[\left(\frac{\partial v_x}{\partial x} \right)^2 + \left(\frac{\partial v_y}{\partial y} \right)^2 + \left(\frac{\partial v_z}{\partial z} \right)^2 \right] + \mu_m \left(\varnothing_x^2 + \varnothing_y^2 + \varnothing_z^2 \right) \quad (5.28)
$$

Energy is dissipated through linear deformation in the first term in brackets of Equation (5.28), and through angular deformation in the second term in brackets of Equation (5.28).

Table 5.5. *Dissipation function χ_D for incompressible Newtonian fluids*

Cartesian

$$\chi_D = \mu_m \left\{ 2 \left[\left(\frac{\partial v_x}{\partial x} \right)^2 + \left(\frac{\partial v_y}{\partial y} \right)^2 + \left(\frac{\partial v_z}{\partial z} \right)^2 \right] \right.$$

$$\left. + \left[\frac{\partial v_y}{\partial x} + \frac{\partial v_x}{\partial y} \right]^2 + \left[\frac{\partial v_z}{\partial y} + \frac{\partial v_y}{\partial z} \right]^2 + \left[\frac{\partial v_x}{\partial z} + \frac{\partial v_z}{\partial x} \right]^2 \right\}$$

Cylindrical

$$\chi_D = \mu_m \left\{ 2 \left[\left(\frac{\partial v_r}{\partial r} \right)^2 + \left(\frac{1}{r} \frac{\partial v_\theta}{\partial \theta} + \frac{v_r}{r} \right)^2 + \left(\frac{\partial v_z}{\partial z} \right)^2 \right] \right.$$

$$\left. + \left[r \frac{\partial}{\partial r} \left(\frac{v_\theta}{r} \right) + \frac{1}{r} \frac{\partial v_r}{\partial \theta} \right]^2 + \left[\frac{1}{r} \frac{\partial v_z}{\partial \theta} + \frac{\partial v_\theta}{\partial z} \right]^2 + \left[\frac{\partial v_r}{\partial z} + \frac{\partial v_z}{\partial r} \right]^2 \right\}$$

Spherical

$$\chi_D = \mu_m \left\{ 2 \left[\left(\frac{\partial v_r}{\partial r} \right)^2 + \left(\frac{1}{r} \frac{\partial v_\theta}{\partial \theta} + \frac{v_r}{r} \right)^2 + \left(\frac{1}{r \sin \theta} \frac{\partial v_\varphi}{\partial \varphi} + \frac{v_r}{r} + \frac{v_\theta \cot \theta}{r} \right)^2 \right] \right.$$

$$+ \left[r \frac{\partial}{\partial r} \left(\frac{v_\theta}{r} \right) + \frac{1}{r} \frac{\partial v_r}{\partial \theta} \right]^2 + \left[\frac{\sin \theta}{r} \frac{\partial}{\partial \theta} \left(\frac{v_\varphi}{\sin \theta} \right) + \frac{1}{r \sin \theta} \frac{\partial v_\theta}{\partial \varphi} \right]^2$$

$$\left. + \left[\frac{1}{r \sin \theta} \frac{\partial v_r}{\partial \varphi} + r \frac{\partial}{\partial r} \left(\frac{v_\varphi}{r} \right) \right]^2 \right\}$$

Example 5.2 Viscous energy dissipation around a sphere

Calculate the total rate of energy dissipation χ_D for flow around a sphere of diameter d_s in creeping motion at a velocity u_∞ in an infinite mass of fluid. The dissipation function in spherical coordinates χ_D is integrated outside a sphere of radius $R = d_s/2$ on the elemental volume $dr \, (r \, d\theta) \, r \sin \theta \, d\varphi$:

$$X_D = \int_0^{2\pi} \int_0^{\pi} \int_{d_s/2}^{r} \chi_D r^2 dr \sin \theta \, d\theta \, d\varphi \qquad \text{(E-5.2.1)}$$

For incompressible flow around the sphere $(v_\varphi = 0)$ and $\partial/\partial \varphi = 0$ the dissipation function in spherical coordinates from Table 5.5 reduces to

$$X_D = 2\pi \mu_m \int_0^{\pi} \int_{d_s/2}^{r} \left\{ 2 \left(\frac{\partial v_r}{\partial r} \right)^2 + 2 \left(\frac{1}{r} \frac{\partial v_\theta}{\partial \theta} + \frac{v_r}{r} \right)^2 + 2 \left(\frac{v_r}{r} + \frac{v_\theta \cot \theta}{r} \right)^2 \right.$$

$$\left. + \left[r \frac{\partial}{\partial r} \left(\frac{v_\theta}{r} \right) + \frac{1}{r} \frac{\partial v_r}{\partial \theta} \right]^2 \right\} r^2 \sin \theta \, dr \, d\theta$$

which from the velocity profile in Equation (5.8) yields

$$X_D = 3\pi \mu_m u_\infty^2 d_s \left(1 - \frac{3d_s}{4r} + \frac{d_s^3}{8r^3} - \frac{d_s^5}{64r^5} \right)$$

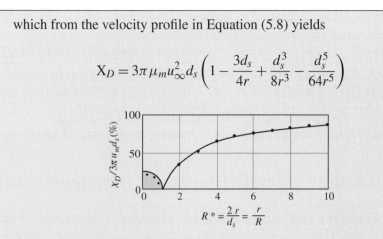

Figure E-5.2.1 Total energy dissipated around a sphere

This function plotted on Figure E-5.2.1 shows that 50% of the energy is dissipated within three times the particle diameter. The total energy dissipated at $r = \infty$ is,

$$X_D = 3\pi \mu_m u_\infty^2 d_s = F_D u_\infty$$

The total rate of energy dissipated per unit volume around the sphere ∀ is

$$\frac{X_D}{\forall} = \frac{18\pi \mu_m u_\infty^2 d_s}{\pi d_s^3} = 18\mu_m \left(\frac{u_\infty}{d_s} \right)^2 \qquad \text{(E-5.2.2)}$$

This equation indicates that, for creeping motion at constant velocity u_∞, the total rate of energy dissipation per unit volume of the sphere is inversely proportional to d_s^2. On the other hand, when u_∞ corresponds to the fall velocity ω, the rate of energy dissipation becomes proportional to d_s^2 for particles finer than silts from Equation (5.21) and inversely proportional to d_s for particles coarser than gravels from Equation (5.22). A maximum is obtained for sand sizes. Sand-sized particles are thus found to be most effective for dissipating viscous energy through settling.

5.6 Laboratory measurements of particle size

Two principal functions of a sediment laboratory are to determine: (1) the particle size distribution of suspended sediment and bed material; and (2) the concentration of suspended sediment. Other functions include the determination of roundness and shape of individual grains and their mineral composition, the amount of

Table 5.6. *Recommended methods for particle size distribution*

Method of analysis	Size range (mm)	Concentration (mg/l)	Sediment mass (g)
Sieves	0.062–32	–	1 g to 10 kg
VA tube	0.062–2.0	–	0.05–15.0
Pipette	0.002–0.062	2,000–5,000	1.0–5.0
BW tube	0.002–0.062	1,000–3,500	0.5–1.8

organic matter, the specific gravity of sediment particles, and the specific weight of deposits.

There are essentially two ways to determine particle size distributions in the laboratory: (1) direct measurement; and (2) sedimentation methods. The direct methods, also discussed in Section 2.3, include immersion and displacement volume measurements, some direct measurements of circumference or diameter, and semi-direct measurements of particle diameter using sieves. Sedimentation methods relate fall velocity measurements to particle size. Standard procedures include the visual accumulation tube (VAT), the bottom withdrawal tube (BWT), the pipette, and the hydrometer. The VAT, used only for sands, operates as a stratified system where particles start from a common source at the top and deposit at the bottom of the tube, according to settling velocities. The pipette and BWT, used only for silts and clays, operate as dispersed systems where particles begin to settle from an initially uniform dispersion. For the BWT method, the distribution is obtained from the quantity of sediment remaining in suspension after various settling times when the coarser sizes and heavier concentrations are withdrawn at the bottom of the tube. Table 5.6 indicates recommended size ranges, analysis concentration, and weight of sediment for the sieves, VAT, BWT, and pipette methods.

Extraneous organic materials should be removed from samples by adding about 5 ml of a 6% solution of hydrogen peroxide for each gram of dry sample in 40 ml of water. The solution must be stirred thoroughly and covered for about 10 min. Large fragments of organic material may then be skimmed off when they are free of sediment particles. If oxidation is slow, or after it has slowed, the mixture is heated to 93°C, stirred occasionally, and more hydrogen peroxide solution added as needed. After the reaction has stopped, the sediment must be carefully washed two or three times with distilled water.

To ensure complete deflocculation of silts and clays when using the pipette, BWT, and hydrometer methods, 1 ml of dispersing agent should be used for each 100 ml sample. Adding 35.7 g/l of sodium hexametaphosphate and 7.99 g/l of sodium carbonate is recommended to prevent flocculation.

The following indirect methods of particle size measurement involve liquid suspensions and fall velocity: the visual accumulation tube (VAT), the bottom withdrawal tube (BWT), the pipette, and the hydrometer.

5.6.1 Visual accumulation tube method (VAT)

The VAT is a fast, economical, and reasonably accurate method of determining the sand size distribution based on fall velocity measurements. Silts finer than 0.062 mm are removed by either wet sieving or by sedimentation methods, and analyzed separately using either the pipette or the BWT method. In some instances, sieving must be employed to remove particles coarser than 2.0 mm.

The equipment for the VAT method of analysis consists primarily of the special settling tube and a recording mechanism shown in Figure 5.5a. The VAT analysis results in a continuous pen trace of sediment accumulation as a function of time. The chart is calibrated at a given temperature to give the relative amount of each size in terms of fall diameter and percentage finer than a given size. The sediment size distribution is corrected for the finer or coarser fractions removed prior to the

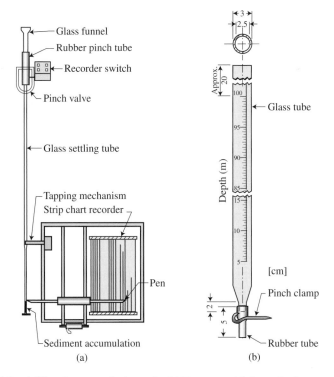

(a)   (b)

Figure 5.5. a) Visual accumulation tube b) Bottom withdrawal tube

VAT analysis. For instance, if 30% of the original sample, finer than 0.062 mm was removed from the VAT analysis, the percentage finer scale of the VAT analysis is corrected to start at 30%.

5.6.2 Bottom withdrawal tube method (BWT)

The bottom withdrawal tube (Fig. 5.5b) is used where the sample contains a very small quantity of fine sands and silts. The sands should be removed and analyzed separately using the VAT. A 100 ml suspension is poured into a 1 m-high settling tube and 10 mℓ samples are withdrawn at the lower end of the tube following the schedule given in Table 5.7. The samples are then poured into evaporating dishes and the sample containers washed with distilled water. The previously weighed evaporating dishes are placed in the oven to dry at a temperature just below the boiling point, to avoid splattering by boiling. When the evaporating dishes or flasks are visibly dry, the temperature is raised to 110°C for 1 hr, after which the containers are transferred from the oven to a desiccator and allowed to cool before weight measurements. The sediment size distribution is then calculated using the Oden curves method of Section 5.4.4.

Table 5.7. *BWT withdrawal time in minutes*

Temperature (°C)	Particle diameter (mm)							
	0.25	0.125	0.0625	0.0312	0.0156	0.0078	0.0039	0.00195
18	0.522	1.48	5.02	20.1	80.5	322	1,288	5,154
19	0.515	1.45	4.88	19.6	78.5	314	1,256	5,026
20	0.508	1.41	4.77	19.2	76.6	306	1,225	4,904
21	0.503	1.39	4.67	18.7	74.9	299	1,198	4,794
22	0.497	1.37	4.55	18.3	73.0	292	1,168	4,675
23	0.488	1.34	4.45	17.8	71.3	285	1,141	4,566
24	0.485	1.32	4.33	17.4	69.6	279	1,114	4,461
25	0.478	1.30	4.25	17.0	68.1	273	1,090	4,361
26	0.472	1.28	4.15	16.7	66.6	266	1,065	4,263
27	0.467	1.26	4.05	16.3	65.1	260	1,042	4,169
28	0.462	1.24	3.97	15.9	63.7	255	1,019	4,079
29	0.455	1.22	3.88	15.6	62.3	249	997	3,991
30	0.450	1.20	3.80	15.3	61.0	244	976	3,907
31	0.445	1.18	3.71	14.9	59.7	239	956	3,825
32	0.442	1.17	3.65	14.6	58.5	234	936	3,747
33	0.438	1.15	3.58	14.2	57.3	229	917	3,671
34	0.435	1.13	3.51	13.9	56.1	244	898	3,494

Note: Time in minutes required for spheres having a specific gravity of 2.65 to fall 1m in water at varying temperatures.

5.6.3 Pipette method

The pipette method is a reliable indirect method to determine the particle size distribution of silts and coarse clays ($d_s < 0.062$ mm). The concentration of a quiescent suspension is measured at a predetermined depth (Fig. 5.6) as a function of settling time. Particles having a settling velocity greater than a given size settle below the point of withdrawal after a certain time determined in Table 5.8. The

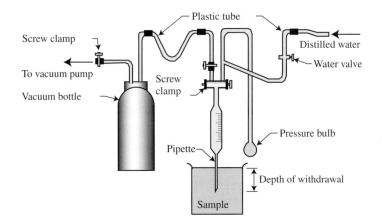

Figure 5.6. Pipette

Table 5.8. *Pipette withdrawal time*

	Diameter of particles (mm); depth of withdrawal (cm)					
	0.062	0.031	0.016	0.008	0.004	0.002
Temperature (°C)	15	15	10	10	5	5
20	44 s	2 m 52 s	7 m 40 s	30 m 40 s	61 m 19 s	4 h 5 m
21	42	2 48	7 29	29 58	59 50	4 0
22	41	2 45	7 18	29 13	28 22	3 54
23	40	2 41	7 8	28 34	57 5	3 48
24	39	2 38	6 58	27 52	55 41	3 43
25	38	2 34	6 48	27 14	54 25	3 38
26	37	2 30	6 39	26 38	53 12	3 33
27	36	2 27	6 31	26 2	52 2	3 28
28	36	2 23	6 22	25 28	50 52	3 24
29	35	2 19	6 13	24 53	49 42	3 19
30	34	2 16	6 6	24 22	48 42	3 15

Note: The values in this table are based on particles of assumed spherical shape with an average specific gravity of 2.65, the constant of acceleration due to gravity $= 980$ cm/s^2, and viscosity varying from 0.010087 cm^2/s at 20°C to 0.008004 cm^2/s at 30°C.

time and depth of withdrawal are predetermined on the basis of the Stokes law in Equation (5.21) at a given water temperature. Samples are dried, weighed, and the Oden curve method is applied to calculate the particle size distribution.

5.6.4 Hydrometer method

The hydrometer measures the change in immersed volume of a floating object in a dilute suspension. The buoyancy force from Chapter 3 being equal to the weight of the object, the time change in submerged volume corresponds to the time change in specific weight of the mixture due to settling of particles in suspension.

Exercises

5.1 Demonstrate that $\tau_{xy} = \tau_{yx}$ from the sum of moments about the center of an infinitesimal element.

5.2 Derive the x-component of the Navier–Stokes equations in Table 5.1 from the equation of motion in Equation (3.14a) and the stress tensor components for incompressible Newtonian fluids in Equations (5.2) and (5.3).

♦ 5.3 Differentiate Equations (5.4a) with respect to y and (5.4b) with respect to x, and subtract them to derive Equation (5.5c).

♦5.4 Determine the shear stress component $\tau_{r\theta}$ in Equation (5.9c) from the tensor $\tau_{r\theta}$ in spherical coordinates (Table 5.3) and the velocity relationships in Equations (5.8a and b).

♦♦5.5 (a) Integrate the shear stress distribution from Equation (5.9c) to determine the surface drag in Equation (5.11) from Equation (5.10); and (b) integrate the dynamic pressure distribution from Equation (5.9b) to obtain the form drag in Equation (5.13) from Equation (5.12).

♦5.6 Derive Rubey's fall velocity equation in Equations (5.23a and b) from combining Equation (5.22b) and $C_D= 2 + 24/\mathrm{Re}_p$.

5.7 Substitute the appropriate stress tensor components for the flow of Newtonian fluids in Cartesian coordinates in Equations (5.2) and (5.3) into the last four terms in parentheses of Eq. (5.26), to obtain the energy dissipation function in Eq. (5.27).

5.8 Describe each member of the dissipation function in Equation (5.28) in terms of the fundamental modes of deformation (translation, linear deformation, etc.).

5.9 Substitute the velocity profile around a sphere from Equation (5.8) into Equation (E-5.2.1) to find the energy dissipation function in Figure E-5.2.1.

5.10 From Example 5.1, assume that the critical velocity u_c is the flow velocity at the top of the sphere. Assume $\tau_c = \mu_m u_c/R$ and calculate the corresponding Shields parameter.

(*Answer*: $\tau_{*c} = \tau_c/\gamma_s - \gamma_m)d_s = 4/9$, this critical shear stress practically corresponds to the beginning of motion of fine particles (Chapter 7)).

Problems
♦♦*Problem 5.1*

Plot the velocity, dynamic pressure, and shear stress distributions around the surface of a sphere for creeping motion given by Stokes' law in Equations (5.8) and (5.9) and compare with irrotational flow without circulation (Problem 4.3).

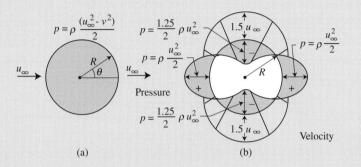

(a) (b)

Figure P-5.1 Flow around a sphere

[*Answer:* (a) Creeping motion in a viscous fluid: $v = 0$ everywhere at the sphere surface, $\tau \neq 0$ everywhere at the sphere surface except at $\theta = 0°$ and $180°$; $p_d \neq 0$ everywhere at the sphere surface except at $\theta = 90°$, (b) Irrotational flow: $\tau = 0$ everywhere at the sphere surface, $v \neq 0$ everywhere at the sphere surface except at $\theta = 0°$ and $180°$, and $p_d \neq 0$ everywhere at the sphere surface except at $\theta = 41.8°$ and $138.2°$]

♦*Problem 5.2*

Plot Rubey's relationship for the drag coefficient C_D on Figure 5.2. How does it compare with the experimental measurements? At a given Re_p, which of Equations (5.17) and (5.19) induces larger settling velocities?

♦♦*Problem 5.3*

Evaluate the dissipation function χ_D from Table 5.5 for a vertical axis Rankine vortex described in cylindrical coordinates by:

(a) forced vortex $v_\theta = \frac{\Gamma_v r}{2\pi r_o^2}, v_z = v_r = 0$ (rotational flow for $r < r_o$)
and
(b) free vortex $v_\theta = \frac{\Gamma_v}{2\pi r}, v_z = v_r = 0$ (irrotational flow for $r > r_o$)

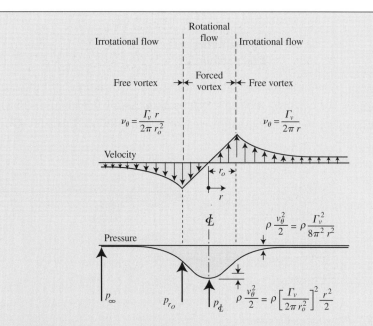

Figure P-5.3 Rankine vortex

(*Answer:* $\chi_D = \mu_m \left(r \frac{\partial}{\partial r} \left(\frac{v_\theta}{r} \right) \right)^2$, (a) $\chi_D = 0$, where the flow is rotational for $r < r_0$; and

(b) $\chi_D = \frac{\mu_m \Gamma_v^2}{\pi^2 r^4} \neq 0$, where the flow is irrotational for $r > r_0$.)

♦♦*Problem 5.4*

The sediment size distribution of a 1200 mg sample is to be determined using the BWT. If the water temperature is 24°C and the solid weight for each 10 ml withdrawal is given, use Table 5.7 to determine the sampling times. Plot the Oden curve, complete the table below and determine the particle size distribution.

Particle diameter (m)	Withdrawal time[a] (min)	Sample volume (ml)	Dry weight of sediment (mg)	Cumulative dry weight (mg)	Percent settled	Percent Finer[b] (%)
0.25	0.485	10	144	144	12	
0.125		10	72	216	18	
0.0625		10	204	420	35	78
0.0312		10	264	684	57	58
0.0156		10	252			
0.0078		10	84			
0.0039		10	48	1,068		
0.00195	4,461	10	45			

[a] Withdrawal times are obtained from Table 5.7
[b] See Figure 5.4b

♦♦*Problem 5.5*

A gravel particle is 32 mm long, 16 mm wide, and 8 mm thick. Determine the volume, sphericity, and Corey shape factor of the particle. Estimate the settling velocity of this particle from the results in Figure 5.2a. Compare with the settling velocity of a sphere with equivalent volume.

♦*Problem 5.6*

Check the BWT withdrawal time in minutes from Table 5.7 using: (1) Rubey's equation (5.23a); and (2) Equation (5.23d).

♦*Problem 5.7*

Check the pipette withdrawal times in Table 5.8 using: (1) Rubey's equation (5.23a); and (2) Equation (5.23d).

♦♦*Problem 5.8*

Determine the particle size distribution of fine sediment from the San Luis Canal in California from the data below. The settling height is 20 cm and the experiment was carried out at room temperature.

Time (min)	% Settled
0.03	10
0.17	20
0.33	30
1	40
2	50
5	60
9	70
30	80
210	90
108,000	100

♦♦*Computer problem 5.1*

Write a simple computer program to determine the particle size d_s, the fall velocity ω, the flocculated fall velocity ω_f, the particle Reynolds number Re_p, the dimensionless particle diameter d_*, and the time of settling per meter for water at 5°C, and complete the table below.

Class name	d_s (mm)	ω (cm/s)	ω_f (cm/s)	Re$_p$	d_*	Settling time
Medium clay						
Medium silt						
Medium sand						
Medium gravel						
Small cobble						
Medium boulder						

6

Turbulent velocity profiles

Most open-channel flows are characterized by irregular velocity fluctuations indicating turbulence. The turbulent fluctuation superimposed on the principal motion is complex in its detail and still poses difficulties for mathematical treatment. This chapter outlines the fundamentals of turbulence with emphasis on turbulent velocity profiles (Section 6.1), turbulent flow along smooth and rough boundaries (Section 6.2), resistance to flow (Section 6.3), departure from logarithmic velocity profiles (Section 6.4), and open-channel flow measurements in Section 6.5.

When describing turbulent flow in mathematical terms, it is convenient to separate the mean motion (notation with overbar) from the fluctuation (notation with superscript +) as sketched in Figure 6.1. Denoting a fluctuating parameter $\hat{v}_x$ of time, average value $\hat{v}_x$, and fluctuation $\hat{v}_x$, the pressure and the velocity components can be rewritten respectively as:

$$\hat{p} = \bar{p} + p^+ \tag{6.1a}$$

$$\hat{v}_x = \bar{v}_x + v_x^+ \tag{6.1b}$$

$$\hat{\tau} = \bar{\tau} + \tau^+ \tag{6.1c}$$

the time-averaged values at a fixed point in space are given by

$$v_x = \bar{v}_x = \frac{1}{t_1} \int_{t_0}^{t_0+t_1} \hat{v}_x dt \tag{6.2}$$

Taking the mean values over a sufficiently long time interval t_1, the time-averaged values of the fluctuations equal zero, thus $\overline{v_x^+} = \overline{v_y^+} = \overline{v_z^+} = \overline{p^+} = 0$. Likewise, the time-averaged values of the derivatives of velocity fluctuations,

113

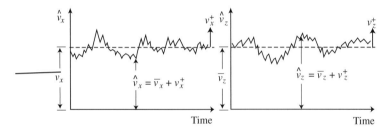

Figure 6.1. Velocity measurements versus time

such as $\overline{\partial v_x^+}/\partial x, \overline{\partial^2 v_x^+}/\partial x^2, \overline{\partial \overline{v}_x v_x^+}/\partial x^2$, also vanish owing to Equation (6.2). The quadratic terms arising from the products of cross-velocity fluctuations like $\overline{v_x^+ v_x^+}, \overline{v_x^+ v_y^+}, \overline{\partial v_x^+ v_y^+}/\partial x$, however, do not vanish. The overbar of simple time-averaged parameters is omitted for notational convenience.

It is seen that both the time-averaged velocity components and the fluctuating components satisfy the equation of continuity, thus for incompressible fluids,

$$\frac{\partial v_x}{\partial x} + \frac{\partial v_y}{\partial y} + \frac{\partial v_z}{\partial z} = 0 \tag{6.3a}$$

$$\frac{\partial v_x^+}{\partial x} + \frac{\partial v_y^+}{\partial y} + \frac{\partial v_z^+}{\partial z} = 0 \tag{6.3b}$$

This formulation of continuity indicates that the magnitude of the fluctuations in $\left|v_x^+\right|, \left|v_y^+\right|$, and $\left|v_z^+\right|$ should remain in similar proportions.

The velocity and pressure terms from Equation (6.1) are substituted into the Navier–Stokes equation (Table 5.1a) to give the following acceleration terms

$$\frac{\partial v_x}{\partial t} + v_x \frac{\partial v_x}{\partial x} + v_y \frac{\partial v_x}{\partial y} + v_z \frac{\partial v_x}{\partial z} = g_x - \frac{1}{\rho} \frac{\partial p}{\partial x} + v_m \nabla^2 v_x$$
$$- \left(\frac{\overline{\partial v_x^+ v_x^+}}{\partial x} + \frac{\overline{\partial v_y^+ v_x^+}}{\partial y} + \frac{\overline{\partial v_z^+ v_x^+}}{\partial z} \right) \tag{6.4a}$$

$$\frac{\partial v_y}{\partial t} + v_x \frac{\partial v_y}{\partial x} + v_y \frac{\partial v_y}{\partial y} + v_z \frac{\partial v_y}{\partial z} = g_y - \frac{1}{\rho} \frac{\partial p}{\partial y} + v_m \nabla^2 v_y$$
$$- \left(\frac{\overline{\partial v_x^+ v_y^+}}{\partial x} + \frac{\overline{\partial v_y^+ v_y^+}}{\partial y} + \frac{\overline{\partial v_z^+ v_y^+}}{\partial z} \right) \tag{6.4b}$$

$$\underbrace{\frac{\partial v_z}{\partial t}}_{local} + \underbrace{v_x \frac{\partial v_z}{\partial x} + v_y \frac{\partial v_z}{\partial y} + v_z \frac{\partial v_z}{\partial z}}_{convective} = \underbrace{g_z}_{gravitational} - \underbrace{\frac{1}{\rho}\frac{\partial p}{\partial z}}_{\substack{pressure \\ gradient}} + \underbrace{v_m \nabla^2 v_z}_{viscous}$$

$$\underbrace{-\left(\frac{\partial \overline{v_x^+ v_z^+}}{\partial x} + \frac{\partial \overline{v_y^+ v_z^+}}{\partial y} + \frac{\partial \overline{v_z^+ v_z^+}}{\partial z} \right)}_{turbulent\ fluctuations} \tag{6.4c}$$

In addition to the terms found in the Navier–Stokes equations, three additional cross-products of velocity fluctuations are obtained from the convective acceleration terms on the left-hand side of Equation (6.4). This formulation is equivalent to Equation (3.14) and Table 5.1, except that shear stresses are now composed of viscous and turbulent terms:

$$\tau_{xx} = 2\mu_m \frac{\partial v_x}{\partial x} - \overline{\rho v_x^+ v_x^+} \tag{6.5a}$$

$$\tau_{yy} = 2\mu_m \frac{\partial v_y}{\partial y} - \overline{\rho v_y^+ v_y^+} \tag{6.5b}$$

$$\tau_{zz} = 2\mu_m \frac{\partial v_z}{\partial z} - \overline{\rho v_z^+ v_z^+} \tag{6.5c}$$

and

$$\tau_{xy} = \tau_{yx} = \mu_m \left(\frac{\partial v_x}{\partial y} + \frac{\partial v_y}{\partial x} \right) - \overline{\rho v_x^+ v_y^+} \tag{6.5d}$$

$$\tau_{xz} = \tau_{zx} = \mu_m \left(\frac{\partial v_x}{\partial z} + \frac{\partial v_z}{\partial x} \right) - \overline{\rho v_x^+ v_z^+} \tag{6.5e}$$

$$\tau_{yz} = \tau_{zy} = \underbrace{\mu_m \left(\frac{\partial v_y}{\partial z} + \frac{\partial v_z}{\partial y} \right)}_{viscous} - \underbrace{\overline{\rho v_y^+ v_z^+}}_{turbulent} \tag{6.5f}$$

These turbulent acceleration terms provide additional stresses called Reynolds stresses, or apparent stresses, which are usually added to the right-hand side of Equation (6.4). As sketched in Figure 6.2, bed shear stress τ_{zx} increases where v_x^+ and v_z^+ have opposite signs.

6.1 Logarithmic velocity profiles

Consider a thin flat plate set parallel to the main flow direction x. We are interested in describing the time-averaged velocity profile v_x as a function of the distance z away from the plate. It is first considered from continuity that the magnitude of

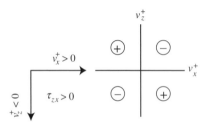

Figure 6.2. Turbulent shear stresses

the transverse velocity fluctuation $\overline{|v_z^+|}$ is the same as that of $\overline{|v_x^+|}$. Drawing an analogy with the mean free path in the kinetic theory of gases, Prandtl imagined the mixing length concept. He hypothesized that the magnitude of velocity fluctuations is proportional to the product of a mixing length and the velocity gradient in the form

$$\overline{|v_x^+|} \sim \overline{|v_z^+|} \sim l_m \frac{dv_x}{dz} \tag{6.6a}$$

in which the proportionality constant l_m denotes the Prandtl mixing length. Near a plate, it is also clear that a positive value for v_x^+ is correlated with a negative value for v_z^+. The average products of velocity fluctuations were then formulated in terms of the mixing length with the aid of Equation (6.6a)

$$\overline{v_x^+ v_z^+} \sim -\overline{|v_x^+||v_z^+|} \tag{6.6b}$$

$$\overline{v_x^+ v_z^+} \sim -l_m^2 \left(\frac{dv_x}{dz}\right)^2 \tag{6.6c}$$

Accordingly, the turbulent shear stress depends on the magnitude of the velocity gradient and the mixing length, or

$$\overline{\tau_{zx}^+} = -\rho_m \overline{v_x^+ v_z^+} = \rho_m l_m^2 \left(\frac{dv_x}{dz}\right)^2 \tag{6.7a}$$

The turbulent shear stress can alternatively be written as a function of the Boussinesq eddy viscosity ε_m:

$$\overline{\tau_{zx}} \cong \rho_m \varepsilon_m \frac{dv_x}{dz} = \rho_m l_m^2 \left(\frac{dv_x}{dz}\right)^2 \tag{6.7b}$$

von Kármán later assumed that the mixing length l_m is proportional to the distance z from the boundary,

$$l_m = \kappa z \tag{6.8}$$

in which κ is the von Kármán constant ($\kappa \simeq 0.41$).

After substituting Equations (6.8) into (6.7a) and (6.5e), the viscous and turbulent shear stress components are

$$\tau_{zx} \cong \underbrace{\mu_m \frac{dv_x}{dz}}_{viscous} + \underbrace{\rho \kappa^2 z^2 \left(\frac{dv_x}{dz}\right)^2}_{turbulent} \tag{6.9}$$

The boundary shear stress τ_o defines the shear velocity u_* as follows:

$$\tau_o = \rho u_*^2 \tag{6.10}\blacklozenge\blacklozenge$$

The shear stress τ_{zx} in the region close to the wall is assumed to remain constant and equal to the boundary shear stress $\tau_o = \rho u_*^2$. After neglecting the viscous shear stress in Equation (6.9), the turbulent velocity profile stems from $\tau_z \equiv \tau_o = \rho u_*^2$, or

$$\sqrt{\frac{\tau_o}{\rho_m}} = u_* = \kappa z \left(\frac{dv_x}{dz}\right) \tag{6.11}$$

Since u_* is constant, the variables v_x and z can be separated and integrated to yield the logarithmic average velocity distribution for steady turbulent flow near a flat boundary

$$\frac{v_x}{u_*} = \frac{1}{\kappa} \ln z + c_o \tag{6.12a}$$

in which c_o is an integration constant evaluated at a distance z_o from the flat boundary. The logarithmic velocity v_{xo} hypothetically equals zero at $z = z_o$, hence

$$\frac{v_x}{u_*} = \frac{1}{\kappa} \ln \frac{z}{z_o} \tag{6.12b}$$

The value of z_o has to be determined from laboratory experiments.

6.2 Smooth and rough plane boundaries

Since the fluid does not slip at the boundary, all turbulent fluctuations must vanish at the surface and remain very small in their immediate neighborhood. It is clear from Equation (6.9) that the viscous stress is dominant as $z \to 0$. The laminar velocity profile near a smooth boundary thus becomes linear because $\tau_0 = \tau_{zx} = \rho u_*^2 = \mu_m dv_x/dz$, or

$$\frac{v_x}{u_*} = \frac{u_* z}{\nu_m} \quad \text{as } z \to 0 \tag{6.13a}$$

Farther away from a plane boundary, the turbulent shear stress becomes dominant and the velocity profile becomes logarithmic as described by Equation (6.12).

Laboratory experiments show that the value of z_o for a smooth boundary is approximately $z_o = v_m/9u_*$. The corresponding turbulent velocity profile when z is large on a smooth boundary is

$$\frac{v_x}{u_x} = \frac{1}{\kappa}\ln\left(\frac{9u_*z}{v_m}\right) \equiv \frac{2.3}{\kappa}\log\left(\frac{9u_*z}{v_m}\right) \cong 5.75\log\left(\frac{u_*z}{v_m}\right) + 5.5 \qquad (6.13b)\blacklozenge$$

The thickness of the laminar sublayer where the flow near the boundary is laminar is obtained from simultaneously solving Equations (6.13a) and (6.13b) at $z = \delta$. This defines the laminar sublayer thickness as

$$\delta = \frac{11.6v_m}{u_*} \qquad (6.14)\blacklozenge\blacklozenge$$

With shear velocities of the order of 0.1 m/s, the laminar sub-layer thickness in open-channel flow is typically of the order of 0.1 mm, which is the size of sands. Generally speaking, a plane bed surface is hydraulically smooth for silts and clays. Velocity measurements for $z \approx \delta$ are shown in Figure 6.3. An approximation to the velocity profile in the buffer zone between turbulent flow and the laminar sublayer has been given by Spalding (in White, 1991)

$$\frac{u_*z}{v_m} = \tilde{u} + 0.1108\left(e^{\kappa\tilde{u}} - 1 - \kappa\tilde{u} - \frac{\left(\kappa\tilde{u}^2\right)^2}{2} - \frac{\left(\kappa\tilde{u}^2\right)^3}{6}\right) \qquad (6.15a)$$

where $\tilde{u} = v_x/u_*$.

The Spalding method is unfortunately implicit and requires iterations to find u at a given z, u_*, and v. Alternatively, an explicit formulation has been proposed by

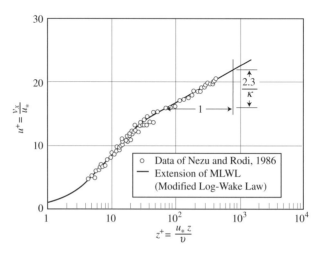

Figure 6.3. Velocity profile in the buffer zone

Guo and Julien (2007)

$$\frac{v_x}{u_*} = 7\tan^{-1}\left(\frac{u_*z}{7v_m}\right) + \frac{7}{3}\tan^{-3}\left(\frac{u_*z}{7v_m}\right) - 0.52\tan^{-4}\left(\frac{u_*z}{7v_m}\right) \qquad (6.15b)$$

This formulation with the argument in radians is useful for $4 < u*z/v < 70$ and the logarithmic velocity profile is valid when $70 < u*z/v < 1000$ offers the advantage to explicitly define the velocity at a given $z, u*$, and v. Figure 6.4 shows a typical velocity profile on a hydraulically smooth surface. Equation (6.15) is applicable when $u*z/v < 70$ and the logarithmic velocity profile is valid when $70 < u*z/v < 1000$.

Natural boundaries are hydraulically smooth when the surface grain roughness $d_s < \delta/3$ or $Re* = u*d_s/v_m < 4$. A transition zone exists where $\delta/3 < d_s < 6\delta$, or $4 < Re* < 70$. Turbulent flows are hydraulically rough when the grain diameter d_s far exceeds the laminar sublayer thickness ($d_s > 6\delta$ or $Re* > 70$). Figure 6.5 illustrates hydraulically smooth and hydraulically rough boundaries.

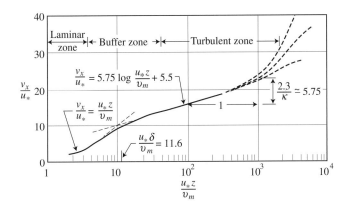

Figure 6.4. Velocity profiles for smooth surfaces

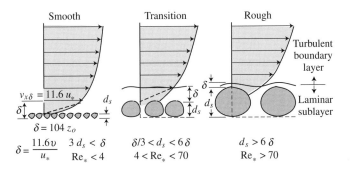

Figure 6.5. Hydraulically smooth and rough boundaries

In early experiments, Nikuradse glued sand particles and measured velocity profiles for turbulent flow over boundaries with grain roughness height k'_s. On rough boundaries, the corresponding value of $z_0 = k'_s/30$.

$$\frac{v_x}{u_*} = \frac{1}{\kappa} \ln\left(\frac{30z}{k'_s}\right) \equiv \frac{2.3}{\kappa} \log\left(\frac{30z}{k'_s}\right) \simeq 5.75 \log\left(\frac{30z}{k'_s}\right) \tag{6.16}\blacklozenge$$

In practice, gravel- and cobble-bed streams are considered hydraulically rough. Compared with the particle size distribution of the bed material, the roughness height has been shown to be approximately $k'_s \cong 3 d_{90}$ or $k'_s \cong 6.8 d_{50}$. It is interesting to notice that z_0 is always less than the surface grain diameter. The flow velocity at an elevation $z = d_{90}$ is equal to $v_x = 5.75 u_*$, and the velocity against a particle is thus roughly $v_p \simeq 6 u_*$. The reference velocity is the velocity at $z = k'_s$ or $v_r = 8.5$ u_*. These two velocities, v_p and v_z, are sometimes used for the design of riprap, as discussed in the next chapter.

6.3 Resistance to flow

In open channels, resistance to flow describes the property of the channel to reduce the mean flow velocity. There are three commonly used parameters that define resistance to flow: (1) the Darcy–Weisbach friction factor f'; (2) the Manning coefficient n'; and (3) the Chézy coefficient C'. Prime is used here to denote resistance to flow on plane surfaces. The respective flow velocity relationships are

$$V = \sqrt{\frac{8}{f'}}\sqrt{g R_h S_f} = \sqrt{\frac{8}{f'}} u_* \tag{6.17a}\blacklozenge$$

$$V = \frac{1}{n'} R_h^{2/3} S_f^{1/2} \text{ in SI, or } V = \frac{1.49}{n'} R_h^{2/3} S_f^{1/2} \text{ in English} \tag{6.17b}\blacklozenge$$

$$V = C' R_h^{1/2} S_f^{1/2} \tag{6.17c}\blacklozenge$$

Where R_h is the hydraulic radius and S_f is the friction slope. Both f' n' describe resistance to flow, but C' describes flow conveyance. It is also interesting to note that only f' is dimensionless. The fundamental dimensions for $C' = L^{1/2}/T$ and $n' = T/L^{1/3}$. However the units of Manning's equation are in a coefficient equal to 1.49 in English units and unity in SI.

The following identities can be defined from $u_* = \sqrt{g R_h S_f}$

$$\frac{V}{u_*} = \frac{C'}{\sqrt{g}} \equiv \sqrt{\frac{8}{f'}} \equiv \frac{R_h^{1/6}}{n'\sqrt{g}} \quad (SI) \equiv 1.49 \frac{R_h^{1/6}}{n'\sqrt{g}} \quad (English) \tag{6.18}$$

The mean velocity of turbulent flows can be determined from the integration of logarithmic velocity profiles (e.g. Eq. (6.12)). The mean flow velocity is found analytically to be at a depth $z = 0.37h$, which is close to $0.6\,h$ from the surface. The mean flow velocity V_x for hydraulically smooth and rough boundaries are, respectively

$$\frac{V}{u_*} \equiv \sqrt{\frac{8}{f'}} \simeq 5.75 \log\left(\frac{u_* h}{v_m}\right) + 3.25 \tag{6.19a}$$

and

$$\frac{V}{u_*} \equiv \sqrt{\frac{8}{f'}} \simeq \frac{2.3}{\kappa} \log\left(\frac{h}{k_s'}\right) + 6.25 \simeq 5.75 \log\left(\frac{12.2h}{k_s'}\right) \tag{6.19b}$$

For sand-bed channels, Kamphuis (1974) suggested $k_s' = 2d_{90}$. For gravel-bed streams, Bray (1982) recommended $k_s' = 3.1\,d_{90}$, $k_s' = 3.5\,d_{84}$, $k_s' = 5.2\,d_{65}$, and $k_s' = 6.8\,d_{50}$. The relationship $k_s' \cong 3\,d_{90}$ appears frequently in the literature. The resulting friction factor for hydraulically rough channels with a plane surface can thus be approximated by

$$\frac{V}{u_*} \equiv \sqrt{\frac{8}{f'}} \simeq 5.75 \log\left(\frac{4h}{d_{90}}\right) \simeq 5.75 \log\left(\frac{2h}{d_{50}}\right) \tag{6.19c}\blacklozenge\blacklozenge$$

Figure 6.6a shows the agreement between this relationship and measurements for plane-bed channels. The Manning relationship also becomes interesting given the Strickler relationship between n and $d_s^{1/6}$ in Table 6.1. For instance, the Manning–Strickler relationship in Figure 6.6a provides satisfactory results except in very rough mountain channels ($h/d_s < 10$) and in deep channels ($h/d_s > 1000$).

Whether or not sediment transport affects resistance to flow cannot be easily answered. On one hand, the energy required to move sediment should increase resistance to flow. On the other hand, the presence of fine sediment between coarser bed particles may effectively reduce the roughness height and decrease resistance to flow. Figure 6.6.b shows the experimental results for an upper-regime plane bed. The results are quite comparable to those of Equation (6.19c).

The grain resistance equation in logarithmic form can be transformed into an equivalent power form in which the exponent b varies with relative submergence h/d_s,

$$\sqrt{\frac{8}{f'}} = a\left(\frac{h}{d_s}\right)^b \equiv \hat{a} \ln\left[\frac{\hat{b}h}{d_s}\right] \tag{6.20}\blacklozenge$$

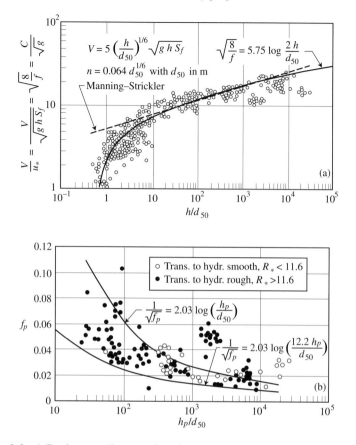

Figure 6.6. a) Resistance diagram for plane bed (from Julien, 2002) b) Darcy–Weisbach friction factor f_p for upper-regime plane bed (from Julien and Raslan, 1998)

under the transformation that imposes both the value and first derivative to be identical:

$$a = \frac{\hat{a}}{b}\left(\frac{d_s}{h}\right)^b \qquad (6.21a)$$

and

$$b = \frac{1}{\ln\left(\frac{\hat{b}h}{d_s}\right)} \qquad (6.21b)$$

The values of the exponent b are plotted in Figure 6.7 as a function of relative submergence h/d_s. It is shown that b gradually decreases to zero as $h/d_s \to \infty$, which implies that the Darcy–Weisbach grain friction factor f' and the grain Chézy

Table 6.1. *Grain resistance and turbulent flow velocity over hydraulically rough plane boundaries* $(C = C' \text{ and } f = f')$

Formulation	Range	Resistance parameter	Velocity[a]
Chézy	$h/d_s \to \infty$	$C = \sqrt{\frac{8g}{f}} \text{ constant}$	$V = C R_h^{1/2} S_f^{1/2}$
Manning	$h/d_s > 100$	$C \cong a\left(\frac{R_h}{d_s}\right)^{1/6} \cong \frac{R_h^{1/6}}{n} \text{(SI)}$	$V = \frac{1}{n} R_h^{2/3} S_f^{1/2} \text{(SI)}$ $n \cong 0.062\, d_{50}^{1/6} \ (d_{50} \text{ in m})$ $n \cong 0.046\, d_{75}^{1/6} \ (d_{75} \text{ in m})$ $n \cong 0.038\, d_{90}^{1/6} \ (d_{90} \text{ in m})$
Logarithmic		$\frac{C}{\sqrt{g}} = \sqrt{\frac{8}{f}} = 5.75 \log\left(\frac{12.2 R_h}{k_{s'}}\right)$	$V = \left(5.75 \log\frac{12.2 R_h}{k_s'}\right)\sqrt{g R_h S_f}$ $k_s' \cong 3\, d_{90} \qquad k_s' \cong 5.2\, d_{65}$ $k_s' \cong 3.5\, d_{84} \qquad k_s' \cong 6.8\, d_{50}$

[a] The hydraulic radius $R_h = A/P$ is used where A is the cross-sectional area and P is the wetted perimeter; the friction slope S_f is the slope of the energy grade line.

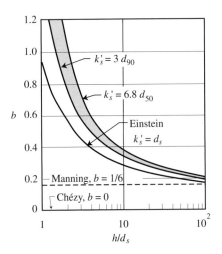

Figure 6.7. Exponent b of the grain resistance equation

coefficient $C' = \sqrt{\frac{8g}{f'}}$ are constant for very large values of h/d_s. At values of $h/d_s > 100$, the exponent b is roughly comparable to 1/6 which corresponds to the Manning–Strickler approximation $(n \sim d_s^{1/6})$. At lower values of the relative submergence $h/d_s < 100$ as in gravel-bed streams, the exponent b of the power form varies with h/d_s and the logarithmic formulation is preferred.

Example 6.1 Turbulent velocity profile

Consider the velocity profile of the Matamek River at a discharge $Q = 462\,\text{ft}^3/\text{s}$, a width $W = 170\,\text{ft}$. The friction slope is about $10\,\text{cm/km}$ and the bed material is very coarse gravel. The cross-sectional area $A = 170 \times 3.7 = 629\,\text{ft}^2$ and the wetted perimeter $P = 170 + 2 \times 3.7 = 177.4\,\text{ft}$, the hydraulic radius $R_h = A/P = 3.54\,\text{ft}$. Assume steady uniform turbulent flow in a wide-rectangular channel, $R_h = h$ (Fig. E-6.1.1). From two points 1 and 2 near the bed, and Equation (6.12),

$$v_1 = \frac{u*}{\kappa} \ln \frac{z_1}{z_0}$$

$$v_2 = \frac{u*}{\kappa} \ln \frac{z_2}{z_0}$$

and estimate the following parameters assuming $\kappa = 0.4$:

Step 1. Shear velocity; from v_1 and v_2

$$u_* = \frac{\kappa(v_2 - v_1)}{\ln\left(\dfrac{z_2}{z_1}\right)} = \frac{0.4(0.85 - 0.55)}{\ln\left(\dfrac{1.5}{0.5}\right)} = 0.11\,\text{ft/s} = 0.0335\,\text{m/s}$$

and the velocity equation is $v = 0.28 \ln z + 0.74$ as shown in Figure E-6.1.1.

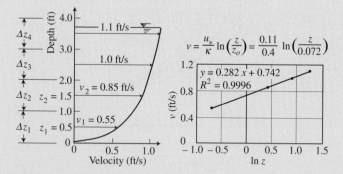

Figure E-6.1.1 Measured velocity profile

Step 2. Boundary shear stress (Equation (6.10)):

$$\tau_o = \rho_m u_*^2 = \frac{1.92\,\text{slugs}}{\text{ft}^3}(0.11)^2 \frac{\text{ft}^2}{\text{s}^2} = 0.023 \frac{\text{lb}}{\text{ft}^2} = 1.1\,\text{Pa}$$

Step 3. Laminar sublayer thickness from Equation (6.14):

$$\delta = \frac{11.6\nu_m}{u_*} = \frac{11.6 \times 1 \times 10^{-5}\text{ft}^2\text{s}}{\text{s} \times 0.11\,\text{ft}} = 0.001\,\text{ft} = 0.32\,\text{mm}$$

Notice that δ is usually less than 1 mm.

Step 4. Elevation z_0 and k_s' from the regression equation

$$0.74 = -\frac{u_*}{\kappa}\ln z_0 \quad \text{or}$$

$$z_0 = e^{-0.74\kappa/u_*} = e^{\frac{-0.74}{0.28}} = 0.07\,\text{ft} = 0.02\,\text{m}$$

$$k_s' = 30z_0 \simeq 2.0\,\text{ft} = 0.6\,\text{m}$$

The flow is hydraulically rough.

Step 5. Mean flow velocity:

$$V_x = \frac{1}{h}\sum_{i=1}^{N} v_i \Delta z_i = \frac{1}{3.7\,\text{ft}}(0.55 + 0.85 + 1.0 + (1.1 \times 0.7))\,\frac{\text{ft}^2}{\text{s}} =$$

$$0.85\,\text{ft/s} \simeq 0.23\,\text{m/s}$$

Step 6. Froude number:

$$\text{Fr} = \frac{V}{\sqrt{gh}} = \frac{0.85\,\text{ft/s}}{\sqrt{32.2 \times 3.7\frac{\text{ft}^2}{\text{s}^2}}} = 0.078$$

Step 7. Friction slope (from Step 2):

$$S_f = \frac{\tau_o}{\gamma_m R_h} \cong \frac{\tau_o}{\gamma_m h} = \frac{0.023\,\text{lb ft}^3}{\text{ft}^2 62.4\,\text{lb} \times 3.7\,\text{ft}} = 1 \times 10^{-4}$$

Step 8. Darcy–Weisbach factor:

$$f = \frac{8S_f}{\text{Fr}^2} = \frac{8 \times 10^{-4}}{0.078^2} = 0.13$$

Step 9. Manning coefficient:

$$n = \frac{1.49}{V}R_h^{2/3}S_f^{1/2} = \frac{1.49}{0.85}(3.7\,\text{ft})^{2/3}\left(1 \times 10^{-4}\right)^{1/2} = 0.042$$

(Note that the units of the Manning equation are in the coefficient 1 in SI or 1.49 in English units.)

Step 10. Chézy coefficient:

$$C = \sqrt{\frac{8g}{f}} = \sqrt{\frac{8 \times 32.2}{0.13}} = 44.5\,\mathrm{ft}^{1/2}/\mathrm{s} = 24.6\frac{\mathrm{m}^{1/2}}{\mathrm{s}}$$

Step 11. Momentum correction factor (Equation (E-3.7.1)):

$$\beta_m = \frac{1}{AV_x^2}\int_A v_x^2\,dA \cong \frac{1}{hV_x^2}\sum_i v_{xi}^2\,dh_i$$

$$\beta_m \cong \frac{1}{3.7\mathrm{ft}}\frac{\mathrm{s}^2}{(0.85)^2\mathrm{ft}^2}\left(0.55^2 + 0.85^2 + 1.0^2 + \left(1.1^2 \times 0.7\right)\right)\frac{\mathrm{ft}^3}{\mathrm{s}^2} = 1.07$$

Step 12. Energy correction factor (Equation (E-3.9.2)):

$$\alpha_e \cong \frac{1}{AV_x^3}\int_A v_x^3\,dA \cong \frac{1}{hV_x^3}\sum_i v_{xi}^3\,dh_i$$

$$\alpha_e \cong \frac{1}{3.7\mathrm{ft}}\frac{\mathrm{s}^3}{(0.85)^3\mathrm{ft}^3}\left(0.55^3 + 0.85^3 + 1.0^3 + \left(1.1^3 \times 0.7\right)\right)\frac{\mathrm{ft}^4}{\mathrm{s}^3} = 1.19$$

6.4 Deviations from logarithmic velocity profiles

Two types of deviations from the logarithmic velocity profiles are considered: (1) the modified log-wake law (Sections 6.4.1 and 6.4.2); and (2) sidewall correction in narrow channels (Section 6.4.3).

6.4.1 Log-wake law

Departure from logarithmic velocity profiles are observed as the distance from the boundary increases (see dotted lines $u_* z/v_m$ on Figure 6.4). The reason for this is essentially related to the invalidity of the following assumptions: (1) constant shear stress throughout the fluid; and (2) mixing length approximation $l_m = \kappa z$.

A more complete description of the velocity distribution v_x is possible after including the law of the wake for steady turbulent open-channel flow as suggested by Coles (1956):

$$\frac{v_x}{u_*} = \underbrace{\left[\frac{2.3}{\kappa}\log\left(\frac{u_* z}{v_m}\right) + 5.5\right]}_{\text{law of the wall}} - \underbrace{\frac{\Delta v_x}{u_*}}_{\substack{\text{roughness}\\\text{function}}} + \underbrace{\frac{2\Pi_w}{\kappa}\sin^2\left(\frac{\pi z}{2h}\right)}_{\text{wake flow function}} \qquad (6.22)$$

where h is the total flow depth, Δv_x represents the velocity reduction due to boundary roughness, and Π_w is the wake strength coefficient.

The terms in brackets depict the original logarithmic law of the wall for smooth boundaries from Equation (6.13b). The last two terms have been added to describe the entire boundary layer velocity profile outside of the thin laminar sublayer. The term $\Delta v_x / u*$ is the channel roughness velocity reduction function.

The last term describes the velocity increase in the wake region as described by the wake strength coefficient Π_w. The wake flow function equals zero near the boundary and increases gradually towards $2\Pi_w/\kappa$ at the upper surface ($z = h$). With $v_x = v_{xm}$ at $z = h$, the upper limit of the velocity profile is

$$\frac{v_{xm}}{u_*} = \frac{2.3}{\kappa} \log\left(\frac{u_* h}{v_m}\right) + 5.5 - \frac{\Delta v_x}{u*} + \frac{2\Pi_w}{\kappa} \tag{6.23}$$

The velocity defect law obtained after subtracting Equation (6.22) from Equation (6.23) gives

$$\frac{v_{xm} - v_x}{u_*} = \left\{\frac{2\Pi_w}{\kappa} - \left[\frac{2.3}{\kappa} \log\frac{z}{h}\right]\right\} - \frac{2\Pi_w}{\kappa} \sin^2\left(\frac{\pi z}{2h}\right) \tag{6.24}\blacklozenge$$

In this form, the term in brackets is the original velocity defect equation for the logarithmic law. The wake flow term vanishes as z approaches zero and the velocity defect asymptotically reaches the term in braces in Equation (6.24) as z/h diminishes. This means that the von Kármán constant κ must be defined from the slope of the logarithmic part in the lower portion (lower 15%) of the velocity profile as shown in Figure 6.8. The wake strength coefficient Π_w is then determined by projecting the straight line, fitting in the lower portion of the velocity profile, to

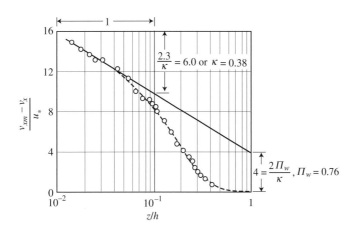

Figure 6.8. Evaluation of κ and Π_w from the velocity defect law

$z/h = 1$, and calculating Π_w from

$$\Pi_w = \frac{\kappa}{2}\left[\frac{v_{xm} - v_x}{u_*}\right] \quad at\ z/h = 1 \tag{6.25}$$

This procedure, illustrated in Problem 6.1, generally shows that Π_w increases with sediment concentration while the von Kármán κ remains constant around 0.4.

6.4.2 Modified log-wake law

In several channels, the maximum flow velocity is observed below the free surface, which makes the velocity profiles difficult to plot graphically (e.g. Figure 6.9). Guo and Julien (2003) suggested a modification to the log-wake law by adding a correction term for the upper boundary condition. The modified log-wake law (MLWL) reads

$$\frac{v_x}{u_*} = \left[\left(\frac{1}{\kappa}\ln\frac{zu_*}{v_m} + B\right) + \frac{2\Pi_w}{\kappa}\sin^2\frac{\pi\xi}{2}\right] - \frac{\xi^3}{3\kappa} \tag{6.26a}$$

where $v_x =$ the time-averaged velocity in the flow direction, $u_* =$ shear velocity, $\kappa =$ von Kármán constant, $z =$ distance from the wall, $v_m =$ kinematic viscosity of the fluid, $B =$ additive constant that relates to the wall roughness, $\Pi_w =$ Coles wake strength, and $\xi = z/z_m$ normalized distance relative to the dip position z_m. The terms in parentheses describe the log law, the terms in square brackets define the law of the wake (Equation (6.22)), and the cubic function is the modification for the flow condition at the upper boundary.

In velocity-defect form, this MLWL becomes a function of the maximum flow velocity v_{xm} at $z = z_m$

$$\frac{v_{xm} - v_x}{u_*} = -\frac{1}{\kappa}\ln\xi + \frac{2\Pi_w}{\kappa}\cos^2\frac{\pi\xi}{2} - \frac{1-\xi^3}{3\kappa} \tag{6.26b}$$

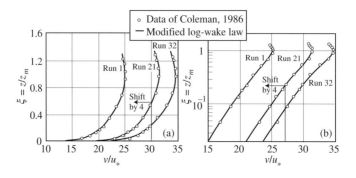

Figure 6.9. Modified log-wake law comparison (after Guo and Julien, 2008)

The MLWL assumes $\kappa = 0.41$ thus leaving four parameters to be determined from a measured velocity profile. These parameters are $z_m, \Pi_w, u*$, and either v_{xm} or B. These four parameters can be simultaneously optimized from the measured velocity profile (e.g. from the function "lsq curve fit" in MATLAB). The procedure detailed in Guo and Julien (2008) is shown in Figure 6.9 for the laboratory data of Coleman (1986).

In some large alluvial rivers, the flow depth is difficult to determine because of the presence of bedforms. In this case, the average bed elevation z_o can be determined from the velocity profile as

$$v_x = \frac{u_*}{\kappa}\left[\ln\left(\frac{z}{z_o}\right) - \frac{1}{3}\left(\frac{z - z_o}{(z_m - z_o)}\right)^3\right] + \frac{2\Pi_w u_*}{\kappa}\sin^2\frac{\pi\,(z - z_o)}{2\,(z_m - z_o)} \qquad (6.27a)$$

in which z_o is the arbitrary mean bed elevation in the presence of bedforms. Note that the constant B in Equation (6.26a) has now been incorporated into z_o. Given that $\kappa = 0.41$, there are four fitting parameters $z_o, z_m, u*$, and Π_w or v_{xm} in Equation (6.27), which can be determined using a non-linear optimization program (e.g. Guo and Julien, 2008). Once the four fitting parameters have been determined, the depth-averaged flow velocity V can be obtained from

$$V = \frac{u_*}{\kappa}\left[\left(\frac{h}{h - z_o}\ln\frac{h}{z}\right) - \frac{\Pi_w}{\pi}\left(\frac{z_m - z_o}{h - z_o}\right)\sin\pi\left(\frac{h - z_o}{z_m - z_o}\right)\right.$$
$$\left. - \frac{1}{12}\left(\frac{h - z_o}{z_m - z_o}\right)^3 + \Pi_w - 1\right] \qquad (6.27b)$$

For instance, the velocity profile of the Mississippi River in Figure 6.10 shows a pronounced velocity dip and a value of $z_o = 0.33$ m, which is substantially larger

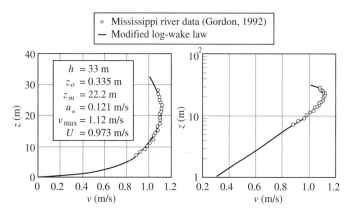

Figure 6.10. Velocity profile of the Mississippi River (after Guo and Julien, 2008)

than the median grain size of the bed material. This high value for z_o reflects the presence of bedforms on the bed of the Mississippi River.

6.4.3 Sidewall correction method

The sidewall correction method is essential for experiments in narrow laboratory channels with smooth metal or plexiglass sidewalls. Consider steady-uniform flow in a narrow open channel at a discharge Q measured from a calibrated orifice and friction slope S_f. When the flume width W is less than five times the flow depth, the sidewall resistance is different from the bed resistance. The Vanoni–Brooks correction method can be applied to determine the bed shear stress, $\tau_b = \rho_m u_{*b}^2$. For a rectangular channel, the hydraulic radius, $R_h = \frac{Wh}{W+2h}$, and the Reynolds number $\mathrm{Re} = \frac{4VR_h}{\nu_m}$, are calculated given the average velocity $V = \frac{Q}{Wh}$. The average shear velocity $u_* = \sqrt{g R_h S_f}$ computed from the slope S_f is then used to calculate the Darcy–Weisbach friction factor $f = \frac{8u_*^2}{V^2}$. The wall friction factor f_w for turbulent flow over a smooth boundary $10^5 < \mathrm{Re}/f < 10^8$ can be calculated from

$$f_w = 0.0026 \left\{ \log \left(\frac{\mathrm{Re}}{f} \right) \right\}^2 - 0.0428 \log \left(\frac{\mathrm{Re}}{f} \right) + 0.1884 \qquad (6.28a)$$

The bed friction factor f_b is then obtained from

$$f_b = f + \frac{2h}{W}(f - f_w) \qquad (6.28b)$$

The hydraulic radius related to the bed $R_b = R_h f_b / f$ is then used to calculate the bed shear stress τ_b from $\tau_b = \gamma_m R_b S_f$. Example 6.2 provides the details of the calculation procedure.

Example 6.2 Application of the sidewall correction method

Consider a discharge $Q = 1.2 \, \mathrm{ft}^3/\mathrm{s}$ in a 4 ft-wide flume inclined at a 0.001 slope. Calculate the bed shear stress τ_b given the normal flow depth of 0.27 ft. The measured water temperature is 70°F.

Step 1. $Q = 1.2 \, \mathrm{ft}^3/\mathrm{s}$, $S_o = S_f = 0.001$, $h_n = 0.27 \, \mathrm{ft}$, $\nu_m \cong 1 \times 10^{-5} \, \mathrm{ft}^2/\mathrm{s}$

Step 2. Hydraulic radius:

$$R_h = \frac{W h_n}{W + 2h_n} = \frac{4\,\text{ft} \times 0.27\,\text{ft}}{(4 + 2 \times 0.27)\,\text{ft}} = 0.238\,\text{ft}$$

Step 3. Flow velocity:

$$V = \frac{Q}{W h_n} = \frac{1.2\,\text{ft}^3}{\text{s}4\,\text{ft} \times 0.27\,\text{ft}} = 1.11\,\text{ft/s}$$

Step 4. Reynolds number:

$$\text{Re} = \frac{4 V R_h}{\nu_m} = 1.06 \times 10^5$$

Step 5. Shear velocity:

$$u* = \sqrt{g R_h S_f} = \sqrt{32.2\frac{\text{ft}}{\text{s}^2} \times 0.238\,\text{ft} \times 0.001} = 0.087\,\text{ft/s}$$

Step 6: Darcy–Weisbach factor:

$$f = \frac{8u*^2}{V^2} = \frac{8(0.087)^2}{(1.11)^2\text{ft}^2}\frac{\text{ft}^2}{\text{s}^2}\text{s}^2 = 0.049$$

Step 7. Wall friction factor:

$$f_w = 0.0026\left[\log\frac{\text{Re}}{f}\right]^2 - 0.0428\left[\log\frac{\text{Re}}{f}\right] + 0.1884 = 0.021$$

Step 8. Bed friction factor:

$$f_b = f + \frac{2h_n}{W}(f - f_w)$$

$$f_b = 0.049 + \frac{2 \times 0.27\text{ft}}{4\text{ft}}(0.049 - 0.021) = 0.0527$$

Step 9. Bed hydraulic radius:

$$R_b = \frac{f_b}{f} R_h = \frac{0.0527}{0.049} \times 0.238\,\text{ft} = 0.255\,\text{ft}$$

Step 10. Bed shear stress:

$$\tau_b = \gamma_m R_b S_f = \frac{62.3\,\text{lb}}{\text{ft}^3} \times 0.255\,\text{ft} \times 0.001 = 0.016\frac{\text{lb}}{\text{ft}^2}$$

6.5 Open-channel flow measurements

Open-channel flow measurements normally include stage and flow velocity measurements.

6.5.1 Stage measurements

Stage measurements determine water surface elevation with reference to a horizontal datum such as the mean sea level, a local datum, or an arbitrary datum below the elevation of zero flow. For instance, the Low Water Reference Plane (LWRP) which corresponds to a stage exceeded 97% of the time cannot be used as datum because the LWRP is not horizontal. Simple non-recording gages require frequent readings to develop continuous water level records. Non-recording gages can either be directly read or can provide measurements of the water surface elevation at a fixed point.

- Staff gages are usually vertical boards or rods precisely graduated with reference to a datum. Staff gages are not effective in cold regions because of ice.
- Point gages consist of mechanical or electromagnetic devices to locate and measure the water surface elevation. Point gages are commonly used in hydraulic laboratories. Measurements can be taken from a graduated rod, drum, or steel tape housed in a small box mounted on a rigid structure directly above the water surface. Electromagnetic devices or Lidars can be used for field measurements.
- Float gages are used primarily with an analog water stage recorder. The gage consists of a float and counterweight connected by a graduated steel tape which passes over a pulley assembly. A relatively large float and counterweight are required for stability, sensitivity, and accuracy – like a 10-in. copper float and a 2-lb lead counterweight.
- Pressure-type gages use water pressure transmitted through a tube to a manometer inside a gage shelter to measure stage. Stage can also be measured by gas bubbling freely into a stream from a submerged tube set at a fixed elevation; the gage pressure in the tube equals the piezometric head at the open end of the tube.
- Crest-stage gages measure maximum flood stage from granulated cork stored inside a 2-in. galvanized pipe. As a flood wave passes, the granulated cork floats as the water rises in the pipe. When the water recedes, the cork adheres to the pipe, marking the crest stage.
- Analog recorders provide a continuous visual record of stage, useful for graphical presentation. Analog recorders are useful in streams with flashy hydrographs.
- Digital recorders provide data in digitally coded form suitable for digital computer processing. Digital recorders store or print out gage heights at preselected time intervals. In many cases, digital measurements are possible at very short time intervals, e.g. 15 minutes.
- Telemetering systems using telephone, radio, or satellite communication are desirable when current information on stage is frequently needed from remote locations. Some telemetering systems continuously indicate or record stage at a given site, others report instantaneous gage readings on request.

6.5.2 *Velocity measurements*

Velocity measuring devices include floats, drag bodies, tracers, velocity-head methods, rotating-element current meters, deflection vanes, optical devices, laser, electromagnetic, and ultrasonic devices. Mechanical meters are limited at both high and low velocities while electromagnetic meters are less so. They are easier to use, have no moving parts, are generally more accurate, and can indicate both velocity and direction and provide electronic readout with averaging. Acoustic and electromagnetic devices have become increasingly popular in recent years.

- Rotating current meters are based on the proportionality between the angular velocity of the rotating device and the flow velocity. By counting the number of revolutions of the rotor in a measured time interval, point velocity is determined. A vertical-axis current meter measures the differential drag on two sides of cups in relative motion in a fluid. The rotation speed of these devices, such as Price current meters, is calibrated against the fluid velocity. Horizontal-axis current meters act as propellers in a moving fluid. Common horizontal-axis meters include the Ott and the Neyrpic current meters. Rotating current meters are useful to determine time-averaged flow velocities.
- Acoustic (ultrasonic) velocity meters measure velocity by determining the travel time of sound pulses transmitted and backscattered from small particles moving with the fluid. Ultrasonic Doppler velocimeters measure the phase shift between the signal emitted along the upstream path and the scattered signal received in the opposite direction. Acoustic Doppler Current Profilers (ADCP) or Acoustic Doppler Velocimeters (ADV) provide instantaneous velocity profiles in a cone from the point of measurement. The accuracy is not as good very near or very far from the instrument. They provide a digital signal that can be easily integrated as the boat crosses a given stream. Despite the ease of providing discharge measurements, instantaneous velocity profiles can deviate significantly from the time-averaged velocity profile along a given vertical.
- Electromagnetic flow meters are based on Faraday's induction law stating that voltage is induced by the motion of a conductor (fluid) perpendicular to a magnetic field. Hand-held electromagnetic flow meters are quite popular on small streams.
- Hot film and hot wire anemometers are electrically heated sensors being cooled by advection. The heat loss being a function of the flow velocity, this laboratory instrument is calibrated to measure fluctuating velocities with high spatial resolution and high frequency response. Hot wires are not very effective in sediment-laden flows.
- Laser Doppler anemometers measure the Doppler frequency shift of light-scattering particles moving with the fluid. The frequency shift provides very accurate flow velocity measurements in the laboratory without flow disturbance.

- The depth-averaged velocity is normally obtained from a time-averaged velocity profile. The following approximate methods for turbulent flows can be used to determine the depth-averaged flow velocity from point velocity measurements:

  (1) The one-point method (at 60% of the total depth measured down from the water surface) uses the observed time-averaged velocity at 0.6 h as the mean velocity in the vertical. This method gives reliable results in uniform cross-sections;
  (2) The two-point method (at 20% and 80% of the total depth measured down from the water surface) averages the two velocity measurements;
  (3) The three-point method (at 20%, 60%, and 80% of the total depth measured down from the water surface) averages the one-point and the two-point methods. Velocities at 0.2 h and 0.8 h are averaged, which value is then averaged with the 0.6 depth velocity measurement to obtain fairly accurate values of depth-averaged flow velocity; and
  (4) The surface method (limited use) assumes a coefficient (usually about 0.85) to convert the surface velocity measured with a float to the depth-averaged velocity. This method is not very accurate. Lee and Julien (2006) used an electromagnetic wave velocimeter for the measurement of surface velocities.

Exercises

6.1. Substitute Equation (6.1) into the Navier–Stokes equations (Table 5.1) to obtain Equation (6.4a).

♦♦6.2. Demonstrate that Equation (6.16) is obtained from Equation (6.12b) when $z_o = k'_s/30$.

♦♦6.3. Demonstrate that Equation (6.13b) is obtained from Equation (6.12b) when $z_o = v_m/9u_*$.

♦♦6.4. Derive Equation (6.14) from Equations (6.13a) and (6.13b) at $z = \delta$.

♦6.5. Define the maximum velocity from Equation (6.26a) at $z = z_m$ and derive the velocity defect form of the MLWL in Equation (6.26b).

♦6.6. Plot the measurement from Example 6.1 on the resistance diagram Figure 6.6a and b. Compare with the expected value for V/u_* and f' from Equations (6.19a–c).

Problems
♦♦*Problem 6.1*

Consider the clear-water and sediment-laden velocity profiles measured in a smooth laboratory flume at a constant discharge by Coleman (1986). Notice the changes in the velocity profiles due to the presence of sediments. Determine the von Kármán constant κ from Equation (6.12) for the two velocity profiles below, given $u_* = 0.041$ m/s, $d_s = 0.105$ mm, $Q = 0.064$ m³/s, $h \cong 0.17$ m, $S_f = 0.002$, and $W = 0.356$ m.

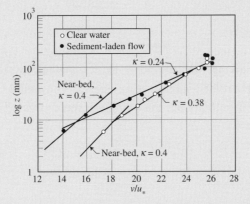

Figure P-6.1 Logarithmic velocity profiles (after Coleman, 1986)

Elevation[a] (mm)	Clear-water flow velocity (m/s)	Sediment-laden flow velocity (m/s)	Concentration by volume
6	0.709	0.576	2.1×10^{-2}
12	0.773	0.649	1.2×10^{-2}
18	0.823	0.743	7.7×10^{-3}
24	0.849	0.798	5.9×10^{-3}
30	0.884	0.838	4.8×10^{-3}
46	0.927	0.916	3.2×10^{-3}
69	0.981	0.976	2.5×10^{-3}
91	1.026	1.047	1.6×10^{-3}
122	1.054	1.07	8.0×10^{-4}
137	1.053	1.07	4.4×10^{-4}
152	1.048	1.057	2.2×10^{-4}
162	1.039	1.048	1.6×10^{-4}

[a] Elevation above the bed.

(*Answer:* The von Kármán constant κ remains close to 0.4 when considering the lowest portion of both velocity profiles. When considering the main portion of the velocity profiles, κ reduces significantly for sediment-laden flow. Also, the presence of sediment reduces the near-bed flow velocity, but increases the wake-strength coefficient.)

Problem 6.2

(a) In turbulent flows, determine the elevation at which the local velocity v_x is equal to the depth-averaged velocity V_x. (*Hint:* $V_x = \frac{1}{(h-z_0)} \int_{z_o}^h v_x dz$ and $\int \ln z\, dz = z \ln z - z$.)

(b) Determine the elevation at which the local velocity v_x equals the shear velocity u_*.

◆◆Problem 6.3

From turbulent velocity measurements at two elevations (v_1 at z_1 and v_2 at z_2) in a wide-rectangular channel, eliminate the constant in Equation 6.12 to determine the shear velocity u_*; the boundary shear stress τ_0; and the laminar sublayer thickness δ.

(*Answer:* $u_* = \kappa(v_1 - v_2)\Big/\ln(z_1/z_2), \tau_0 = \rho_m u_*^2 = \rho_m \kappa^2 [(v_1 - v_2)\Big/$ $\ln(z_1/z_2)]^2$, and $\delta = 11.6 v_m/u_* = 11.6 v_m \ln(z_1/z_2)/\kappa(v_1 - v_2)$

◆◆Problem 6.4

With reference to Problem 6.1: (a) calculate the laminar sublayer thickness δ; (b) compare the flow depth to the hydraulic radius; (c) determine the Darcy–Weisbach friction factor f; and find the mean flow velocity from: (d) the velocity profile; (e) the integral of the log law; (f) the one-point method; (g) the two-point method; (h) the three-point method; and (i) the surface velocity.

◆◆Problem 6.5

Apply the sidewall correction method in Example 6.2 to the data of Problem 6.1.

◆Problem 6.6

With reference to Problem 6.1

(a) evaluate the parameters κ and Π_w, from the velocity defect formulation in Figure 6.8 and Equation (6.25). Compare the value of κ with the value obtained previously (Problem 6.1) from the simple logarithmic law.

(b) Plot the experimental velocity profiles in velocity defect form (like Figure 6.8) with and without sediment transport.

◆Problem 6.7

Consider the velocity and concentration profiles in the table below for the Yangtze River at Feng-Jie from Guo (1998). The flow depth is 42.2 m and the river slope is 5.8 cm/km. Plot the logarithmic velocity profile and determine the hydraulic parameters.

z (m)	v (m/s)	C (mg/l)
42.1	2.82	1,410
33.8	2.82	1,700
16.9	2.46	1,830
8.4	2.22	1,930
4.2	2.03	2,260
0.5	1.63	2,260
0.1	1.43	4,930

♦♦Problem 6.8

Consider the velocity and concentration profiles in the table below for the Missouri River. The flow depth is 7.8 ft, the slope is 12 cm/km and the width is 800 ft. The suspended sand concentration is for the fraction passing the 0.105 mm sieve and retained on a 0.074 mm sieve. Plot the logarithmic velocity profile and determine the hydraulic parameters.

z (ft)	v (ft/s)	C (mg/l)
0.7	4.3	411
0.9	4.5	380
1.2	4.64	305
1.4	4.77	299
1.7	4.83	277
2.2	5.12	238
2.7	5.30	217
2.9	5.40	–
3.2	5.42	196
3.4	5.42	–
3.7	5.50	184
4.2	5.60	–
4.8	5.60	148
5.8	5.70	130
6.8	5.95	–

♦♦Problem 6.9

Consider the velocity and concentration profiles of the Low Flow Conveyance Channel, New Mexico in June 1999, from Baird (2004). The flow discharge is 625 ft^3/s, the flow depth is 5.6 ft, the hydraulic radius is 4.03 ft, the top width is 50.1 ft and the friction slope is 38 cm/km. The bed particle size is fairly uniform sand at $d_{50} = 0.15$ mm. Follow Example 6.1, plot the logarithmic velocity profile and determine the hydraulic characteristics.

Plane bed, June 1999

z (ft)	v (ft/s)	C (mg/l)
0.1	2.65	
0.17	3.16	
0.27	3.44	1,493
0.3	–	
0.37	3.27	1,194
0.47	3.62	975
0.5	–	853
0.6	3.51	
0.7	3.70	914
0.8	3.9	
0.9	4.07	
1	3.91	776
1.1	4.12	
1.2	4.22	648
1.3	4.05	
1.4	4.31	463
1.5	4.34	
1.6	4.45	459
1.7	4.41	
1.8	4.58	283
1.9	4.79	
2	4.73	271
2.2	4.61	190
2.4	4.74	
2.6	4.83	223
2.7	–	
2.8	4.97	
3	5.07	
3.2	4.78	
3.4	5.01	
3.6	5.07	
3.7	–	
3.8	5.03	
4	4.99	
4.2	5.06	
4.4	4.98	
4.5	–	
4.6	4.85	
4.8	4.78	
4.9	–	
5	4.56	
5.2	4.25	
5.4	4.20	
5.5	4.09	
5.5	4.41	

Dunes, May, 2001

z (m)	v (m/s)	C (mg/l)
0.03	0.325	
0.06	0.698	
0.09	0.705	738
0.12	0.729	
0.15	0.746	
0.18	0.756	
0.21	0.755	654
0.24	0.743	
0.27	0.812	
0.30	0.787	533
0.36	0.742	
0.42	0.840	462
0.45	–	
0.48	0.904	401
0.54	0.864	
0.60	0.922	491
0.70	0.969	
0.73	–	238
0.79	0.899	
0.88	0.975	337
0.97	1.023	
1.03	–	
1.06	1.021	
1.15	0.955	
1.24	1.070	
1.31	–	
1.34	1.058	
1.43	1.081	
1.52	1.077	
1.61	1.091	
1.70	1.074	
1.79	1.058	
1.88	0.989	
1.92	–	
1.98	1.011	
2.07	0.975	
2.13	1.029	
2.16	–	

♦♦*Problem 6.10*

Consider the velocity and concentration profile of the Rhine River on November 3, 1998 in the table below (from the Dutch Rijkswaterstaat, also in Julien, 2002). The flow depth is 9.9 m, the navigable channel width is 260 m, the discharge is 9,464 m^3/s, the friction slope is 13.12 cm/km and the particle size distribution of the bed material is $d_{10} = 0.34$ mm, $d_{16} = 0.4$ mm, $d_{35} = 0.71$ mm, $d_{50} = 1.2$ mm, $d_{65} = 3$ mm, $d_{84} = 9.9$ mm and $d_{90} = 12.2$ mm. The bed was covered with dunes approximately 0.87 m high and 19.8 m long. Follow Example 6.1 and plot the logarithmic velocity profile. Determine the main flow characteristics and discuss the results.

z (m)	v (m/s)	C (mg/l)
0	–	–
0.3	0.74	494
0.3	0.81	488
0.3	0.72	498
0.4	0.47	432
0.5	0.84	398
0.8	1.22	293
0.9	1.34	185
1.2	1.38	132
1.3	1.47	134
2.2	1.63	83
3.5	1.92	49
4.0	1.85	43
4.1	1.86	43
5.3	1.99	35
6.0	1.98	33
7.3	2.08	26
8.0	2.04	25
9.0	1.9	23
9.9	–	–

♦*Problem 6.11*

Consider the time-averaged velocity profile of the Mississippi River from Gordon (1992), at a flow depth of 33 m. Assume $\kappa = 0.4$, plot the velocity profile and determine the shear velocity and z_0. Compare the results with Figure 6.10.

Elevation (m)	Velocity (m/s)
7.32	0.88
8.47	0.91
9.38	0.96
10.30	0.98
11.45	1.00
12.59	1.03
13.74	1.04
14.65	1.06
15.34	1.07
16.48	1.07
17.63	1.09
18.55	1.10
19.69	1.10
20.61	1.10
21.52	1.11
22.44	1.11
23.58	1.11
24.73	1.10
25.42	1.09
26.33	1.09
27.25	1.08
28.39	1.08

♦♦*Problem 6.12*

Consider the velocity profile of the Mississippi River in the table below, from Akalin (2002). The instantaneous velocity profile was measured with an Acoustic Doppler Current Profiler (ADCP) at Union Point at vertical 1648 on April 17, 1998. The flow depth is 32.2 m, the hydraulic radius is 30.4 m, the channel width is 1,100 m, the friction slope is $S_f = 3.78$ cm/km and the particle size distribution of the bed material is $d_{10} = 0.17$ mm, $d_{50} = 0.25$ mm, and $d_{90} = 0.45$ mm.

Depth (ft)	v (ft/s)	C (mg/l)	C_{sand} (mg/l)	Suspended sand % finer than			
				0.425 (mm)	0.25 (mm)	0.125 (mm)	0.062 (mm)
7.05	7.16	303	44	100	95.8	72	3.3
10.3	6.86						
13.6	6.84						
16.9	7.01	241	19	100	94.3	56.6	16.6
20.1	6.93						
21.3	6.67						
23.4	6.39	405	111	99.2	96.2	20.8	1.2
26.7	6.58						
30.0	6.69						
33.3	6.49	380	112	98.6	93.8	24.8	3.1
35.5	6.43						
36.5	6.49						
39.8	6.49	504	207	97.6	94.7	16.7	1.7
43.1	6.31	486	231	98.7	93.5	17.2	0.6
46.4	6.01						
49.7	5.78						
52.9	5.75						
56.2	5.35						
59.5	5.31						
62.8	5.28						
63.9	4.07						
66.1	3.64						
69.3	3.37						
71.0	2.74						

7

Incipient motion

The threshold conditions between erosion and sedimentation of a single particle describe incipient motion. The stability of granular material in air is first examined to define the angle of repose in Section 7.1. The following sections cover submerged particles. In Section 7.2, the simplified particle equilibrium conditions on near-horizontal surfaces are discussed for uniform grain sizes, bed sediment mixtures, and cohesive material. The equilibrium of particles under tridimensional moments of forces is detailed in Section 7.3. A simplified force balance is presented in Section 7.4. Two examples of particle stability analysis and stable channel design conclude this chapter.

7.1 Angle of repose

The stability of a single particle on a plane horizontal surface is first considered in Figure 7.1a for simple two-dimensional particle shapes. The threshold condition is obtained when the particle center of mass G is vertically above the point of contact C. The critical angle at which motion occurs is called the angle of repose ϕ and equals 180° divided by the number of sides of the polygons. For instance, the angle of repose ϕ of an equilateral triangle is $\phi = 180°/3 = 60°$, a square is $\phi = 180°/4 = 45°$, and the angle of repose of a sphere is $\phi = 180°/\infty = 0°$. It is concluded that the angle of repose of particles on a flat surface increases with angularity.

As shown in Figure 7.2, a given particle does not necessarily have a unique value of angle of repose. For instance, it is easy to demonstrate that a simple cylinder has an angle of repose $\tan \phi = $ diameter/height when standing. It can also roll freely on a table at $\tan \phi = 0$. The angle of repose thus depends on the position of the particle relative to the contact surface.

Long cylinders standing piled on top of each other (Figure 7.1b) rest at an angle of repose $\phi = 30°$. A sphere standing on spheres of equal diameter (Figure 7.1c)

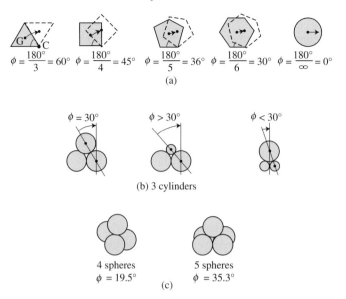

(a)

(b) 3 cylinders

(c)

Figure 7.1. Angle of repose for simple shapes

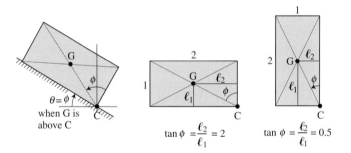

Figure 7.2. Angle of repose depends on particle orientation

reaches threshold of motion at $\phi = 19.5°$ for three points of contact, and at $\phi = 35.3°$ for four points of contact. The angle of repose is thus not solely a function of the particle. It depends on the interaction of the particle with the contact surface.

In the case of material with different diameter particles, consider a sphere of diameter d_2 resting on top of four identical spheres of diameter d_1, as sketched in Figure 7.3. Geometrically, the angle of repose is given by:

$$\tan \phi = \frac{d_1}{\sqrt{(d_1 + d_2)^2 - 2d_1^2}} = \sqrt{\frac{1}{\left(1 + \dfrac{d_2}{d_1}\right)^2 - 2}} \qquad (7.1a)$$

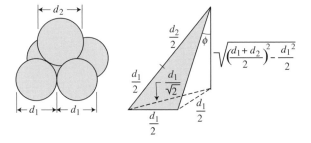

Figure 7.3. Angle of repose for spheres of different diameter

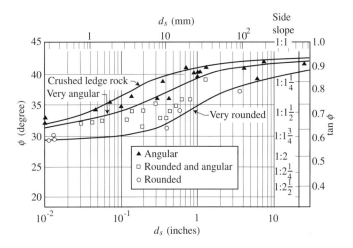

Figure 7.4. Angle of repose for granular material (after Simons, 1957)

The angle of repose ϕ is $35.3°$ when $d_2 = d_1$; it decreases when $d_2 > d_1$ and increases for fine particles $d_2 < d_1$ until $d_2 = 0.41\, d_1$. Surface particles of diameter d_1 can act as a filter to pass particles finer than $0.41\, d_1$ between bed particles. The sub-surface layer of sediment mixtures may therefore tend to become well graded. For granular material, the angle of repose empirically varies with grain size and angularity of the material, as shown in Figure 7.4.

Figure 7.5 shows the application of a drag force F_D passing through the center of gravity of a particle of weight W on a horizontal surface. Incipient motion is defined as the magnitude of the horizontal force $F_H = F_D$ that would set the particle at the beginning of motion obtained when the sum of moments, with $F_V = W$, equals zero. Accordingly, it is important to notice that the ratio of forces thus becomes equal to $\tan \phi$.

A more compact base material composed of equal diameter spheres in a staggered pattern leads to the analysis of a sphere of diameter d_2 resting on top of three identical

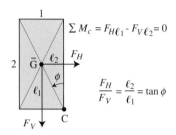

Figure 7.5. Sum of moments

spheres of diameter d_1. From Exercise 7.1,

$$\tan\phi = \frac{1}{\sqrt{3\left(1+\dfrac{d_2}{d_1}\right)^2 - 4}} \qquad (7.1b)$$

Same size particles give $\phi = 19.5°$, and Figure 7.1c shows that compact bed surfaces result in low angles of repose.

7.2 Submerged incipient motion

Fluid flow around sediment particles exerts forces which tend to initiate particle motion. The resisting force of non-cohesive material is the particle weight. Threshold conditions occur when the hydrodynamic moments of forces acting on a single particle balance the resisting moments of force. The particle is then impeding incipient motion.

The forces acting on a non-cohesive sediment particle sketched in Figure 7.6 are the particle weight F_W, buoyancy force F_B, lift force F_L, drag force F_D, and resisting force F_R. As a first approximation it is assumed here that the lift force is negligible. A more refined analysis is presented in Section 7.3. Further assuming that the bed surface slope is horizontal $S_0 = \tan\theta \cong 0$, and the water surface is almost horizontal, the buoyancy force F_B acts in a vertical direction, opposite to the particle weight F_W. The submerged weight of the particle $F_S = F_W - F_B$ is therefore the passive vertical force. The active horizontal force is $F_H = F_D \cong \tau_o d_s^2 = \rho_m u_*^2 d_s^2$. The passive vertical force is $F_V = F_S \sim (\gamma_s - \gamma_m)d_s^3$. The ratio of forces defines the dimensionless shear stress τ_*, called the Shields parameter:

$$\tau_* = \frac{\tau_o}{(\gamma_s - \gamma_m)d_s} = \frac{\rho_m u_*^2}{(\gamma_s - \gamma_m)d_s} \qquad (7.2a)$$

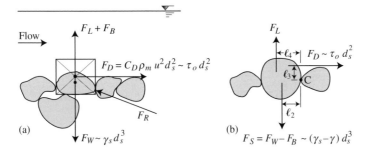

Figure 7.6. Force diagram under steady uniform flow

where: $\tau_o =$ boundary shear stress
$u_* =$ shear velocity
$\gamma_s =$ specific weight of a sediment particle
$\gamma_m =$ specific weight of the fluid mixture
$d_s =$ particle size

It is interesting to consider the effect of a lift force on incipient motion. With reference to Figure 7.6b, with moment arms l_2, l_3, and l_4 for the submerged weight, drag and lift forces respectively, the sum of moments about the point of contact C gives

$$F_D l_3 + F_L l_4 = F_S l_2 \tag{7.2b}$$

Given that the Shields parameter expresses the ratio of drag force to submerged weight, the lift force will decrease the critical Shields parameter in proportion to $F_L l_4 / F_D l_3$, which corresponds to the ratio of lift to drag moments.

$$\tau_{*c} \sim \frac{F_D}{F_S} = \frac{l_2}{l_3} \frac{1}{\left(1 + \dfrac{F_L l_4}{F_D l_3}\right)} \tag{7.2c}$$

In Section 7.3, this ratio will be defined as M/N, and $\Pi_{ld} = F_L/F_D$ defines the lift-drag force ratio. It will be found that $l_4/l_3 \approx 2.6$ in Equation (7.14).

7.2.1 Uniform grain size

Owing to the analysis in Figure 7.5, the critical value of the Shields parameter τ_{*c} corresponding to the beginning of motion ($\tau_o = \tau_c$) depends on $\tan \phi$. Besides the angle of repose, one should consider the ratio of sediment size to the laminar sublayer thickness expressed either as d_s/δ, or the grain shear Reynolds number $Re_* = u_* d_s / \nu_m$ (because $\delta = 11.6 \nu_m / u_*$), the shape of the particle and the lift to

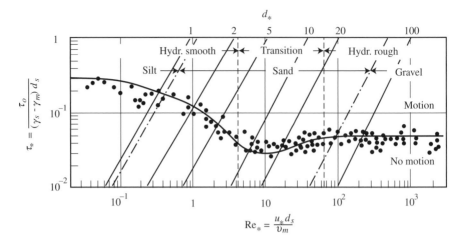

Figure 7.7. Shields diagram for granular material (modified after Yalin and Karahan, 1979)

drag ratio should also be considered.

$$\tau_{*c} = \frac{\tau_c}{(\gamma_s - \gamma_m)d_s} = f\left(\tan\phi, \frac{d_s}{\delta}, \frac{\text{lift}}{\text{drag}}, \text{shape}\ldots\right) \qquad (7.3a)$$

It is interesting that Duboys (1879) first derived the Shields parameter, and defined its proportionality with tan ϕ. Shields (1936) defined the relationship to Re$_*$. He determined the threshold condition by measuring sediment transport for values of τ_* at least twice as large as the critical value and then extrapolated to the point of vanishing sediment transport. His laboratory experiments and those of Yalin and Karahan (1979) and Whitehouse *et al.* (2000) using the median grain size for d_s led to the Shields diagram shown in Figure 7.7. At high values of d_s/δ, or Re$_* = u_*d_s/\nu$, one obtains a constant value of $\tau_{*c} \approx 0.047$, or

$$\tau_{*c} \cong 0.06\tan\phi; \quad \text{when} \frac{u_*d_s}{\nu_m} > 50 \qquad (7.3b)$$

The value of $\tau_{*c} \approx 0.047$ has been widely used for single-size particles. The formulation with tanϕ may be preferable to describe sediment mixtures. An exact value of the critical Shields parameter remains very difficult to define with great accuracy because the definition of beginning of motion is subjective, and because tanϕ varies for the reasons discussed in Section 7.1.

Since the shear velocity u_* appears both in the Shields parameter $\tau_* = \frac{\rho_m u_*^2}{(\gamma_s - \gamma)d_s}$ and in the grain shear Reynolds number Re$_* = \frac{u_*d_s}{\nu_m}$, an iterative procedure is required to solve the Shields diagram. It is practical to replace the abscissa of the Shields diagram after eliminating the shear velocity from Re$_*$ and defining the

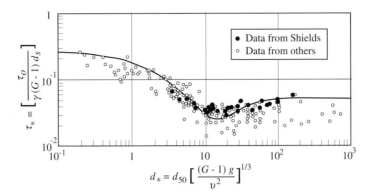

Figure 7.8. Modified Shields diagram

dimensionless particle diameter d_* from $d_*^3 = \dfrac{\text{Re}_*^2}{\tau_*}$. Thus, the abscissa of the Shields diagram can be replaced by the dimensionless particle diameter d_* resulting in the modified Shields diagram in Figure 7.8. The critical values of the Shields parameter τ_{*c} can be approximated as follows:

$$\tau_{*c} \approx 0.3 e^{\frac{-d_*}{3}} + 0.06 \tan\phi \left(1 - e^{\frac{-d_*}{20}}\right) \tag{7.4}$$

where

$$d_* = d_s \left(\frac{(G-1)g}{v_m^2}\right)^{1/3}$$

For turbulent flows over rough boundaries, large Re_* or d_*, Equation (7.4) reduces to Equation (7.3b). The critical shear stress becomes proportional to the sediment size since the threshold value of the Shields parameter remains constant. Based on a comparison of data from the Highway Research Board (1970), a graphical relationship between critical shear stress τ_c and mean grain size diameter d_{50} on a flat horizontal surface is shown in Figure 7.9 with equivalent values for granular material in Table 7.1. It is interesting to notice that the critical shear stress for beginning of motion is approximately 1 Pa per mm of grain size.

7.2.2 Sediment mixtures

The particle size distribution of bed material can vary with time depending on the magnitude of the applied shear stress in relation to the mobility of sediment particles of different sizes. Consider a gravel-bed mixture, identical for the three cases illustrated in Figure 7.10, for which the applied shear stress can move: (a) none of the particles (low shear stress); (b) the fine particles only; (c) all the particles

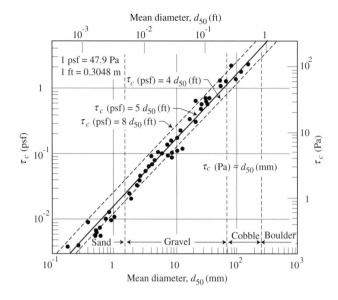

Figure 7.9. Critical shear stress on a horizontal surface

Table 7.1. *Threshold conditions for uniform material at 20°C*

Class name	d_s (mm)	d_*	ϕ (deg)	τ_{*c}	τ_c (Pa)	u_{*c} (m/s)
Boulder						
Very large	> 2,048	51,800	42	0.054	1,790	1.33
Large	> 1,024	25,900	42	0.054	895	0.94
Medium	> 512	12,950	42	0.054	447	0.67
Small	> 256	6,475	42	0.054	223	0.47
Cobble						
Large	> 128	3,235	42	0.054	111	0.33
Small	> 64	1,620	41	0.052	53	0.23
Gravel						
Very coarse	> 32	810	40	0.05	26	0.16
Coarse	> 16	404	38	0.047	12	0.11
Medium	> 8	202	36	0.044	5.7	0.074
Fine	> 4	101	35	0.042	2.71	0.052
Very fine	> 2	50	33	0.039	1.26	0.036
Sand						
Very coarse	> 1	25	32	0.029	0.47	0.0216
Coarse	> 0.5	12.5	31	0.033	0.27	0.0164
Medium	> 0.25	6.3	30	0.048	0.194	0.0139
Fine	>0.125	3.2	30	0.072	0.145	0.0120
Very fine	> 0.0625	1.6	30	0.109	0.110	0.0105
Silt						
Coarse	> 0.031	0.8	30	0.165	0.083	0.0091
Medium	> 0.016	0.4	30	0.25	0.065	0.0080

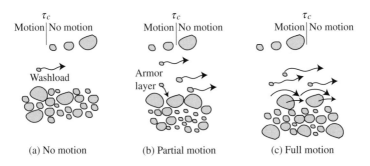

(a) No motion   (b) Partial motion   (c) Full motion

Figure 7.10. Bed surface for sediment mixtures

(high shear stress). It is observed that without motion of the fines, the bed surface is comprised of the original bed material.

With motion of the finer fractions at a shear stress insufficiently high to displace the coarse particles, the bed surface is coarsened to form an armor layer while the fractions in motion are finer than those of the bed surface. Finally, as the shear stress becomes sufficiently large to break the coarse armor layer, all size fractions are brought into motion.

Gravel- and cobble-bed streams tend to have well-graded sediment mixtures. Gradation coefficients $\sigma_g > 3$ are common and particle sizes can range from a few mm to hundreds of mm. Figure 7.11 provides an example of the particle size distribution of the bed material at Little Granite Creek (Weinhold, 2002). The distribution of the sub-surface material shows nearly equal volumes of sediment for each size class between 1–180 mm. The thickness of the bed-surface (or pavement)

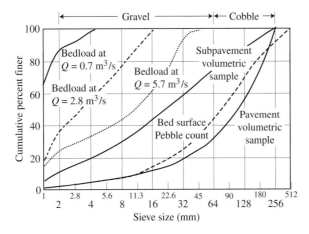

Figure 7.11. Particle size distributions of a cobble–gravel bed stream (modified after Weinhold, 2002)

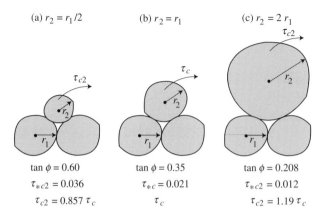

(a) $r_2 = r_1/2$ (b) $r_2 = r_1$ (c) $r_2 = 2\,r_1$

$\tan\phi = 0.60$ $\tan\phi = 0.35$ $\tan\phi = 0.208$

$\tau_{*c2} = 0.036$ $\tau_{*c} = 0.021$ $\tau_{*c2} = 0.012$

$\tau_{c2} = 0.857\,\tau_c$ τ_c $\tau_{c2} = 1.19\,\tau_c$

Figure 7.12. Relative stability for different particle sizes

layer is typically $d_{max} \cong 2-3d_{50}$ of the surface layer. The surface material distribution is skewed, the median grain diameter is approximately 100 mm and the ratio of d_{84}/d_{50} is much smaller than d_{50}/d_{16}. The surface layer is usually called an armor layer when it can be mobilized relatively frequently, every year or so. The surface layer will be paved when only an extreme event can break up the surface layer. Bunte and Abt (2001) provide ample details on measurement techniques for gravel-bed streams.

The mobility of the surface layer becomes interesting because incipient motion depends on $\tan\phi$ in Equation (7.3), and $\tan\phi$ varies with the relative size of sediment particles. For instance, the stability analysis of different particle sizes from Equations (7.1b) and (7.3) is shown in Figure 7.12. The effect of the change in angle of repose from Equation (7.1b) is to make it relatively easy to mobilize particles larger than the bed material. This example shows that these three different sized particles can be mobilized at nearly the same critical shear stress ($0.857\,\tau_c$ versus $1.19\,\tau_c$). This concept has been expanded to define near-equal mobility.

Armor layers can only form in graded-bed sediment mixtures. The analysis of armor layers (Fig. 7.10) centers around the coarser fractions of the mixture which are not moving; finer fractions will be present in the mixture as long as they are shielded by the stable coarser particles. All particles will enter motion as soon as the shear stress exceeds the threshold of motion of coarse particles. This concept is referred to as the equal mobility concept for which all fractions of sediment enter motion at the same value of applied shear stress. Equal mobility of gravel beds implies that all size fractions of a mixture will be mobilized at the same shear stress. It is intuitively obvious that once an armor layer can be mobilized, all size fractions of the underlying material will also reach incipient motion at the same shear stress. When considering size fraction d_i of a mixture of median grain

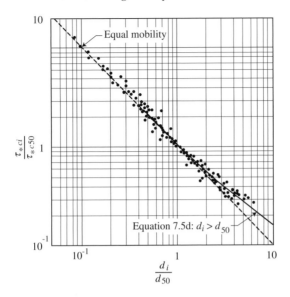

Figure 7.13. Equal mobility diagram (modified after Patel and Ranga Raju, 1999)

diameter d_{50}, the incipient motion of this size is defined by the Shields parameter τ_{*c50}, and the corresponding Shields parameter of fraction d_i is τ_{*ci}. The ratio of τ_{*c50}/τ_{*ci} can be plotted as a function of d_i/d_{50} to give the equal mobility diagram shown in Figure 7.13 with experimental data from Patel and Ranga Raju (1999). On this diagram, equal mobility is obtained when the slope of the line is -1. This implies that all size fractions would reach incipient motion at the same shear stress.

Near-equal mobility recognizes inherent limitations of equal mobility. Even if equal mobility implies that all size fractions found in the armor layer would move, the fact remains that boulders will not move in a gravel-bed stream even at very high flows. Also, Figure 7.11 clearly shows the size fractions mobilized as bedload, and all size fractions do not move at all discharges. Obviously, these discrepancies must be reconciled. Near-equal mobility has often been defined by exponents of the equal mobility diagram.

$$\frac{\tau_{*ci}}{\tau_{*c50}} = \frac{\tau_{ci}}{(G-1)\,\gamma\,d_i}\frac{(G-1)\gamma\,d_{50}}{\tau_{c50}} = \left(\frac{d_i}{d_{50}}\right)^{-x}$$

or

$$\tau_{ci} = \tau_{c50}\left(\frac{d_i}{d_{50}}\right)^{1-x} \tag{7.5a}$$

where x is close to, but slightly less than, unity.

Other methods for near-equal mobility have been available for several decades, e.g. Egiazaroff (1965), Hayashi *et al.* (1980), and many others since: Parker *et al.* (1982) and Klaassen *et al.* (1988);

$$\tau_{*ci} = \tau_{*c50} \left(\frac{\log 19}{\log (19 d_i / d_{50})} \right)^2 \tag{7.5b}$$

$$\tau_{*ci} = \tau_{*c50} \left(\frac{\log 8}{\log (8 d_i / d_{50})} \right)^2 \quad \text{for } d_i > d_{50} \tag{7.5c}$$

and $\tau_{*ci} = \tau_{*c50} \left(\frac{d_i}{d_{50}} \right)^{-1}$ for $d_i < d_{50}$

Equations (7.1b) and (7.3b) can also be combined to give the following theoretical relationship:

$$\tau_{*ci} = \tau_{*c50} \sqrt{\frac{8}{3 (1 + d_i / d_{50})^2 - 4}}; \text{when } d_i / d_{50} > 0.25 \tag{7.5d}$$

In practice, armor layers can be brought into motion when the coarse fractions of the bed-surface material reach incipient motion. This is when $\tau_{*c50} \approx 0.047$ based on d_{50} of the armor or pavement layer, and $\tan \phi \simeq 0.8$. Higher values of τ_{*c50} are required when the d_{50} of the sub-surface material is considered. When fitting straight lines through the equal mobility diagram on Figure 7.13, values of the near-equal mobility exponent x are typically around 0.9–1. It is important to remember that near-equal mobility only applies to size fractions found in the bed material. In practice this is about 5–10 times larger than d_{50}. Also, field measurements can provide the useful site calibration of these parameters (Bakke *et al.*, 1999 and Weinhold, 2002).

7.2.3 Cohesive material

Incipient motion conditions for cohesive material are more difficult to determine because it depends on clay mineralogy, water quality, and other chemical interactions, site-specific critical shear stress values can be determined from laboratory tests.

For design purposes, values of permissible velocity V_c in canals are listed in Table 7.2 and Table 7.3, with information from Fortier and Scobey (1926) and Mirtskhoulava (1988) for several unusual materials and for cohesive sediment.

Table 7.2. *Maximum permissible velocities for canals with h < 1m*

Soil type	Manning n	Clear water, no detritus (m/s)	Water transporting colloidal silt (m/s)	Water transporting silts, sands, or gravels (m/s)
Stiff clay (very colloidal)	0.025	1.14	1.52	0.91
Alluvial silt when colloidal	0.025	1.14	1.52	0.91
Alluvial silt when noncolloidal	0.02	0.61	1.07	0.61
Volcanic ash	0.02	0.76	1.07	0.61
Silt loam (noncolloidal)	0.02	0.61	0.91	0.61
Ordinary firm loam	0.02	0.76	1.07	0.69
Sandy loam (noncolloidal)	0.02	0.53	0.76	0.61
Fine sand (colloidal)	0.02	0.46	0.76	0.46
Fine gravel	0.02	0.76	1.52	1.14
Graded, loam to cobbles, when noncolloidal	0.03	1.14	1.52	1.52
Graded, silt to cobbles, when colloidal	0.03	1.22	1.68	1.52
Coarse gravel (noncolloidal)	0.025	1.22	1.83	1.98
Cobbles and shingles	0.035	1.52	1.68	1.98
Shales and hard pans	0.025	1.83	1.83	1.52

Source: Fortier and Scobey (1926).

7.3 Moment stability analysis

Layers of large stones, commonly called riprap, are used to protect embankment slopes against erosion. The stability of riprap depends on the stability of individual particles subjected to hydrodynamic forces under various embankment configurations and stone properties.

Figure 7.14 illustrates the forces acting on a cohesionless particle resting on an embankment inclined at a sideslope angle Θ_1, and a downstream angle Θ_0. Those are the lift force F_L, the drag force F_D, the buoyancy force F_B, and the weight of the particle F_W (Stevens and Simons, 1971). As long as the water surface slope angle in the downstream direction is small, the buoyancy force can be subtracted from the

Table 7.3. *Maximum permissible velocities for cohesive channels*

ρ_{md} (kg/m³)	ρ_m (kg/m³)	Soil type	Loamy sand[a] (m/s)	Non-plastic clay (m/s)	Clay[a] (m/s)	Heavy clayey soil (m/s)	Loamy clay[a] (m/s)	Maximum velocity (m/s)
320	1,200	Fine sandy loamy clay	$0.1 \log 8.8\, h/k_s$	–	$0.12 \log 8.8\, h/k_s$	–	$0.12 \log 8.8\, h/k_s$	0.45–0.91
320–1040	1,200–1,650	Alluvial mud	$0.15 \log 8.8\, h/k_s$	–	$0.3 \log 8.8\, h/k_s$	–	$0.25 \log 8.8\, h/k_s$	0.61–0.84
8.70	1,544	Alluvial loamy clay	–	0.32	0.35	0.4	0.45	0.76–0.84
1040–1664	1,650–2,040	Hard loamy clay	–	–	$0.45 \log 8.8h/k_s$	–	$0.4 \log 8.8\, h/k_s$	0.91–1.14
1190	1,742	Hard clay	–	0.7	0.8	0.85	0.9	0.76–1.52
1664	2,040	Rigid clay	–	1.05	1.2	1.25	1.3	1.22–1.52
1664–1824	2,040–2,140	Clayey shale	–	–	$0.65 \log 8.8h/k_s$	–	$0.6 \log 8.8\, h/k_s$	0.76–2.13
2030	2,270	Hard rock	–	1.35	1.65	1.70	1.8	3–4.5

[a] h is the flow depth and k_s the boundary roughness height.
Source: Modified after Etcheverry (1916); Fortier and Scobey (1926); and Mirtskhoulava (1988).

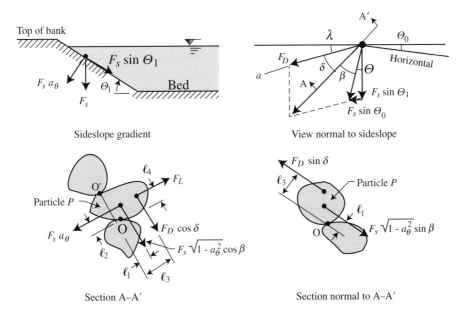

Figure 7.14. Moment stability analysis of a particle

weight of the particle to give the submerged weight of the particle $F_S = F_W - F_B$. The lift force is defined as the fluid force normal to the embankment plane and the drag force is acting along the plane in the same direction as the velocity field surrounding the particle.

Geometrically, the projection of the submerged weight on the embankment plane gives $\tan\Theta = \cos\Theta_1 \sin\Theta_0 / \cos\Theta_0 \sin\Theta_1$. Similarly, the submerged weight projection into the embankment plane is $a_\Theta = \sqrt{1 - \cos^2\Theta_0 \sin^2\Theta_1 - \cos^2\Theta_1 \sin^2\Theta_0}$. For small angles Θ_0 ($\Theta_0 < 20°$), these expressions can thus be approximated as $a_\Theta \approx \sqrt{\cos^2\Theta_1 - \sin^2\Theta_0}$ and this corresponds to $\tan\Theta \simeq \sin\Theta_0 / \sin\Theta_1$. The submerged weight has one sideslope component $F_S \sin\Theta_1$, one downslope component $F_S \sin\Theta_0$, and a component normal to the plane $F_S a_\Theta$. The streamline deviates from the downstream direction at an angle λ along the embankment plane (λ is defined positive downward). Once in motion, the particle follows a direction at an angle β from the downward direction (projection of a vertical on the embankment plane).

Stability against rotation of a particle determines incipient conditions of motion when the equilibrium of moments about the point of rotation 0 is satisfied (see Figure 7.14 Section A–A′).

$$l_2 F_S a_\Theta = l_1 F_S \sqrt{1 - a_\Theta^2} \cos\beta + l_3 F_D \cos\delta + l_4 F_L \tag{7.6}$$

the angles δ and β, and the moment arms l_1, l_2, l_3 and l_4 are shown on Figure 7.14.

The first two terms on the right-hand side of Equation (7.6) determine about which pivot point particle P is to rotate. Two stability factors SF_0 and SF_{01}, against rotation about points 0 and $0'$ are respectively defined as the ratio of the resisting moments to the moments generating motion.

$$SF_0 = \frac{l_2 F_S a_{\Theta}}{l_1 F_S \sqrt{1 - a_{\Theta}^2} \cos\beta + l_3 F_D \cos\delta + l_4 F_L} \tag{7.7a}$$

$$SF_{01} = \frac{l_2 F_S a_{\Theta} + l_1 F_S \sqrt{1 - a_{\Theta}^2} \cos\beta}{l_3 F_D \cos\delta + l_4 F_L} \tag{7.7b}$$

Note that Equation (7.7b) is only used when $F_D \cos\delta$ is applied upslope (large negative λ value). Each term in Equation (7.7) must be positive because the formulation describes the ratio of positive stabilizing moment to positive destabilizing moments.

Because the stability factor SF_o equals unity when the angle Θ equals the angle of repose ϕ under static fluid conditions ($F_D = F_L = 0$), it is then found that $\tan\phi = l_2/l_1$. Dividing both the numerator and the denominator by $l_1 F_S$, transforms Equation 7.7 to:

$$SF_0 = \frac{a_{\Theta} \tan\phi}{\eta_1 \tan\phi + \sqrt{1 - a_{\Theta}^2} \cos\beta} \tag{7.8a} \blacklozenge\blacklozenge$$

$$SF_{01} = \frac{a_{\Theta} \tan\phi + \sqrt{1 - a_{\Theta}^2} \cos\beta}{\eta_1 \tan\phi} \tag{7.8b}$$

in which, after defining $M = l_4 F_L / l_2 F_S$; and $N = l_3 F_D / l_2 F_S$:

$$\eta_1 = M + N \cos\delta \tag{7.9}$$

The variable η_1 relates to the stability number $\eta_0 = M + N$ for particles on a plane horizontal surface ($\Theta_0 = \Theta_1 = \delta = 0$) after considering $\lambda + \delta + \beta + \Theta = 90°$, or

$$\eta_1 = \eta_0 \left(\frac{(M/N) + \sin(\lambda + \beta + \Theta)}{1 + (M/N)} \right) \tag{7.10} \blacklozenge$$

When the flow is fully turbulent over a hydraulically rough horizontal surface, incipient motion corresponds to $SF_0 = 1$, or:

$$\eta_0 = \frac{\tau_o}{\tau_c} \cong \frac{\tau_o}{(\gamma_s - \gamma_m) d_s \tau_c^*} = \frac{21\tau_0}{(\gamma_s - \gamma_m) d_s} \tag{7.11a} \blacklozenge$$

This normalized form of the Shields parameter shows that $\eta_0 = 1$ when $\tau_{*c} = 0.047$, describing incipient motion of particles on a plane bed under turbulent flow over hydraulically rough boundaries.

Alternative relationships for η_0 are obtained when replacing the boundary shear stress τ_o with the reference velocity v_r, the velocity against the particle v_p or the average flow velocity V. The reference velocity $v_r = 8.5u_*$ is the velocity at a height $z = k'_s$; thus, from Equation (6.16), $\tau_o = \rho_m u_*^2 = 0.0138\rho_m v_r^2$. Taking the velocity against the particle $v_p \cong 0.71\, v_r = 6\, u_*$, $\tau_o = 0.027\rho_m v_p^2$. Finally, from Equation (6.10), the grain shear component on a plane surface has a notation with prime $'$ as

$$\tau_o = \tau'_o = \rho_m u'^2_* = f' \rho_m V^2/8 = \rho_m V^2 \Big/ \left(5.75 \log(12.2h/k'_s)\right)^2$$

$$\tau'_o = \gamma R_h S'_f = 0.0138\rho_m v_r^2 = 0.027\rho_m v_p^2$$

and alternative equations for η_0 as a function of the reference velocity v_r, the velocity against the particle v_p and the mean flow velocity V are respectively:

$$\eta_0 = \frac{0.3v_r^2}{(G-1)gd_s} \tag{7.11b}$$

$$\eta_0 = \frac{0.6v_p^2}{(G-1)gd_s} \tag{7.11c}$$

$$\eta_0 = \frac{V^2}{(G-1)gd_s(5.75 \log(12.2h/k's))^2} \tag{7.11d}$$

The second equilibrium condition given by the direction of the particle along the section normal to A–A′ on Figure 7.14 is:

$$l_3 F_D \sin\delta = l_1 F_S \sqrt{1 - a_\Theta^2}\,\sin\beta \tag{7.12}$$

After writing δ as a function of λ, Θ, and β, and solving for β gives

$$\beta = \tan^{-1}\left\{ \frac{\cos(\lambda+\Theta)}{\frac{(M+N)\sqrt{1-a_\Theta^2}}{N\eta_0 \tan\phi} + \sin(\lambda+\Theta)} \right\} \tag{7.13}◆$$

In summary, the stability factors for particles on sideslopes can be calculated from Equation (7.8) after solving successively Equations (7.11), (7.13), and (7.10), with the use of two geometric relationships, $a_\Theta = \sqrt{\cos^2\Theta_1 - \sin^2\Theta_0}$ and $\tan\Theta = \frac{\sin\Theta_0}{\sin\Theta_1}$. For simplified applications, one can use $M = N$ because the stability factor is not very sensitive to the M/N ratio.

Recent research on the lift-to-drag ratio $\Pi_{ld} = F_L/F_D$ on spherical particles by Lucker and Zanke (2007) shows variability of Π_{ld} relative to the grain shear

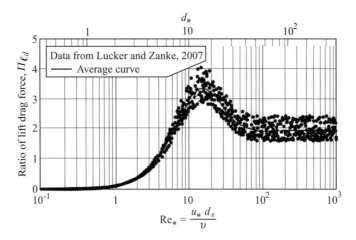

Figure 7.15. Lift-to-drag force ratio versus Re_* for spheres

Reynolds number Re_*. In Figure 7.15, this data can be used to fit an empirical relationship as a function of d_* with $\tan\phi \simeq 0.8$ as

$$\Pi_{ld} = \frac{F_L}{F_D} \approx -0.38 + \left[2.6e^{\frac{-d_*}{3}} + 0.5\tan\phi\left(1 - e^{\frac{-d_*}{20}}\right)\right]^{-1} \qquad (7.14a)$$

From this relationship, it is interesting that the critical value of the Shields parameter for incipient motion on a plane horizontal surface in Equation (7.2c), given Equation (7.9) corresponds to Equation (7.4), or $\tau_{*c} = 0.3/(1 + (M/N))$. The corresponding ratio of moment of forces is

$$\frac{M}{N} \approx \frac{l_4 F_L}{l_3 F_D} = 2.6\Pi_{ld} \approx -1 + \left[e^{\frac{-d_*}{3}} + 0.2\tan\phi\left(1 - e^{\frac{-d_*}{20}}\right)\right]^{-1} \qquad (7.14b)$$

Accordingly, $M/N \approx 5$ when $d_* = d_s\left[(G-1)g/v^2\right]^{1/3} > 100$ and M/N approaches zero when $d_* < 1$, as expected from Chapter 5. From these empirical relationships, it can be noted that the moment arms are in the approximate following proportion: $l_1 \simeq 0.37l_3, l_2 \simeq 0.3l_3, l_4 \simeq 2.6l_3$.

Riprap is safe from failure when $SF_0 > 1$ and particles are expected to move when $SF_0 < 1$. Impending motion corresponds to $SF_0 = 1$. This analysis reduces to the method of Stevens and Simons (1971) for SF_0 when $\Theta_0 = 0$. Example 7.1 provides the detailed calculations of particle stability using this method.

Example 7.1 Application of the moment stability of a particle

A round 5 cm (50 mm) particle stands on the bed of a channel. If the downstream channel slope is 0.05 and the sideslope angle is 20°, calculate:

(1) the stability factor of the particle under an applied shear $\tau_o = 1$ lb/ft^2 when the streamlines are deflected downward at a 20° angle; and
(2) calculate the direction of the path line if the particle in (1) enters motion; and
(3) repeat the calculations for a 5 mm particle and compare the results.

(1) *Stability factor*

Step 1. The particle size is $5 \text{ cm} = 0.164$ ft, and $d_* = 0.05(1.65 \times 9.81 \times 10^{12})^{1/3} = 1,264$;

Step 2. The angle of repose is approximately $\phi = 37°$ from Figure 7.4;

Step 3. The lateral slope angle is $\Theta_1 = 20°$;

Step 4. The downstream slope angle $\Theta_0 = \tan^{-1} 0.05 = 2.86°$;

Step 5. The angle $\Theta = \tan^{-1} (\sin \Theta_0 / \sin \Theta_1) = 8.3°$;

Step 6. The factor $a_\Theta = \sqrt{\cos^2 \Theta_1 - \sin^2 \Theta_0} = 0.938$;

Step 7. The deviation angle $\lambda = 20°$;

Step 8. From Equation (7.11a)

$$\eta_o = \frac{21 \times 1 \, \text{lb}}{\text{ft}^2} \frac{\text{ft}^3}{(1.65) \times 62.4 \, \text{lb} \times 0.164 \, \text{ft}} = 1.24$$

Note that this particle would move in a horizontal plane because $\eta_o > 1$

Step 9. $M/N \approx 5$ from Equation (7.14b);

Step 10. From Equation (7.13),

$$\beta = \tan^{-1} \left\{ \cos(20° + 8.3°) \bigg/ \left(6 \frac{\sqrt{1 - (0.938)^2}}{1.24 \tan 37°} + \sin(20° + 8.3°) \right) \right\} = 18°$$

Step 11. From Equation (7.10),

$$\eta_1 = 1.24 \left\{ \frac{5 + \sin(20° + 18° + 8.3°)}{6} \right\} = 1.19$$

Step 12. From Equation (7.8a), because $\lambda \geq 0$,

$$SF_o = \frac{0.938 \tan 37°}{1.19 \tan 37° + \sqrt{1 - (0.938)^2} \cos 18°} = 0.58$$

The particle is moving because $SF_0 < 1$; and

(2) Direction of the path line
The particle will move along a path line at an angle $\beta = 18°$ from the vertical in the downstream direction.
(3) Calculations for a 5 mm particle

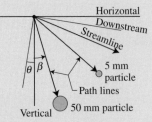

Figure E-7.1 Path lines for different particle sizes

In this case, $\phi = 31°$, $SF_0 = 0.07$ and $\beta = 48°$. There is a 30° difference between the orientation angles of these two particles as sketched in Figure E-7.1.

7.4 Simplified stability analysis

An approximate formulation of particle stability is presented as the ratio of critical shear stress on an embankment slope $\tau_{\Theta c}$ compared to the critical shear stress on a flat surface τ_c. The method is simplified and uses only the magnitude of active and passive forces, regardless of orientation. Accordingly, only the ratio of forces is used. The angle of repose ϕ can be expressed as a function of the resultant destabilizing force F_{R1} to the stabilizing force $F_S a_\Theta$. With reference to Figure 7.16,

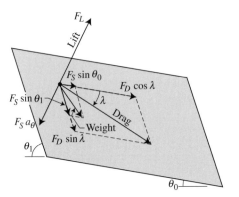

Figure 7.16. Simplified stability of particle

the magnitude of the destabilizing force is equal to the square root of the sum of the three squared orthogonal force components:

$$\tan \phi = \frac{F_{R1}}{F_S a_\Theta} = \frac{\left[(F_D \cos \lambda + F_S \sin \Theta_o)^2 + (F_D \sin \lambda + F_S \sin \Theta_1)^2 + F_L^2 \right]^{1/2}}{F_S a_\Theta}$$

(7.15a)

where the drag force at an angle λ can be expressed as $F_D = c_2 \, \tau_{\Theta c} d_s^2$.

For incipient motion ($\tau_{\Theta c} = \tau_c$) on a horizontal surface ($\Theta_1 = \Theta_0 = \lambda = 0$), the drag force $F_D = c_2 \tau_c d_s^2$, and with a constant value of the lift–drag ratio $\Pi_{ld} = F_L/F_D$, Equation (7.15a) reduces to:

$$\tan \phi = \frac{\left(c_2^2 \tau_c^2 d_s^4 + F_L^2 \right)^{1/2}}{F_S} = \frac{c_2 \tau_c d_s^2}{F_S} \sqrt{1 + \Pi_{ld}^2}$$

(7.15b)

Equations (7.15a and b) are combined and solved for $\tau_{\Theta c}/\tau_c$

$$\frac{\tau_{\Theta c}}{\tau_c} = - \frac{\sin \Theta_1 \sin \lambda + \sin \Theta_0 \cos \lambda}{\sqrt{1 + \Pi_{ld}^2} \tan \phi} +$$

$$\sqrt{ \frac{(\sin \Theta_1 \sin \lambda + \sin \Theta_0 \cos \lambda)^2}{(1 + \Pi_{ld}^2) \tan^2 \phi} + 1 - \left(\frac{\sin^2 \Theta_0 + \sin^2 \Theta_1}{\sin^2 \phi} \right) }$$

(7.16)◆

This equation expresses the ratio of the critical shear stress on the embankment $\tau_{\Theta c}$ to the critical shear stress on a horizontal surface τ_c as a function of: (1) the embankment slopes Θ_0 and Θ_1; (2) the shear stress direction angle λ; (3) the angle of repose ϕ; and (4) the lift–drag ratio Π_{ld}. After expanding the square root term of Equation (7.16) into a power series, it follows that

$$\frac{\tau_{\Theta c}}{\tau_c} = \sqrt{ \left(1 - \frac{\sin^2 \Theta_1}{\sin^2 \phi} - \frac{\sin^2 \Theta_0}{\sin^2 \phi} \right) \left(1 - x + \frac{x^2}{2} + \cdots \right) }$$

(7.17a)

where

$$x = \frac{\cos \phi (\sin \Theta_1 \sin \lambda + \sin \Theta_0 \cos \lambda)}{\sqrt{1 + \Pi_{ld}^2} \sqrt{\sin^2 \phi - \sin^2 \Theta_1 - \sin^2 \Theta_0}}$$

(7.17b)

Consider the particular case when $\Theta_0 = 0$. As the lift forces become negligible ($\Pi_{ld} = 0$), Equation (7.16) reduces to Brooks' (1958) relationship. As the lift–drag ratio goes to infinity ($\Pi_{ld} \to \infty$), or when the streamlines are horizontal ($\lambda = 0$), Equation (7.17) with $x = 0$ reduces to Lane's (1953) relationship:

$$\frac{\tau_{\Theta c}}{\tau_c} = \cos \Theta_1 \sqrt{1 - \left(\frac{\tan^2 \Theta_1}{\tan^2 \phi} \right)} = \sqrt{1 - \left(\frac{\sin^2 \Theta_1}{\sin^2 \phi} \right)} \qquad (7.18)\blacklozenge$$

This simple relationship has become very useful for defining the cross-sectional geometry of straight channels.

Consider steady uniform flow in a straight trapezoidal channel for a base width–depth ratio $W_b/h > 4$, as sketched in Figure 7.17. The bed shear stress $\tau_b \cong 0.97 \gamma_m h S_o$ and the bank shear stress $\tau_s \cong 0.75 \gamma_m h S_o$ impose two conditions for channel stability: (a) the bed particles are stable when $\tau_b < \tau_c$; and (b) the sideslope particles are stable when $\tau_s < \tau_{\Theta c}$ calculated from Equation (7.18). Table 7.4 suggests sideslope angles Θ_1 for a variety of channel bank material. A detailed stable channel design procedure is presented in Example 7.2.

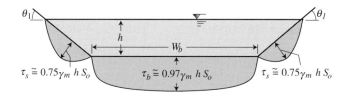

Figure 7.17. Applied shear stress distribution in trapezoidal channels

Table 7.4. *Embankment sideslopes*

Bank material	Θ_1 (deg)	Vertical:horizontal
Rock	78	1:0.2
Smooth or weathered rock, shell	45–63	1:1–1:0.5
Soil (clay, silt, and sand mixtures)	34	1:1.5
Sandy soil	34	1:1.5
Silt and loam (loose sandy earth)	26	1:2
Fine sand	18	1:3
Other very fine material	18	1:3
Compacted clay	34	1:1.5

Example 7.2 Design of a stable trapezoidal channel

Traditional channel design methods can be based on incipient motion; from Lane (1953) and Simons (1957). The regime theory approach of Lacey (1929) and downstream hydraulic geometry approaches, Wargadalan (1993) and Julien (2002) can also be used. Design a stable straight trapezoidal channel made of very uniform rounded gravel $d_{50} = 1.5$ in and $d_{90} = 2$ in. A total clear water discharge $Q = 1000$ cfs is conveyed on a bed slope $S_o = 0.0015$.

Step 1. The angle of repose $\phi = 37°$ is found from Figure 7.4 for rounded material. The flow depth is such that the bed particles are at incipient motion, assuming $R_h = h$. The bed shear stress τ_b equals the critical shear stress τ_c at the following flow depth h:

$$\tau_b = 0.97\gamma h S_o = 0.06(\gamma_s - \gamma)d_{50}\tan\phi$$

$$h = \frac{0.06(G-1)\,d_{50}}{0.97}\,\frac{}{S_o}\tan\phi = \frac{0.06 \times 1.65 \times 1.5\,\text{ft} \times \tan 37°}{0.97 \times 0.0015 \times 12} = 6.3\,\text{ft}$$

Step 2. The sideslope angle Θ_1 is calculated to correspond to beginning of motion, from Equation (7.18). The applied bank shear stress τ_s equals the critical shear stress $\tau_{\Theta c}$ at a side slope angle Θ_1 given as shown in Figure 7.17

$$\tau_s = 0.75\gamma h S_o = 0.06\tan\phi\,(\gamma_s - \gamma)d_{50}\sqrt{1 - \frac{\sin^2\Theta_1}{\sin^2\phi}}$$

$$\sqrt{1 - \frac{\sin^2\Theta_1}{\sin^2 37°}} = \frac{0.75 \times 6.3\,\text{ft} \times 0.0015 \times 12\,\text{in}}{0.06 \times 1.65 \times 1.5\,\text{in ft} \times \tan 37°} = 0.76$$

$$\sin^2\Theta_1 = \sin^2 37°\left(1 - 0.76^2\right), \text{or}\,\Theta_1 = 23°$$

Step 3. The cross-sectional area of a trapezoidal channel is given by $A = W_b h + h^2\cot\Theta_1 = W_b h + 93.5\text{ft}^2$, given the base width W_b. The wetted perimeter $P = W_b + 2h\,\text{cosec}\,\Theta_1 = W_b + 32.2$ ft and the hydraulic radius $R_h = A/P$. Using the Darcy–Weisbach friction factor from Equation (6.19), the velocity is given by

$$V = \frac{Q}{A} = \sqrt{\frac{8g}{f}}R_h^{1/2}S_o^{1/2} = \sqrt{gR_hS_o}\left(5.75\log\left(\frac{12.2h}{3d_{90}}\right)\right) = 2.76\sqrt{R_h}$$

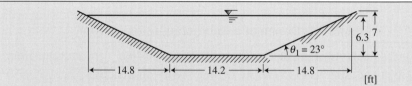

Figure E-7.2.1 Stable trapezoidal channel

As a first approximation, $A = (Q/V) \cong W_b h$ with $R_h \cong h$, gives $W_b \cong 22.9\,\text{ft}$. Through iterations or goal seek functions, the base width should be $W_b = 14.2\,\text{ft}$ for a stable channel. The cross-sectional geometry is plotted in Figure E-7.2.1.

Exercises

◆7.1 Analyze the angle of repose of one sphere of diameter d_2 resting on an equilateral triangle of spheres of diameter d_1. Check the angle of repose in Figure 7.1c and Equation (7.1b).

7.2 Demonstrate that the two formulations on the right-hand side of Equation (7.18) are identical.

7.3 Define d_s/δ as a function of τ_* and d_*. (*Answer*: $d_s/\delta \cong 0.086\,(\tau_* d_*^3)^{0.5}$)

Problems
◆◆*Problem 7.1*

Demonstrate that the critical shear stress on an incline is $\tau_{c\alpha} = \tau_c \sin(\phi+\alpha)/\sin\phi$ from Equation (7.8a). (*Hint*: Consider $\lambda = 90°$, $\Theta_1 = -\alpha$, and $\Theta = \Theta_0$, $\beta = 0$. See application to scour hole formation in a plunge pool in Bormann and Julien (1991).)

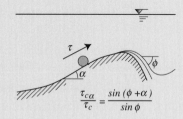

Figure P-7.1 Threshold up an incline

◆*Problem 7.2*

What is the sediment size corresponding to beginning of motion when the shear velocity $u_* = 0.1\,\text{m/s}$? (*Answer*: medium gravel, $d_s \cong 1\,cm$ from Table 7.1)

♦♦Problem 7.3

Calculate the stability factor of 8-in riprap on an embankment inclined at a $1V{:}2H$ sideslope if the shear stress $\tau_o = 1$ lb/ft^2. (*Answer: SF_0=1.3; thus, stable.*)

♦♦Problem 7.4

An angular 10 mm sediment particle is submerged on an embankment inclined at $\Theta_1 = 20°$ and $\Theta_0 = 0°$. Calculate the critical shear stress from the moment stability method when the streamlines near the particle are: (a) $\lambda = 15°$ (deflected downward); (b) $\lambda = 0°$ (horizontal flow); and (c) $\lambda = -15°$ (deflected upward).

♦♦Problem 7.5

Compare the values of critical shear stresses $\tau_{\Theta c}$ from Problem 7.4 with those calculated with Equation (7.16) and with Lane's method (Eq. 7.18), given $\Theta_0 = 0$ and $\Pi_{ld} = 0$. (*Answer: at $\phi = 37°$, and $\tau_{*c} = 0.047$.*)

Angle λ (deg)	Moment stability (N/m^2)	Simplified stability (N/m^2)	Lane's method (N/m^2)
−15° up	6.6	7.2	6.3
0	5.7	6.3	6.3
15° down	5.0	5.4	6.3

♦Problem 7.6

Design a stable channel conveying 14 m^3/s in coarse gravel, $d_{50} = 10$ mm and $d_{90} = 20$ mm, at a slope $S_o = 0.0006$.

♦Problem 7.7

The riverbed of the Rio Grande is composed of a mixture of 0.2 mm sand and 10 mm gravel. At what flow depth would the riverbed armor if the slope is 0.0008?

♦♦Problem 7.8

Repeat the calculations of Example 7.1 for $\lambda = -20°$ for two particle sizes (5 mm and 50 mm). Compare the particle direction angles, discuss why the orientation angles β are different.

◆◆*Problem 7.9*

Consider the gravel transport and particle size distributions at Little Granite Creek, Wyoming. The data were measured with a 7.6 cm-high Helley–Smith sampler in a very steep channel in the table below (Data source from Weinhold, 2002, and Ryan and Emmett, 2002). The channel bed slope is 0.02, the top width is 7.2 m, and the stage–discharge relationship in SI is $h \sim 0.3Q^{0.42}$. Bankfull discharge is about 220 cfs, and the particle size distribution of the bed material and sediment transport are shown in the table below and in Figure 7.11. Determine the following in SI units:

(a) Plot the sediment rating curve in tons per day as a function of discharge.
(b) What is the Shields value of the d_{50} of the surface material at bankfull discharge. Also calculate for the sub-surface material and discuss the results.
(c) On the sediment rating curve, plot a line with a concentration of 1 mg/l.
(d) Plot the particle size distribution of the sediment transport in the Helley–Smith at 50, 100, 200, 300 and 400 cfs.
(e) Determine the threshold condition for each size fraction and plot the results on Figure 7.13.
(f) Based on Figure 7.9 and d_{50} surface, determine at what discharge the bed material would reach incipient motion.

Sediment mass retained in grams

Avg Q cfs	Total transport metric t/d	Pan	Particle diameter (mm)								
			0.3	0.5	1.0	2.0	4.0	8.0	16.0	32	64
25	0.04	0.5	0.5	0.7	0.2	0.5	0.0	0.0	0.0	0	0
41	0.02	0.2	0.2	0.2	0.3	0.2	0.0	0.0	0.0	0	0
51	0.04	0.1	0.3	0.3	0.3	0.2	0.7	0.4	0.0	0	0
68	0.69	0.9	3.0	5.5	9.7	12.5	9.5	2.3	0.0	0	0
86	0.07	0.6	2.9	2.0	1.7	0.8	1.0	0.0	0.0	0	0
96	0.15	0.4	2.7	2.9	2.8	2.9	3.5	2.1	0.6	0	0
111	0.14	0.7	2.2	2.4	2.6	3.2	3.1	2.4	0.0	0	0
125	0.08	0.5	1.6	1.8	3.2	1.9	0.0	0.0	0.0	0	0
142	1.25	4.6	11.8	11.7	13.4	18.3	25.3	32.6	26.6	0	0
149	1.03	9.1	22.6	17.9	17.7	12.8	9.0	9.5	19.3	0	0
156	1.54	13.9	31.9	29.5	31.0	25.3	23.8	16.2	4.5	0	0
166	4.56	20.6	65.2	52.9	56.5	63.2	48.8	59.1	23.6	124	0
190	2.02	12.0	45.2	42.6	52.9	33.7	29.8	5.6	0.0	0	0
200	11.0	19.7	104	155	192	162	118	125	116	199	0
216	47.7	68.6	300	323	359	489	927	1,236	1,015	369	0
249	44.1	45.6	212	317	477	626	629	795	805	571	389
298	27.5	45.3	86.3	73.4	84.9	167	240	485	918	1,075	0
346	60.9	86.4	368	369	469	548	557	1,133	1,990	1,421	0
363	17.8	55.8	270	188	215	257	246	213.0	354	247	0
408	131	342	1,504	1,525	1,562	1,551	1,303	1,317	2,390	3,236	642

◆◆Problem 7.10

Based on Equation (7.2c), consider that $\tau_{*c} = 0.03$ when $F_L/F_D = 0$, then combine F_L/F_D from Figure 7.15 with $l_4/l_3 \approx 2.6$. Compare the values of τ_{*c} that are obtained with the Shields diagram value in Figure 7.8.

◆Problem 7.11

Estimate the maximum permissible velocity of a consolidated stiff clay channel at a density of 2,000 kg/m^3. If the discharge is 2,000 cfs, give approximate dimensions of a conveyance canal in terms of width, depth, velocity, sideslope angle, and downstream slope. (*Hint*: consider possible velocity and embankment sideslope angles to determine cross-section area and geometry.)

8

Bedforms

As soon as sediment particles enter motion, the random patterns of erosion and sedimentation generate very small perturbations of the bed surface elevation. In many instances, these perturbations grow in time until various surface configurations called bedforms cover the entire bed surface. The mechanics of bedforms is presented in Section 8.1 and the classification and geometry of bedforms is covered in Section 8.2. Resistance to flow, (Section 8.3), which depends largely on bedform configuration, directly affects water surface elevation in alluvial channels. Changes in bedform resistance induce shifting of the stage–discharge relationship (Section 8.4) and create problems in the determination of river discharges from water level measurements. Two examples and one case study complete this chapter.

8.1 Mechanics of bedforms

The mechanics of bedforms is rather complex and involves the main flow component as well as the near-bed flow conditions. The main flow characteristics can be deduced from an analysis of the equations of motion in the downstream x and upward z directions (Equations (3.17a and c)) for steady flow conditions

$$\frac{\partial}{\partial x}\left(\frac{p}{\rho_m} + \frac{v^2}{2}\right) = g_x + (v_y \otimes_z - v_z \otimes_y) + \frac{1}{\rho_m}\left(\frac{\partial \tau_{xx}}{\partial x} + \frac{\partial \tau_{yx}}{\partial y} + \frac{\partial \tau_{zx}}{\partial z}\right) \tag{8.1}$$

$$\frac{\partial}{\partial z}\left(\frac{p}{\rho_m} + \frac{v^2}{2}\right) = g_z + (v_x \otimes_y - v_y \otimes_x) + \frac{1}{\rho_m}\left(\frac{\partial \tau_{xz}}{\partial x} + \frac{\partial \tau_{yz}}{\partial y} + \frac{\partial \tau_{zz}}{\partial z}\right) \tag{8.2}$$

Consider two-dimensional flow in a wide open channel: (1) the shear stress components τ_{xx}, τ_{yx}, τ_{yz}, and τ_{zz} can be neglected as a first approximation; and (2) the velocity components v_y and v_z are assumed negligible such that $v_x = v$. With the only rotation component $\otimes_y = \partial v_x / \partial z$ then obtained from Equation (3.4), and the specific energy function $E = \frac{p}{\gamma_m} + \frac{v^2}{2g}$ from Equation (3.23a), Equations (8.1)

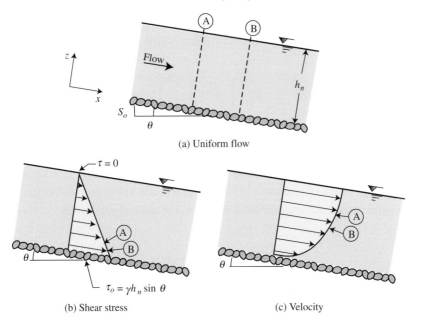

(a) Uniform flow

(b) Shear stress

(c) Velocity

Figure 8.1. Steady uniform flow in alluvial channels a) uniform flow b) shear stress c) velocity

and (8.2) respectively reduce to:

$$g\frac{\partial}{\partial x}\left[\frac{p}{\rho_m g}+\frac{v^2}{2g}\right]=g_x+\frac{1}{\rho_m}\frac{\partial \tau_{zx}}{\partial z} \tag{8.3}$$

$$\frac{\partial}{\partial z}\left[\frac{p}{\rho_m}+\frac{v^2}{2}\right]=g_z+v_x\frac{\partial v_x}{\partial z}+\frac{1}{\rho_m}\frac{\partial \tau_{xz}}{\partial x} \tag{8.4}$$

The pressure distribution remains hydrostatic when $\partial\tau_{xz}/\partial x=0$, and Equation (8.4) can be easily depth-integrated over h_n with $g_z=-g\cos\theta$ to give

$$p=\rho_m g(h_n-z)\cos\theta \tag{8.5}$$

For steady uniform flow sketched in Figure 8.1a, the velocity v and the pressure p remain constant along x and the left-hand side of Equation (8.3) reduces to zero. The shear stress distribution τ_{zx} is then obtained after integrating the right-hand side of Equation (8.3) over the normal depth h_n with $g_x=g\sin\theta$, thus

$$\tau_{zx}=\rho_m g(h_n-z)\sin\theta \tag{8.6}$$

It is noticeable that for steady uniform flow, the shear stress increases linearly from $\tau_{zx}=0$ at the free surface to the boundary shear stress $\tau_o=\tau_{zx}=\rho_m g h_n\sin\theta=\gamma_m h_n S_o$ at the bed.

In the case of nonuniform flow with hydrostatic pressure distribution, the governing equation describing the internal shear stress distribution (Equation (8.3)) can be rewritten as

$$-\underbrace{\frac{1}{\rho_m}\frac{\partial \tau_{zx}}{\partial z}}_{\text{uniform flow}} = g\sin\theta - \underbrace{g\frac{\partial}{\partial x}\left[\frac{p}{\rho_m g} + \frac{v^2}{2g}\right]}_{\text{nonuniform flow perturbation}} \quad \textit{throughout the flow} \quad (8.7)$$

It is becoming clear that the term in brackets of Equation (8.7) is the specific energy *E*. As shown in Figure 8.1c, it should also be considered that $v \to 0$ near the bed and the term in brackets reduces to flow depth *h* in the lower part of the profile. The cases of subcritical and supercritical flow are considered separately.

8.1.1 Lower regime

In the lower regime consider subcritical flow with a small bed perturbation of amplitude Δz, as sketched in Figure 8.2a. Two points C and D are identified where the flow depth is h_n on each side of the perturbation for comparison with shear stresses for steady uniform flow. Considering the entire flow depth, it is shown from the specific energy diagram in Figure 8.2b that a small perturbation Δz causes a decrease in specific energy when approaching the perturbation, thus ($g\ \partial E/\partial x < 0$ at point C). It follows that the gradient of shear stress $\partial \tau_{zx}/\partial z$ on the upstream side becomes larger (thus increasing shear stress) than for the corresponding steady uniform flow condition. Conversely, on the downstream side of the perturbation, the corresponding increase in specific energy ($g\ \partial E/\partial x > 0$) causes a reduction in shear stress (reduced $\partial \tau_{zx}/\partial z$).

Near the bed, however, the velocity term in the brackets of Equation (8.7) can be neglected and the shear stress gradient near the bed reduces to

$$-\frac{1}{\rho}\frac{\partial \tau_{zx}}{\partial z} = g\sin\theta - g\frac{\partial h}{\partial x}; \text{ near the bed} \quad (8.8)$$

On the downstream side of the perturbation, the gradient of shear near the bed becomes positive when $\partial h/\partial x$ exceeds $\sin\theta$, as shown in Figure 8.2c for curve D. Integration over the entire flow depth may result in negative values of bed shear stress τ_0 when $\partial h/\partial x$ becomes large. Separation occurs when $\tau_0 < 0$ on the downstream side of the perturbation as sketched in Figure 8.2d.

As a consequence of increased shear stress on the upstream face of the perturbation in subcritical flows, increased sediment transport causes erosion in converging flow. On the downstream side of the perturbation, the reduced bed shear stress and sediment transport capacity induces sedimentation on the lee side of the perturbation. This mechanism causes bedforms to move downstream. Depending on

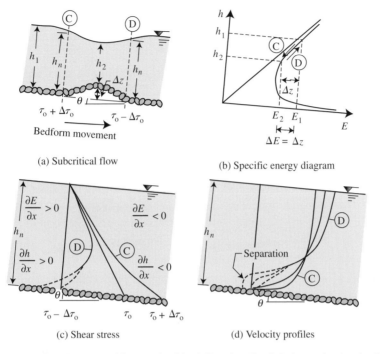

Figure 8.2. Steady non uniform subcritical flow in alluvial channels a) subcritial flow b) specific energy diagram c) shear stress c) velocity profiles

near-bed conditions, these perturbations can amplify until ripples and dunes fully develop.

8.1.2 Upper regime

In the upper regime, consider the case of supercritical flow and a similar analysis is sketched in Figure 8.3. The small perturbation causes an increase in flow depth and a reduction in specific energy near F, upstream of the perturbation. At point F, this results in a steeper shear stress gradient in the upper part of the velocity profile and reverse gradient near the bed (Figure 8.3c), which may induce separation of the velocity profile near the bed (Figure 8.3d). Conversely, the downstream face of the perturbation at G shows an increase in boundary shear stress which increases sediment transport capacity. Because $\partial h/\partial x$ is much larger than $\partial E/\partial x$, this pattern can also be observed at lower values of the Froude number. Although the sediment particles are transported in the downstream direction, bedforms in supercritical flow can migrate upstream in shallow streams as the sediment deposition on the upstream face combines with the rate of erosion on the downstream face of the

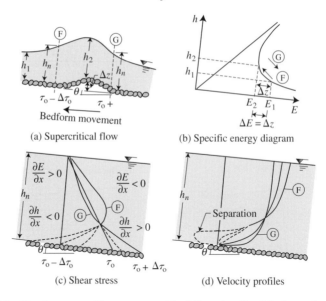

(a) Supercritical flow

(b) Specific energy diagram

(c) Shear stress

(d) Velocity profiles

Figure 8.3. Steady non-uniform supercritical flow in alluvial channels a) supercritical flow b) specific energy diagram c) shear stress d) velocity profiles

perturbation. The term antidune describes bedforms and free surface oscillations migrating upstream.

8.2 Bedform classification and geometry

The various bedform configurations depend on the main flow characteristics of Section 8.1 as well as the bed material characteristics defined as the bed particle size, the fall velocity, and the grain shear Reynolds number.

From extensive laboratory experiments at Colorado State University by Simons and Richardson (1963, 1966), several types of bedforms have been identified. Flat bed, or plane bed, refers to a bed surface without bedforms. With reference to the bedform configurations sketched in Figure 8.4, ripples are small bedforms with wave heights less than a few cm (~ 0.1 ft). Ripple shapes vary from nearly triangular to almost sinusoidal. Dunes are much larger than ripples and are out of phase with the water surface waves. From longitudinal profiles, dunes are often triangular with fairly gentle upstream slopes and downstream slopes approaching the angle of repose of the bed material. The large eddies on the lee side of dunes cause surface boils, clearly visible from bridges and river banks. The lower regime on the left-hand side of Figure 8.4 includes plane bed (without sediment transport), ripples, dunes, and washed-out dunes. The upper regime consists of upper-regime plane

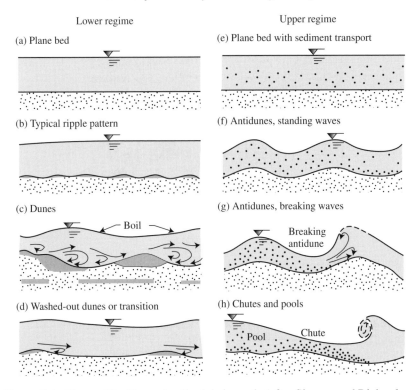

Figure 8.4. Types of bedforms in alluvial channels (after Simons and Richardson, 1966) a) plane bed b) typical ripple pattern c) dunes d) washed-out dunes or transition e) plane bed with sediment transport f) antidunes, standing waves g) antidunes, breaking waves h) chutes and pools

bed (with sediment transport), antidunes, breaking waves, and chutes and pools. In the upper regime, bedforms are in phase with free-surface waves. They grow with increasing Froude number until they become unstable and break. Chutes and pools occur at relatively large slopes and consist of elongated chutes with supercritical flow, connected by pools where the flow is generally subcritical.

To separate lower and upper flow regimes, the Froude number values in the transition zone have been found by Athaullah (1968) to decrease with relative submergence, as shown on Figure 8.5.

8.2.1 Bedform prediction

The prediction of bedform configurations has been the subject of numerous laboratory and field investigations. None provides a definite classification but the following are among the most instructive predictors. Liu (1957) used the ratio u_*/ω

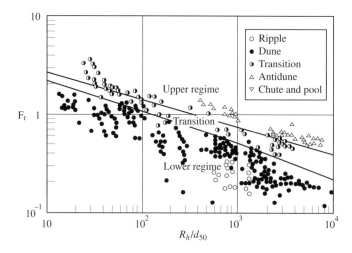

Figure 8.5. Lower- and upper-regime bedform classification (after Athaullah, 1968)

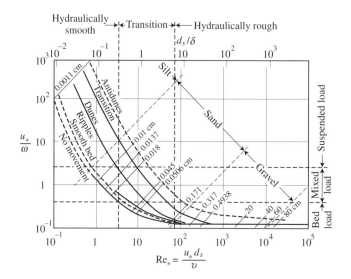

Figure 8.6. Bedform classification (after Liu, 1957)

of the shear velocity u_* to the particle fall velocity ω as a function of the grain shear Reynolds number $\mathrm{Re}_* = \frac{u_* d_s}{\upsilon_m}$. His analysis in Figure 8.6 suggests that ripples and dunes essentially cannot form in gravel-bed channels. Silts also move in suspension rather than bedload, as will be discussed in Chapter 10.

Simons and Richardson (1963, 1966) proposed a bedform predictor encompassing both lower and upper regimes when plotting the stream power $\gamma q S_f$ as a function of particle diameter (Figure 8.7). Accordingly, ripples cannot be found

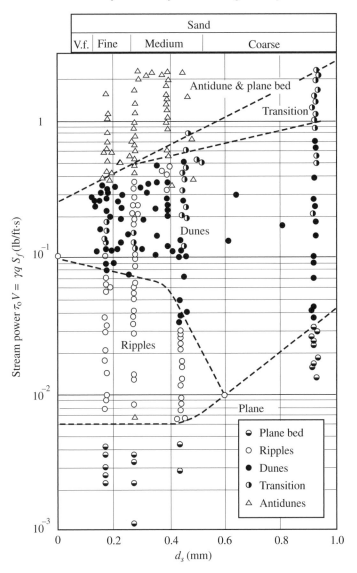

Figure 8.7. Bedform classification (after Simons and Richardson, 1963, 1966)

for $d_s > 0.6$ mm. This bedform predictor is based on extensive laboratory experiments and is quite reliable for shallow streams. However, it deviates from observed bedforms in deep streams. Chabert and Chauvin (1963) proposed a bedform predictor based on the Shields diagram, shown in Figure 8.8. Ripples form when $d_* = d_s \left((G-1)g/v_m^2\right)^{1/3} < 20$, which corresponds to $Re_* < 15$, or transition to hydraulically smooth boundaries.

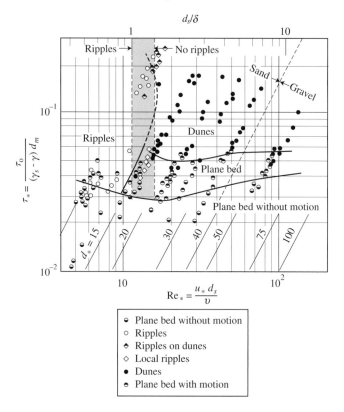

Figure 8.8. Bedform classification (after Chabert and Chauvin, 1963)

van Rijn (1984a,b) proposed a bedform classification based on the dimensionless particle diameter d_* and the transport-stage parameter T, respectively defined as

$$d_* = d_{50} \left[\frac{(G-1)g}{v_m^2} \right]^{1/3}$$ (8.9a)

$$T = \frac{\tau_*' - \tau_{*c}}{\tau_{*c}} = \frac{(u_*')^2 - u_{*c}^2}{u_{*c}^2} = \frac{\rho_m V^2}{\tau_c \left(5.75 \log \frac{4R_b}{d_{90}} \right)^2} - 1$$ (8.9b)

or

$$\tau_*' \approx 0.04 \left(\frac{d_{50}}{h} \right)^{1/3} \frac{V^2}{((G-1)gd_{50})}$$ (8.9c)

in which d_{50} is the mean bed particle diameter (50% passing by weight), G is the particle specific gravity, v_m is the fluid mixture kinematic viscosity, V is the depth-averaged flow velocity, g is the gravitational acceleration, R_b is the hydraulic radius related to the bed obtained from the Vanoni–Brooks method (Section 6.4.3), d_{90} is

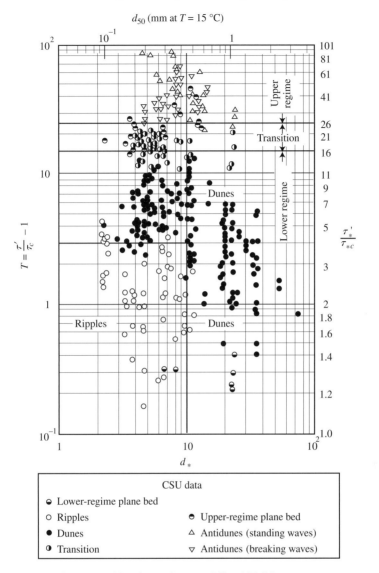

Figure 8.9. Bedform classification (after van Rijn, 1984b)

the 90% passing-bed particle diameter, and τ_c is the critical shear stress obtained from the Shields diagram. The parameters τ'_* and u'_* are expanded upon in Section 8.3. van Rijn (1984b) suggested that ripples form when both $d_* < 10$ and $T < 3$, as shown in Figure 8.10. Dunes are present elsewhere when $T < 15$, dunes washout when $15 < T < 25$, and the upper flow regime starts when $T > 25$.

Julien and Raslan (1998) found that the value of the transport-stage parameter T_P for the upper-regime plane bed increases with relative submergence. As shown in

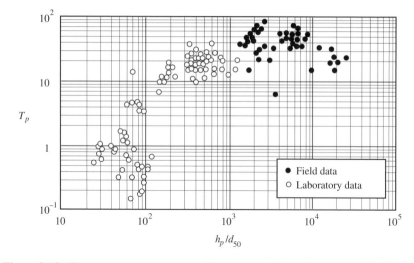

Figure 8.10. Transport-stage parameter T_p versus relative submergence h_p/d_{50} for upper-regime plane bed (after Julien and Raslan, 1998)

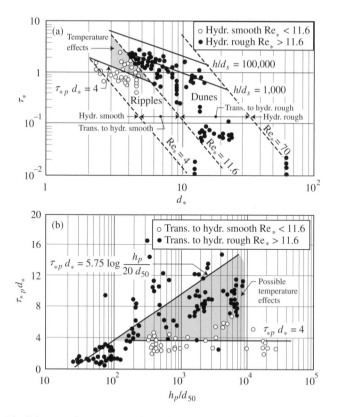

Figure 8.11. Diagram for upper-regime plane bed: a) τ_* versus d_* and b) τ_* versus d_* versus h_p/d_{50} (after Julien and Raslan, 1998)

Figure 8.10, the upper-regime plane bed of rivers is much higher than for laboratory data. The available data for upper regime could be found between $4 < \text{Re}_* < 70$, which corresponds to the range from hydraulically smooth to hydraulically rough. As shown in Figures 8.11a and b, two separate regimes could be identified for the transition to upper regime with:

$$\tau_{*p} d_* \approx 4 \text{ when } \text{Re}_* < 11.6 \tag{8.10a}$$

$$\tau_{*p} d_* = 5.75 \log 0.05 \, h_p/d_{50} \text{ when } \text{Re}_* > 11.6 \tag{8.10b}$$

In summary, ripples form when three conditions are satisfied: (1) $2 < d_* < 6$; (2) $4 < \text{Re}_* < 11.6$; and (3) $\tau_* < 4/d_*$ or $\tau_* < 1$. Dunes form when $3 < d_* < 70$, $11.6 < \text{Re}_* < 70$ and $\tau_*' < (5.75/d_*) \log (h/20 d_{50})$. Temperature effects are possible for the limited range of conditions shown in Figure 8.11a, where $3 < d_* < 6$ and $h/d_{50} > 200$. This corresponds to $\text{Re}_* \approx 11.6$ and $\tau_* \approx 1$. This leads to the possibility that with a slight change in viscosity bedform configurations may be in the lower or upper regime, depending on water temperature, as observed in the field, e.g. Example 8.1.

Example 8.1 Temperature effects on bedforms, Missouri River, Nebraska

Temperature effects are possible for fine or medium sands $2 < d_* < 6$. For instance, Julien and Raslan (1998) present the example of the Missouri River at $h = 3.07$m, $d_{50} = 0.218$mm, and $S = 1.42 \times 10^{-4}$. Accordingly, $\nu = 1.58 \times 10^{-6}$ m^2/s at $T° = 3°C$ and $d_* = 4.06$, $\tau_* = 1.21$ and $\text{Re}_* = 9 < 11.6$. The flow depth for the upper-regime plane bed is 2.5 m and the Missouri River should have a plane bed during the cold winter months. However, at $T° = 20°C$ with $\nu = 1 \times 10^{-6}$m^2/s, the parameters change to $d_* = 5.5$, $\tau_* = 1.21$ and $\text{Re}_* = 14.2 > 11.6$. The corresponding τ_{*p} from Equation (8.10b) becomes, $\tau_{*p} = 2.97$, and, $\tau_* < \tau_{*p}$ is now in the lower regime with dunes in the Missouri River during the summer months. At a given discharge, the bedforms for fine and medium sand-bed rivers can change depending on water temperature with low Manning n in winter and high Manning n in summer.

8.2.2 Bedform geometry

The geometry of dunes is a concern in engineering projects dealing with navigation, flood control, and resistance to flow. In the lower regime, the geometry of bedforms refers to representative dune height Δ and wavelength Λ of dunes as a function of the average flow depth h, average bed particle diameter d_{50}, and other flow parameters such as the transport-stage parameter T and the grain shear Reynolds

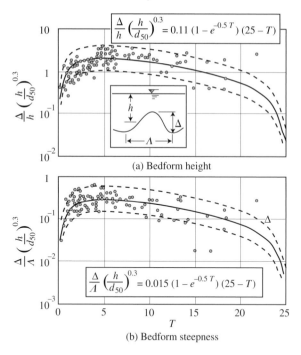

(a) Bedform height

(b) Bedform steepness

Figure 8.12. Laboratory bedform height and steepness (after van Rijn, 1984b) a) bedform height b) bedform steepness

number $Re_* = u_* d_s/v_m$. The dune height and steepness predictors proposed by van Rijn (1984) are:

$$\frac{\Delta}{h} = 0.11 \left(\frac{d_{50}}{h}\right)^{0.3} \left(1 - e^{-0.5T}\right) (25 - T) \qquad (8.11a)$$

and

$$\frac{\Delta}{\Lambda} = 0.015 \left(\frac{d_{50}}{h}\right)^{0.3} \left(1 - e^{-0.5T}\right) (25 - T) \qquad (8.11b)$$

The dune length Λ obtained from dividing these two equations $\Lambda = 7.3h$ is quite close to the theoretical value $\Lambda = 2\pi h$ proposed by Yalin (1964) and revisited by Zhou and Mendoza (2005). The agreement with laboratory data is quite good, as shown in Figure 8.12a,b. However, both curves tend to underestimate the bedform height and steepness of field data as shown in Figure 8.13. The field data from different sources including Adriaanse (1986), Neill (1969), and Peters (1978) has been compiled by Julien and Klaassen (1995). Lower-regime bedforms can be observed at values of T well beyond 25 in very large rivers.

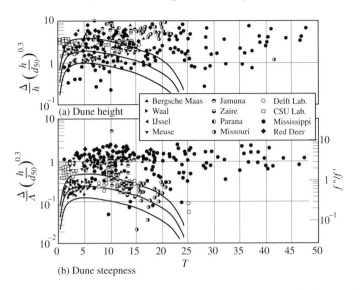

Figure 8.13. Dune height and steepness in rivers (after Julien and Klaassen, 1995)

The variability of the field data is large but Julien and Klaassen (1995) proposed a first approximation for average dune height $\overline{\Delta}$ and wavelength $\overline{\Lambda}$

$$\overline{\Delta} \approx 2.5 \, h^{0.7} d_s^{0.3} \qquad (8.12a) \blacklozenge$$

$$\overline{\Lambda} \approx 6.5 h \qquad (8.12b) \blacklozenge$$

A similar approach has been suggested by Yalin and daSilva (2001), typical values of $\Delta/\Lambda < 0.1$ for laboratory data and $\Delta/\Lambda < 0.06$ for field data, which gives roughly $\Delta/h < 0.36$ for field data.

8.3 Resistance to flow with bedforms

The lower flow regime is generally subcritical (Fr < 1), as expected from Section 8.1, and the water surface undulations are out of phase with the bed waves (Figure 8.4). Surface roughness characteristics are summarized in Table 8.1. Resistance to flow is large because flow separation occurs on the downstream side of the waves. This generates large-scale turbulence dissipating considerable energy. Sediment transport is relatively low because bed sediment particles move primarily in contact with the bed.

In the upper flow regime, resistance to flow is low because grain roughness predominates. However, the energy dissipated by standing waves and the formation of breaking antidunes increases flow resistance. Standing waves and antidunes are common in shallow supercritical flow (Fr > 1). Standing waves are in-phase

Table 8.1. *Typical bedform characteristics*

Bedform	Manning n	Concentration (mg/l)	Roughness	Surface profiles
Lower Regime				
Plane bed	0.014	0	grain	–
Ripples	0.018 – 0.028	10 – 200	form	–
Dunes	0.020 – 0.040	200 – 3,000	form	out of phase
Washed-out dunes	0.014 – 0.025	1,000 – 4,000	variable	out of phase
Upper Regime				
Plane bed	0.010 – 0.013	2,000 – 4,000	grain	–
Antidunes	0.010 – 0.020	2,000 – 5,000	grain	in phase
Chutes and pools	0.018 – 0.035	5,000 – 50,000	variable	in phase

sinusoidal sand and water waves (Figure 8.4) that build up in amplitude from a plane bed and plane water surface. Resistance to flow for breaking antidunes is only slightly larger than for plane bed because they cover only a small portion of channel reach at a given time. Bed material transport in the upper flow regime is high because, except when antidunes are breaking, the contact sediment discharge is almost continuous, and the suspended sediment concentration is large.

The analysis of total resistance is somewhat analogous to the analysis of viscous flow around a spherical particle presented in Section 5.3. The total resistance is separated into: (1) grain resistance accounting for forces acting on individual particles; and (2) form resistance due to bedform configurations. The total bed shear stress is the sum of two components

$$\tau_b = \tau_b' + \tau_b'' \tag{8.13a}$$

where τ_b is the total bed shear stress, τ'_b is the grain shear stress, and τ''_b is the form shear stress.

The corresponding identities using the grain shear velocity u'_*, the grain hydraulic radius R'_h, grain friction slope S'_f, grain Darcy–Weisbach friction factor f' and their corresponding values u''_*, R''_h, S''_f, and f'' for the form resistance are formulated from $\tau = \rho u_*^2 = \gamma R_h S_f = \frac{f}{8}\rho V^2$ as:

$$u_*^2 = u_*'^2 + u_*''^2 \tag{8.13b}$$

$$R_h = R'_h + R''_h \tag{8.13c}$$

$$S_f = S_f' + S_f''$$

(8.13d)

$$f = f' + f''$$

(8.13e)

When written in terms of Shields parameters, Equation (8.13a) gives

$$\tau_* = \tau_*' + \tau_*'', \text{or:} \quad \frac{\tau_b}{(\gamma_s - \gamma_m)\, d_{50}} = \frac{\tau_b'}{(\gamma_s - \gamma_m)\, d_{50}} + \frac{\tau_b''}{(\gamma_s - \gamma_m)\, d_{50}}$$

(8.13f)◆

where τ_* is the total Shields parameter, τ_*' is the grain Shields parameter, and τ_*'' is the form Shields parameter. Notice that for Manning and Chézy coefficients: $n \neq n' + n''$ and $C \neq C' + C''$.

8.3.1 Total and grain resistance with bedforms

The total resistance to flow can be obtained from field measurements of flow depth h, main flow velocity V, and friction slope S_f from

$$f = \frac{8\, g h S_f}{V^2}$$

(8.14)

The roughness height k_s can then be obtained from

$$\sqrt{\frac{8}{f}} = 5.75 \log\left(\frac{12.2\, h}{k_s}\right)$$

(8.15)

In sand-bed channels with bedforms, van Rijn (1984b) showed in Figure 8.14 that the roughness height k_s depends on dune length Λ and can be as large as the dune height Δ. A very crude approximation deserving further testing is

$$k_s \approx 1.1\Delta \left(1 - e^{-25\Delta/\Lambda}\right)$$

(8.16)

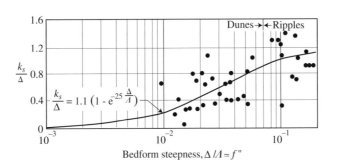

Figure 8.14. Equivalent bedform roughness (after van Rijn, 1984b)

To separate the grain resistance from form resistance, equation (6.19b) see page 121 can be used to define grain resistance, e.g.

$$\sqrt{\frac{8}{f'}} = 5.75 \log \left(\frac{12.2h}{k_s'} \right) \tag{8.17}\blacklozenge$$

where k_s' is the grain roughness such that $k_s' \approx 6.8 \, d_{50}$ or $k_s' \approx 3 \, d_{90}$. When using the Manning–Strickler approximation, one obtains a simple relationship for grain resistance, e.g. Figure 6.6a,

$$\frac{V}{u_*'} = \sqrt{\frac{8}{f'}} \approx 5 \left(\frac{h}{d_s} \right)^{1/6} \tag{8.18a}$$

or

$$f' \cong 0.32 \left(\frac{d_{50}}{h} \right)^{1/3} \tag{8.18b}$$

Subtracting grain resistance f' from the total resistance f has been commonly used to estimate the form resistance f'' from Equations (8.13e, 8.14 and 8.18b).

8.3.2 Bedform resistance

A different approach is discussed in this section. For instance, Engelund (1966) used the Carnot formula to define the head loss $\Delta H''$ from

$$\Delta H'' = C_E \frac{(V_1 - V_2)^2}{2g} \tag{8.19a}$$

where C_E is the expansion loss coefficient, V_1 and V_2 are the depth-averaged velocities at the crest and toe of the dune, respectively. Given the dune height Δ, dune length Λ, mean flow depth h and unit discharge q, one obtains

$$\Delta H'' = \frac{C_E}{2g} \left(\frac{2q}{2h - \Delta} - \frac{2q}{2h + \Delta} \right)^2 \approx C_E \frac{V^2}{2g} \left(\frac{\Delta}{h} \right)^2 \tag{8.19b}$$

or

$$S_f'' = \frac{\Delta H''}{\Lambda} = \frac{C_E}{2} \frac{\Delta^2}{\Lambda h} \mathrm{Fr}^2 \tag{8.19c}$$

Accordingly, the bedform energy slope $S_f'' = \Delta H'' / \Lambda$ depends on the product of dune steepness Δ / Λ and relative dune height Δ / h. However, Engelund (1977)

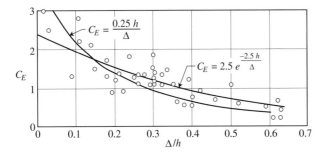

Figure 8.15. Expansion loss coefficient (modified after Engelund, 1977)

later defined C_E as shown in Figure 8.15

$$C_E \approx 2.5\,e^{\frac{-2.5h}{\Delta}} \tag{8.20}$$

It is interesting the C_E can also be approximated by $C_E \approx 0.25h/\Delta$, which can be combined in Equation (8.19c) with $S_f'' = (f''/8)\,\mathrm{Fr}^2$ to give the very simple approximation

$$f'' \approx \frac{\Delta}{\Lambda} \tag{8.21}\blacklozenge$$

When combined with the grain resistance in Equation (8.18b), it is interesting to find

$$\frac{f''}{f'} \approx \frac{3\Delta}{\Lambda}\left(\frac{h}{d_{50}}\right)^{1/3} \tag{8.22}$$

The dune steepness parameter in Figure 8.13b therefore approximately describes the ratio of f'' to f'.

In summary, despite obvious complexity, the analysis of bedform configurations bears remarkable simplicity after considering Equations (8.12, 8.18b and 8.21), or $\bar{\Delta} \approx 2.5\,h^{0.7}d_{50}^{0.3}$, $\bar{\Lambda} \approx 6.5h$, $f'' \approx \Delta/\Lambda$. Example 8.2 provides a detailed procedure to estimate bedform configuration, dune geometry, and resistance to flow.

Example 8.2 Resistance to flow with bedforms

Consider a 46 m-wide canal in Pakistan. The slope is 11.5 cm/km and the bed material size is $d_{50} = 0.4\,\mathrm{mm}$ and $d_{90} = 0.65\,\mathrm{mm}$. Determine the type and geometry of bedform and estimate the discharge when the flow depth is 3 m.

Step 1. $d_* = d_{50}\left(\dfrac{(G-1)g}{\nu^2}\right)^{1/3} = 0.0004\left(\dfrac{1.65 \times 9.8}{\left(1 \times 10^{-6}\right)^2}\right)^{1/3} = 10.1$

Step 2.a $\tau_* = \dfrac{hS}{(G-1)d_{50}} = \dfrac{3\,\mathrm{m} \times 11.5 \times 10^{-5}}{1.65 \times 4 \times 10^{-4}} = 0.52$

Step 2.b From Equation (8.9c) $\tau_*' \approx 0.04 \left(\dfrac{d_{50}}{h}\right)^{1/3} \dfrac{V^2}{((G-1)\,g\,d_{50})}$ when velocity measurements are available.

Step 3.

$\mathrm{Re}_* = \dfrac{u_* d_s}{\nu} = \dfrac{\sqrt{g\,hS}\,d_s}{\nu} = \dfrac{\sqrt{9.81 \times 3 \times 11.5 \times 10^{-5}} \times 4 \times 10^{-4}}{1 \times 10^{-6}} = 23.2$

Step 4. Upper-regime plane bed when $\mathrm{Re}_* > 11.6$ from Equation (8.10b) is

$\tau_{*p} = \left(\dfrac{5.75}{d_*}\right) \log\left(\dfrac{h}{20\,d_{50}}\right) = \left(\dfrac{5.75}{10.1}\right) \log\left(\dfrac{3}{20 \times 0.0004}\right) = 1.46$

$\tau_*' < \tau_{*p}$ and dunes are expected, the average dune height and length are:

Step 5. $\bar{\Delta} \approx 2.5\,h^{0.7} d_{50}^{0.3} = 2.5 \times 3^{0.7} \left(4 \times 10^{-4}\right)^{0.3} = 0.51\,\mathrm{m}$ from Eq. (8.12a)

Step 6. $\bar{\Lambda} \approx 6.5\,h = 6.5 \times 3\,\mathrm{m} = 19.5\,\mathrm{m}$ from Equation (8.12b)

Step 7. $f' \approx 0.32 \left(\dfrac{d_{50}}{h}\right)^{1/3} = 0.32 \times \left(\dfrac{4 \times 10^{-4}}{3}\right)^{1/3} = 0.016$ from Eq. (8.18b)

Step 8. $f'' \approx \dfrac{\bar{\Delta}}{\bar{\Lambda}} = \dfrac{0.51\mathrm{m}}{19.5\mathrm{m}} = 0.027$ from Equation (8.21)

Step 9. $f = f' + f'' = 0.016 + 0.027 = 0.043$ from Equation (8.13e)

Step 10. $V = \sqrt{\dfrac{8}{f}}\sqrt{g hS} = \sqrt{\dfrac{8}{0.043} \times 9.81 \times 3 \times 11.5 \times 10^{-5}} = \dfrac{0.79\mathrm{m}}{\mathrm{s}}$ from Eq. (8.14)

Step 11. $Q = WhV = 46\,m \times 3\,m \times 0.79\,\mathrm{m/s} = 109\,\mathrm{m}^3/\mathrm{s}$

8.4 Field observations of bedforms

Three examples are discussed in this section. The first example is on the Missouri River. There is no unique relationship between channel slope, flow depth, and flow velocity in alluvial sand-bed channels. Changes in bed configuration cause shifts in the stage–discharge relationship. Bed configuration changes affect resistance to flow, flow velocity, and sediment transport. There are many examples of these changes in natural streams. On an approximately 300 mile reach of the Missouri River from Sioux City, Iowa, to Kansas City, Kansas, the bed configurations at

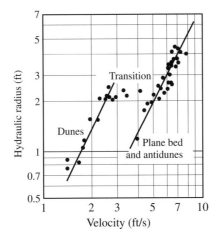

Figure 8.16. Velocity versus hydraulic radius for the Rio Grande (after Nordin, 1964)

a given discharge change from large dunes in the middle of the summer, when the water temperature is about 80°F, to washed-out dunes or plane bed in the fall at water temperature around 45°F. The changes in bedform configuration at constant discharge (approximately 34,000 ft^3/s) decreased Manning n from 0.018 to 0.014 and reduced the average flow depth from approximately 11 ft to approximately 9 ft, with a corresponding increase in average flow velocity from 4.6 ft/sec to 5.5 ft/sec. The reason for these temperature effects has been explained in Example 8.1.

As a second example, the Rio Grande also displays a discontinuous stage–discharge relationship, Nordin (1964) documented the change in hydraulic radius versus mean velocity in Figure 8.16. During runoff events, the bed varies from dunes at low flow, to plane bed, and antidunes at high flow, which causes the discontinuity in the stage–discharge relationship. The change from dunes to plane bed occurs at a larger discharge than the change from plane bed back to dunes.

As a third example, the 1984 flood of the Meuse River resulted in significant changes in bedform configuration. A sequence of bathymetric profiles is shown in Figure 8.17, where the amplitude and wavelength of bedforms change rapidly during floods. Soundings prior to the flood, $Q = 1,434\,\text{m}^3$/s, on February 8 show small amplitude irregularities of the bed profile similar to those after the flood, $Q = 654\,\text{m}^3$/s, on February 20. At higher discharge, the large dunes showed rounded crests and some dunes measured up to 3 m in amplitude. Comparing results of February 8 and February 15, at a similar flow discharge, both the dune height and wavelength are smaller under rising discharge than falling discharge. This loop

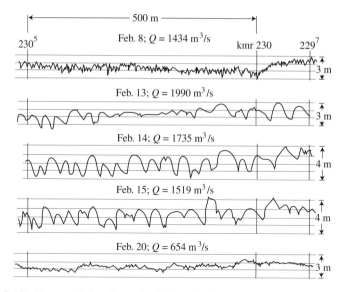

Figure 8.17. Dunes of the Bergsche Maas during the 1984 flood of the Meuse river (kmr denotes river kilometer, after Adriaanse, 1986)

rating effect is expected because the time scale required for the formation of 2 m-high sand dunes is in the order of 1–3 days in the Bergsche Maas. The following case study illustrates the changes in bedform configuration of the Rhine during a major flood.

Case study 8.1 River Rhine, The Netherlands

During the 1998 flood of the Rhine river, the peak discharge reached 9,464 m³/s, and figures among the largest floods. The bedform of a relatively straight reach of the Upper Rhine (Bovenrijn) just upstream of the Pannerdens canal was analyzed by Julien and Klaassen (1995) and Julien *et al.* (2002). As shown in Figure CS-8.1, the flow depth reached 12 m and the peak flow velocity was 1.8 m/s. Bedform data were recorded twice daily during the flood and the primary dune height showed a strong hysteresis effect with discharge. The average dune height was about 1.2 m measured two days after the peak discharge. The roughness height k_s varied from 0.2 to 0.5 m during the flood without hysteresis effect. Similarly, the Darcy–Weisbach f also gradually increased from 0.025 to 0.04 without hysteresis.

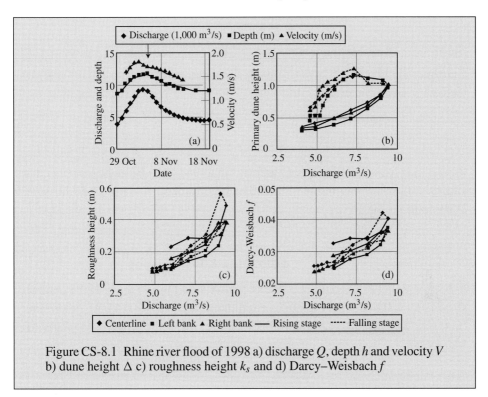

Figure CS-8.1 Rhine river flood of 1998 a) discharge Q, depth h and velocity V b) dune height Δ c) roughness height k_s and d) Darcy–Weisbach f

Exercises

♦8.1 Demonstrate that Equations (8.1) and (8.2) reduce to Equations (8.3) and (8.4).

♦8.2 Demonstrate Equation (8.9c) from $\tau' = \rho u_*'^2 = \frac{f'}{8}\rho V^2$ and $\tau_*' = \tau'/(G-1)\gamma d_s$ from Equations (8.18a and b) and (8.13f).

Problems
♦ Problem 8.1

Estimate the dune height of the Missouri River in Example 8.1 during the summer months.

♦ Problem 8.2

Determine the flow regime and type of bedform in the Rio Grande conveyance channel given: the mean velocity $V = 0.5$ m/s, the flow depth $h = 0.40$ m, the bed slope $S_o = 52$ cm/km and the grain size distributions $d_{50} = 0.24$ mm and $d_{65} = 0.35$ mm.

Problem 8.3

Check the type of bedform in a 200 ft-wide channel conveying 8,500 cfs in a channel sloping at 9.6×10^{-5} given the mean velocity $V = 3.6\,\text{ft/s}$ and the median grain diameter $d_m = 0.213\,\text{mm}$.

♦Problem 8.4

Predict the type and geometry of bedforms in a sand-bed channel $d_{35} = 0.35\,\text{mm}$ and $d_{65} = 0.42\,\text{mm}$ sloping at $S_o = 0.001$ with flow depth $h = 1\,\text{m}$ when the water temperature is 40°F.

♦♦Problem 8.5

A 20 m-wide alluvial channel conveys a discharge $Q = 45\,\text{m}^3\text{/s}$. If the channel slope is $S_o = 0.0003$ and the median sediment size is $d_m = 0.4\,\text{mm}$, determine: (a) the flow depth; (b) the type of bedform; (c) the bedform geometry; and (d) Manning n.

♦♦Problem 8.6

Dunes as high as 6.4 m and as long as 518 m were measured in the Mississippi River. Given the 38.7 m flow depth, the river energy slope 7.5 cm/km, the water temperature 3°C, the depth-averaged flow velocity 2.6 m/s, and the grain size $d_{50} = 0.25\,\text{mm}$ and $d_{90} = 0.59\,\text{mm}$: (a) check the type of bedform and compare with the predicted average dune geometry; (b) estimate S'' from Equation (8.19c); and (c) calculate k_s from field data using Equation (8.15), and plot the results on Figure 8.14. Could bedforms change during the summer months?

♦♦Problem 8.7

Measurements on the Zaïre River from Peters (1978) show dunes of 1.2–1.9 m in amplitude and 95–400 m in length. At a flow depth of 13.2 m, the velocity is 1.3 m/s and the river slope is 4.83 cm/km. The bed material is $d_{50} = 0.34\,\text{mm}$ and $d_{90} = 0.54\,\text{mm}$, and the water temperature is 27°C. Determine the following: (a) compare the bedform type and geometry with all bedform predictors; (b) estimate f''/f'; and (c) plot the results on Figure 8.13.

♦♦*Problem 8.8*

Data from the Jamuna River in Bangladesh from Klaassen (also listed in Julien, 1992) show dunes at a discharge of 10,000 m³/s, flow depth of 11 m, velocity $\overline{V} = 1.5$ m/s, slope $S = 7$ cm/km and $d_{50} = 200\mu$m. Measured dune heights and length are typically 2 m and 50 m, respectively. Compare the bedform type and geometry, calculate T and plot the measurement on relevant graphs in this chapter.

♦*Problem 8.9*

The slope of the Amazon River at Obidos is 1.38 cm/km. If the width is 2,200 m, depth $h = 48.5$ m and $d_{50} = 0.12$ mm, estimate the following: (a) height and length of bedforms; (b) grain and form resistance f' and f''; and (c) flow velocity V and discharge Q.

♦*Computer problem 8.1*

From the Bergsche Maas bedform data given in the table below: (a) calculate the grain Chézy coefficient C' and the sediment transport parameter T; (b) plot the

Q (m³/s)	S_f (cm/km)	h (m)	V (m/s)	d_s (μm)	Δ (m)	Λ (m)	T	C' (m$^{1/2}$/s)
2,160	12.5	8.6	1.35	480	1.5	22	9.09	81.6
2,160	12.5	8.0	1.35	410	1.0	14	9.05	84.2
2,160	12.5	10.5	1.30	300	1.5	30		
2,160	12.5	10.0	1.40	500	1.6	32		
2,160	12.5	7.6	1.40	520	1.4	21		
2,160	12.5	8.4	1.40	380	1.5	22		
2,160	12.5	8.7	1.70	300	1.5	30		
2,160	12.5	7.5	1.55	250	2.5	50		

Source: data from Adriaanse (1986), also in Julien (1992).

data on van Rijn's dune height and dune steepness diagrams: Figure 8.12 page 182; and (c) calculate k_s from field measurements and plot on Figure 8.14.

♦♦*Computer problem 8.2*

Examine bedforms and resistance to flow in the backwater profile analyzed in Computer problem 3.1. Assume that the rigid boundary is replaced with uniform 1 mm-sand. Select an appropriate bedform predictor and determine the type of bedform to be expected along the 25 km reach of the channel using previously calculated hydraulic parameters. (Check your results with a second bedform predictor.) Also determine the corresponding resistance to flow along the channel and recalculate the backwater profile. Briefly discuss the methods, assumptions, and results. Three sketches should be provided along the 25 km of the reach: (a) type of bedforms; (b) Manning n or Darcy–Weisbach f; and (c) water surface elevation.

9

Bedload

Non-cohesive bed particles enter motion as soon as the shear stress applied on the bed material exceeds the critical shear stress. Generally, silt and clay particles enter suspension (see Chapter 10), and sand and gravel particles roll and slide in a thin layer near the bed called the bed layer. The bed layer thickness is typically less than 1 mm in sand-bed channels, up to tens of centimeters in gravel-bed streams. Note that the bed layer thickness should not be mistaken for the laminar sublayer thickness defined in Chapter 6. As sketched in Figure 9.1, the bed layer thickness a is a few grain diameters thick, and $a = 2\,d_s$ has been commonly used. Bedload, or contact load, refers to the transport of sediment particles which frequently maintain contact with the bed. Bedload transport can be treated either as a deterministic or a probabilistic problem. Deterministic methods have been proposed by DuBoys and Meyer-Peter Müller; probabilistic methods were developed by Kalinske and Einstein. Both approaches yield satisfactory estimates of bedload discharge, as discussed in Section 9.1. The characteristics of the bed layer are described in Section 9.2. Bed material sampling is discussed in Section 9.3 and bed sediment discharge measurement techniques are summarized in Section 9.4.

Bedload L_b refers to a quantity of sediment that is moving in the bed layer, which can be measured by volume, mass, or weight. In SI, it is usual to measure bedload by mass in metric tons (1,000 kg), but the English system of units measures bedload by weight in tons (2,000 lb). Conversions from volume to mass involve the mass density of sediment ρ_s such that $L_{bm} = \rho_s \times L_{bv}$. Similarly, conversions from mass to weight involve the gravitational acceleration g, such that $L_{bw} = gL_{bm} = \gamma_s L_{bv}$.

The bedload discharge Q_b is the flux of sediment moving in the bed layer. The bedload L_b thus corresponds to a time integration $L_b = \int_0^T Q_b \, dt$ on a daily, monthly, or annual basis. The fundamental dimensions of Q_b by volume, mass, or weight are summarized in Table 9.1. The unit bedload discharge q_b is the flux of sediment per unit width and per unit time moving in the bed layer. The unit bedload discharge can be measured by weight (M/T^3), mass (M/LT) or volume (L^2/T). The

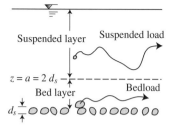

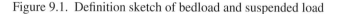

Figure 9.1. Definition sketch of bedload and suspended load

Table 9.1. *Fundamental dimensions of bedload*

	by volume	by mass	by weight
Bedload	$L_{bv}(L^3)$	$L_{bm}(M)$	$L_{bw}\ (ML/T^2)$
Bedload discharge	$Q_{bv}(L^3/T)$	$Q_{bm}(M/T)$	$Q_{bw}\ (ML/T^3)$
Unit bedload discharge	$q_{bv}(L^2/T)$	$q_{bm}\ (M/LT)$	$q_{bw}(M/T^3)$

Note: $L_{bw} = gL_{bm} = \gamma_s L_{bv}$, $L_{bm} = \rho_s L_{bv}$,
$Q_b = \int_0^W q_b dW$, $L_b = \int_0^T Q_b dt$, or $L_b = \int_0^T \int_0^W q_b dW\, dt$

bedload discharge is the integral over a channel width of the unit bedload discharge, or $Q_b = \int_0^W q_b\, dW$.

Table 9.1 summarizes the relationships between bedload, bedload discharge, and unit bedload discharge. The fundamental dimensions for measurements by volume, mass, or weight are also given in this table. In terms of notation, the first subscript *b* refers to bedload and the second subscript refers to volume, mass, or weight, such that q_{bm} is the unit bedload discharge by mass in *M/LT*, e.g. in kg/ms in SI units. Finally, it must be noted that only the volume of solids is considered in the conversions.

9.1 Bedload equations

Three bedload equations are first described in this section. In the presence of bedforms, only the grain shear stress should be considered and for simplicity in the notation, the shear stress in this section refers to the grain shear stress. Bedload transport by size fractions is covered in Section 9.1.4.

9.1.1 DuBoys' equation

The pioneering contribution of M.P. DuBoys (1879) is based on the concept that sediment moves in thin layers along the bed. The applied bed shear stress τ_0 must exceed the critical shear stress τ_c to initiate motion. The volume of gravel material in motion per unit width and time q_{bv} in ft^2/s is calculated from:

$$q_{bv} = \frac{0.173}{d_s^{3/4}} \tau_o(\tau_o - 0.0125 - 0.019\,d_s) \tag{9.1}$$

where d_s is the particle size in millimeters and τ_0 is the boundary grain shear stress in lb/ft^2. Note that the critical shear stress ($\tau_c = 0.0125 + 0.019d_s$; τ_c in lb/ft^2) is quite compatible with Figure 7.9.

9.1.2 Meyer-Peter Müller's equation

Meyer-Peter and Müller (1948) developed a complex bedload formula for gravels based on the median sediment size d_{50} of the surface layer of the bed material. Chien (1956) demonstrated that the elaborate original formulation can be reduced in the following simple form:

$$\frac{q_{bv}}{\sqrt{(G-1)g\,d_s^3}} = 8(\tau_* - 0.047)^{3/2} \tag{9.2a}\blacklozenge$$

This formulation is most appropriate for channels with large width–depth ratios. The corresponding dimensional formulation for q_{bv} with dimensions of L^2/T is:

$$q_{bv} \approx \frac{12.9}{\gamma_s\sqrt{\rho}}(\tau_o - \tau_c)^{1.5} \tag{9.2b}$$

The complete formulation for composite channel configurations can be found in Simons and Senturk (1977) and in Richardson *et al.* (1990).

9.1.3 Einstein–Brown's equation

H.A. Einstein (1942) made the seminal contribution to bedload sediment transport. He introduced the idea that grains move in steps proportional to their size. He defined the bed layer thickness as twice the particle diameter. He extensively used probability concepts to formulate a relationship for contact sediment discharge. The gravel sediment discharge q_{bv} in volume of sediment per unit width and time (q_{bv} in L^2/T) is transformed, using Rubey's clear-water fall velocity ω_o from Equation (5.23b), into a dimensionless volumetric unit sediment discharge q_{bv*} as:

$$q_{bv*} = \frac{q_{bv}}{\omega_o\,d_s} = \frac{q_{bv}}{\sqrt{(G-1)gd_s^3}\left\{\sqrt{\dfrac{2}{3} + \dfrac{36v^2}{(G-1)gd_s^3}} - \sqrt{\dfrac{36v^2}{(G-1)gd_s^3}}\right\}} \tag{9.3a}$$

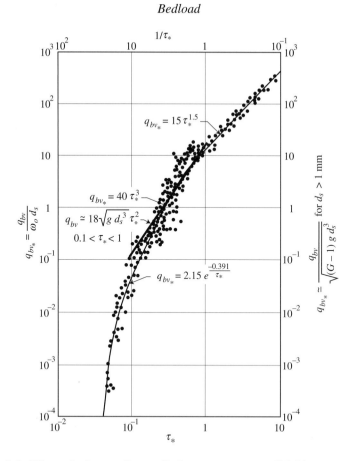

Figure 9.2. Dimensionless sediment discharge q_{bv*} versus Shields parameter τ_*

For very coarse sands and gravels, the fall velocity $\omega_o \simeq \sqrt{(G-1)g\,d_s}$, and

$$q_{bv*} \simeq \frac{q_{bv}}{\sqrt{(G-1)g\,d_s^3}} \quad \text{when } d_s > 1\,\text{mm} \qquad (9.3b)$$

The dimensionless rate of sediment transport q_{bv*} is shown on Figure 9.2 as a function of the Shields parameter $\tau* = \tau_o/(\gamma_s - \gamma)\,d_s$, with measurements from Gilbert (1914), Bogardi (1974) and Wilson (1966). Brown (1950) suggested the following two relationships:

$$q_{bv*} = 2.15e^{-0.391/\tau*}; \quad \text{when } \tau* < 0.18 \qquad (9.4a)$$

and

$$q_{bv*} = 40\tau_*^3; \quad \text{when } 0.52 > \tau* > 0.18 \qquad (9.4b)$$

Considering sediment transport data at high shear rates $\tau_* > 0.52$ one obtains

$$q_{bv*} = 15\tau_*^{1.5}; \text{ when } \tau* > 0.52 \qquad (9.4c)$$

At such high shear rates this third approximation is not very accurate, however, because large quantities of sediment will move in suspension as discussed in Chapters 10 and 11.

The slope of the sediment-rating curve is usually so steep that bedload sediment transport rapidly becomes negligible at low flow. In the domain where $0.1 < \tau_* < 1$, the sediment transport rates increase rapidly and Julien (2002) suggested the following approximation for the unit sediment discharge by volume.

$$q_{bv} \simeq 18\sqrt{g\, d_s^3 \tau_*^2} \quad 0.1 < \tau_* < 1 \qquad (9.4d)$$

After using the resistance relationship $V/u_* \simeq 5\,(h/d_{50})^{1/6}$ from Figure 6.6a, this transport equation reduces to

$$q_{bv} \simeq 0.06\, h u_* \, \text{Fr}^3 \qquad (9.4e)$$

This relationship may give the impression that sediment transport solely depends on hydraulic parameters. This may be misleading because it should be considered that the Froude number depends on resistance to flow and grain size.

In practice, daily transport rates below 1 metric ton per day ($\simeq 10$ g/s) are considered very small and difficult to measure.

9.1.4 Bedload transport by size fractions

Sediment transport calculations by size fractions are obtained as follows:

$$q_b = \sum \Delta p_i\, q_{bi} \qquad (9.5)\blacklozenge$$

where Δp_i is the fraction by weight of sediment particles of the fraction i found in the bed, and q_{bi} is the unit sediment discharge of the fraction i. Notice that $\sum_i \Delta p_i = 1$, fractions in percentage are divided by 100.

In gravel-bed streams, the bedload transport rates depend on the ability of the flow to mobilize partial areas of the bed. It is usually considered that the surface layer (armor or surface layer) is coarser than the substrate (sub-surface layer). Accordingly, when a small portion of the armor layer breaks, similar proportions of finer material will also be released and transported downstream. This concept is also somewhat related to the near-equal mobility concept discussed in Chapter 7. When the main source of sediment is coming from an armored bed, the transport rates of sand and gravel sizes will often be in near-equal quantities. For instance, the gravel transport data from Problem 7.9 at Little Granite Creek, Wyoming, illustrates this, and site-calibrations of sediment transport with the methods of Bakke *et al.* (1999) and Weinhold (2002) can be useful.

   In the case of gravel sediment transport of paved surfaces, the sediment transport calculations by size fractions in Equation (9.5) will largely overestimate transport rates. Instead, calculations based on d_{50} of the armor layer are recommended because transport rates depend on the transport rates of the armor layer rather than on the transport capacity of each size fraction.

Example 9.1 Bedload transport on dunes

Consider dunes formed essentially by bedload sediment transport as sketched in Figure E-9.1.1. The unit bedload discharge by volume q_{bv} causes the downstream

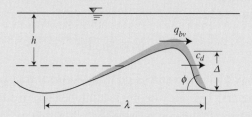

Figure E-9.1.1 Sketch of bedload transport on a dune

migration of dunes of wavelength Λ and amplitude Δ at a celerity of the dune c_Δ. Consider that for large dunes, the sediment transport slides on the downstream side of the dune at the angle of repose ϕ. Assuming a triangular dune, the volume per unit width is $\Lambda\Delta/2$ and the period T for transport is given by $T = \Lambda/c_\Delta$. The following relationship between the bedload transport rate and bedform geometry is the following

$$q_{bv} \simeq \frac{c_\Delta \Delta}{2} \tag{E-9.1.1}$$

From which an estimate for the celerity of dunes can be obtained as

$$c_\Delta \simeq \frac{2q_{bv}}{\Delta} \tag{E-9.1.2}$$

As a first approximation, when combining with the sediment transport Equation (9.4e), constant $\Delta/h \cong 0.3$ and $V \simeq 20u_*$, the following relationship has been proposed by Kopaliani (2002)

$$c_\Delta \simeq 0.02\,V\,\mathrm{Fr}^3 \simeq 0.4\,u_*\,\mathrm{Fr}^3 \tag{E-9.1.3}$$

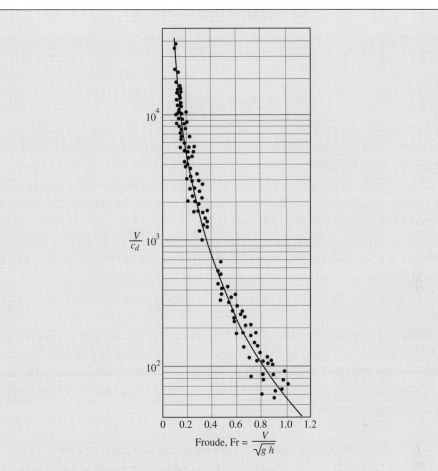

Figure E-9.1.2 Dune celerity versus Froude number

The comparison with field and laboratory measurements is shown in Figure E-9.1.2. Corresponding relationships for dune height from sediment transport can then be defined, as shown by Kopaliani (2007):

$$\Delta \simeq \frac{q_{bv}}{0.01 \, V \, \mathrm{Fr}^3} \qquad (\text{E-9.1.4})$$

or

$$q_{bv} \simeq 0.01 \, V \, \Delta Fr^3 \qquad (\text{E-9.1.5})$$

It is interesting to note that at a given sediment transport rate higher dunes will decrease velocity and Froude number. Conversely, at a given velocity and Froude number, the higher dunes are the result of a local increase in sediment transport.

9.2 Bed layer characteristics

This section covers bedload particle velocity (Section 9.2.1), near-bed sediment concentration and pick-up rate (Section 9.2.2).

9.2.1 Bedload particle velocity

The velocity of bedload particles in the bed layer can be estimated from the velocity profile. As a first approximation, the velocity profile for hydraulically rough boundaries from Equation (6.16) with $k'_s = 6.8\,d_{50}$ and $z = 2\,d_{50}$ gives a bedload velocity v_a in the bed layer $v_a \simeq 5.5u_*$. This is comparable to Einstein's assumption that $k'_s = d_{50}$ and $v_a \simeq 10u_*$. A comparison with the laboratory measurement of Bounvilay (2003) is shown in Figure 9.3a.

When the moving bedload particle has a different diameter d_s than the stationary bed particle of diameter k_s, the boundary Shields parameter $\tau_{*ks} = \frac{\tau_0}{(\gamma_s - \gamma)k_s}$ can be defined, and the bedload particle velocity v_a can be estimated as

$$v_a \simeq (3.3\ln\tau_{*ks} + 17.5)\,u_* \qquad (9.6)\blacklozenge$$

A comparison of this relationship with the experimental measurements of Bounvilay (2003) is shown in Figure 9.3b. It is interesting to notice that coarser particles placed on top of a rigid bed of smaller particles can move when $\tau_{*ks} < 0.06$. This relationship can estimate the bedload particle velocity but should not be used to define threshold conditions of motion.

9.2.2 Bed layer sediment concentration and pick-up rate

Based on the Colorado State University laboratory data (Guy *et al.* 1966), the near-bed volumetric sediment concentration C_{av} has been analyzed by Zyserman and Fredsøe (1994). The relationship shown in Figure 9.4 can be approximated with

$$C_{av} \simeq 0.025\,q_{bv*} = 0.2\,(\tau_* - \tau_{*c})^{1.5} \quad \text{for } \tau_* < 2 \qquad (9.7)\blacklozenge$$

In the case of bedforms, only the grain Shields parameter should be considered. The maximum concentration $C_{av} = 0.54$ should be assumed when $\tau_* > 2$.

The pick-up rate of sediment E_b is the quantity of sediment per unit area per unit time that moves out of the bed layer into the suspended layer. The pick-up rate can be written by volume $E_{bv}(L/T)$, by mass $E_{bm}(M/L^2T)$ or by weight E_{bw} (M/LT^3). Of course, conversions are obtained from $E_{bw} = gE_{bm} = \rho_s g E_{bv} = \gamma_s E_{bv}$. Under equilibrium conditions, the pick-up rate is equal and opposite to the settling flux per unit area as sketched in Figure 9.5. Accordingly, $E_{bv} = \omega C_a$ and from Equation (9.7)

$$E_{bv} = \omega C_{av} = 0.2\omega\,(\tau_* - \tau_{*c})^{1.5} \qquad (9.8)$$

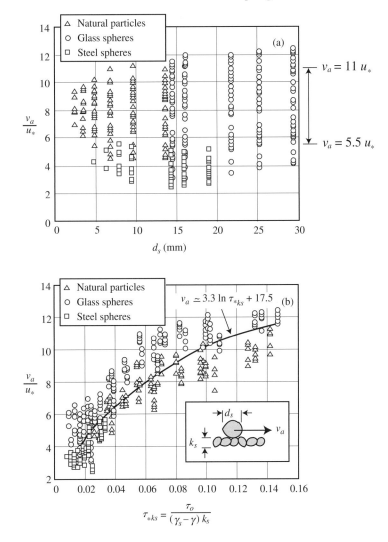

Figure 9.3. a) Ratio of v_a/u_* as a function of particle diameter d_s (modified after Bounvilay, 2003) b) Bedload particle velocity versus Shields parameter (modified after Bounvilay, 2003)

The pick-up rate function can be plotted as $E_{bv}/\omega = C_{av}$ as shown in Figure 9.4.

9.3 Bed sediment sampling

Bed samples are usually collected at, or slightly below, the bed surface to determine the particle size distribution and density of sediment particles available for transport.

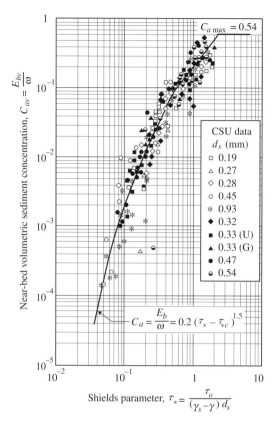

Figure 9.4. Near-bed sediment concentration (modified after Zyserman and Fredsøe, 1994)

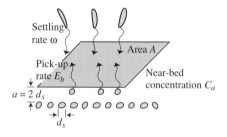

Figure 9.5. Equilibrium pick-up rate

To obtain satisfactory submerged bed material samples, the samplers should enclose a volume of sediment and then isolate the sample from currents while the sampler is being lifted to the surface. The ease with which the sample can be transferred to a suitable container is also important.

Shallow samples

Samplers for obtaining bed material are generally one of the following types: (1) drag bucket or scoop; (2) grab bucket or clamshell; (3) vertical pipe or core; (4) piston core; and (5) rotating bucket.

With the drag bucket or scoop, some of the sample material may wash away, and the clamshell and grab bucket do not always close properly if the sample contains gravel or clays. Accordingly, when these samplers must be used, special control is required to ensure that the samples are representative of the bed material.

Vertical pipe or core samplers are essentially tubes which are forced into the streambed; the sample is retained inside the cylinder by creating a partial vacuum above the sample. Penetration in fine-grained sediment is easy, but penetration in sand usually is limited to about 0.5 m or less. These samplers generally yield good-quality samples and are inexpensive and simple to maintain.

The US BMH-53 is a piston core sampler consisting of a 9 in.-long, 2 in.-diameter brass or stainless steel pipe with cutting edge and suction piston attached to a control rod. The piston is retracted as the cutting cylinder is forced into the streambed. The partial vacuum in the sampling chamber, which develops as the piston is withdrawn, is of assistance in collecting and holding the sample in the cylinder. This sampler can be used only in streams that are shallow enough to be waded in.

Deep samples

The US BM-54, US BM-60 and Shipek are rotating bucket samplers designed for sand-bed streams. The US BM-54 weighs 100 lb and is designed to be suspended from a cable and to scoop up a sample of the bed sediment that is 3 in. in width and about 2 in. in maximum depth. When the sampler contacts the stream bed with the bucket completely retracted, the tension in the suspension cable is released and a heavy coil spring quickly rotates the bucket through 180° to scoop up the sample. A rubber stop prevents any sediment from being lost.

The US BM-60 bed material sampler is similar to the US BM-54 and was developed for both handline and cable suspension. The sampler weighs 30 lb if made of aluminum, and 40 lb if made of brass. It is used to collect samples in streams with low velocities but with depths beyond the range of the US BMH-53 sampler.

Coarse particle samples

Materials coarser than gravel and cobble are extremely difficult to sample effectively because penetration is difficult and large quantities of material must be collected as shown in Figure 2.3. Strictly speaking, hundreds to thousands of pounds of bed material are required for an accurate determination of the median diameter and the particle size distribution. Manual collection and measurement is necessary to determine a representative sample of such material. In armored and paved gravel-bed streams, the particle size distribution of the surface and sub-surface layers are different, as shown in Figure 7.11, and must be sampled separately.

    The recommended method for wadeable streams is to use a grid pattern to locate sampling points. The particle at each grid point is retrieved and its intermediate diameter measured and recorded. Where a grid point is over sand or finer material, a small volume (about 15 ml) is collected and combined with samples from other such points for sieve analysis. At least 200 points should be sampled for the relative quantity of some of the coarser sizes to be accurate. In sand-bed channels, the measurement procedures of Edwards and Glysson (1988) are recommended. In gravel-bed streams, the report of Bunte and Abt (2001) is perhaps the most comprehensive.

9.4 Bedload measurements

Direct bedload measurements are usually possible with samplers or other devices, including sediment traps, bedload samplers, and vortex tubes, or other techniques, such as tracer techniques or measurement of the migration of bedforms. Bunte *et al.* (2004) provides a detailed review of bedload measurement techniques for gravel- and cobble-bed streams. In sand-bed channels, bedload is small compared to the suspended load, and measurement techniques for suspended sediment are discussed in Chapter 10. Box and basket samplers and sediment traps can sometimes be installed in small streams at a reasonable cost. These direct measuring devices measure the volume of bed material in motion near the bed during major events. The total volume or weight of sediment accumulated in the trap can be determined after each event, although very little information on the rate of sediment transport at a given discharge can be obtained.

    Tracer techniques can be applied in coarse bed material streams by painting, staining, or radio tracking coarse particles from the bed. The position of the particles after a major event indicates the distance traveled during the flood and reflects sediment transport. The use of radioactive tracers is discouraged, however, for environmental reasons.

    In sand-bed channels, the rate of bedload transport depends largely on the motion of large-size bedforms such as dunes. Since large sediment volumes are contained within dunes, their motion can be monitored and the bedload discharge corresponds to the dune volume divided by the time required during a full wavelength migration.

    Bedload samplers are most useful in providing the sediment size distribution and qualitative information on the rates of sediment transport in the layer extending 0.3 ft above the bed. The efficiency of bedload samplers such as the Helley–Smith sketched in Figure 9.6 depends on the size fractions: very coarse material ($d_s >$ 80 mm) will undoubtedly not enter the sampler, and very fine material ($d_s < 0.5$ mm) will be washed through the sample bag. Sampling over a long period of time may cause clogging of the sample bag and bias the measurements. Also, when positioned over gravel- and cobble-bed streams, substantial amounts of finer particles (sand particles) will be transported underneath the sampler. The Helley–Smith sampler seems best suited to coarse sand to fine gravel-bed streams, given the primary advantages of low cost and great mobility, Bunte *et al.* (2004) and Bunte and Abt (2005).

Another type of sampler is the vortex tube (Figure 9.7), which has proven to be effective in the removal of bedload in narrow open channels. Some vortex tubes have been effective in removing coarse bed material up to gravel and cobble size in laboratories, irrigation canals, and mountain streams. The main feature of the vortex tube is a vented circular tube with an opening along the top side mounted flush with the bed elevation. As water flows over the tube, the shearing action

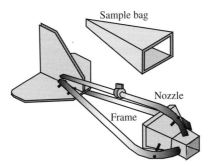

Figure 9.6. Helley–Smith bedload sampler (from Emmett, 1979)

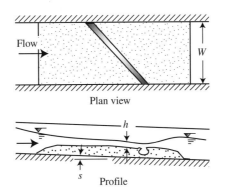

Figure 9.7. Vortex tube sediment ejector

across the opening sets up vortex motion within and along the tube. This whirling action pulls the sediment particles passing over the lip of the opening and carries the particles to the outlet of the vortex tube (e.g. Atkinson, 1994). Case study 9.1 provides detailed bedload calculations using the methods described in Section 9.1. Case study 9.2 then looks at sediment transport calculations at Little Granite Creek to follow up on the discussion of the data presented in Problem 7.9.

Case study 9.1 Mountain Creek, United States

This case study illustrates sediment transport calculations when bedload is dominant and $u_*/\omega < 1$. Mountain Creek near Greenville, South Carolina, is a small 14 ft-wide sand-bed stream. The geometric mean sediment size is 0.86 mm with standard deviation $\sigma_g = 1.8$, $d_{35} = 0.68$ mm, $d_{50} = 0.86$ mm, $d_{65} = 1.08$ mm, and $d_{90} = 1.88$ mm. The complete sieve analysis and sediment-rating measurements are given in the following tabulation:

Particle size distribution		Sediment-rating measurements		
Sieve (mm)	% Finer	Flow depth h (ft)	Unit discharge q (ft^2/s)	q_{bw} (lb/ft $\cdot$ s)
0.074	0.07	0.16	0.21	–
0.125	0.33	0.18	0.23	–
0.246	1.70	0.22	0.32	0.004
0.351	6.20	0.25	0.40	0.006
0.495	19.0	0.27	0.48	0.007
0.701	37.3	0.29	0.52	0.009
0.991	60.5	0.32	0.60	0.012
1.400	79.4	0.40	0.87	0.030
1.980	90.4	0.43	1.10	0.039
3.960	99.3	0.60	1.50	–

The water surface slope of Mountain Creek varied between 0.00155 and 0.0016 during the measurements. The unit discharge increased from 0.2 to 1.1 ft^2/s at flow depths corresponding to 0.16 to 0.43 ft, as indicated on the above table. Assuming water temperature at 78°F: (1) calculate bedload from the methods of DuBoys, Meyer-Peter and Müller, and Einstein–Brown using the median grain size; and (2) compare calculations by size fractions with field measurements.

Calculations based on the median grain size.

Bedload calculations for $d_{50} = 0.86$ mm $= 0.00282$ ft, $S_f = S_o = 0.0016$, $h = 0.6$ ft, $\tau_o = \gamma h\, S_f = 0.06$ lb/ft^2, $T° = 78°$F and $\nu = 1 \times 10^{-5}$ ft^2/s follow:

(a) DuBoys' equation:

$$q_{bv} = \frac{0.173}{d_s^{3/4}} \tau_o (\tau_o - 0.0125 - 0.019 d_s)$$

$$= \frac{0.173}{(0.86)^{3/4}} 0.06(0.06 - 0.0125 - 0.019 \times 0.86)$$

$$= 3.6 \times 10^{-4}\ \text{ft}^2/\text{s}$$

$$q_{bw} = \gamma_s\, q_{bv} = 0.0595\ \text{lb/ft s}$$

(b) Meyer-Peter Müller equation:

$$\tau_* = \frac{\tau_o}{(\gamma_s - \gamma)d_s} = \frac{0.06\ \text{lb ft}^3}{\text{ft}^2 \times 1.65 \times 62.4\text{lb} \times 0.00282\ \text{ft}} = 0.206$$

$$q_{hv} = \sqrt{(G-1)gd_s^3}\ 8(\tau_* - 0.047)^{3/2}$$

$$= \sqrt{1.65 \times \frac{32.2\ \text{ft}}{\text{s}^2} \times (0.00282)^3\ \text{ft}^3 \times 8 \times (0.206 - 0.047)^{3/2}}$$

$$= 5.5 \times 10^{-4}\ \text{ft}^2/\text{s}$$

$$q_{bw} = \gamma_s\, q_{bv} = 0.091\ \text{lb/ft s}$$

(c) Einstein–Brown equation:

$$\tau_* = 0.206 > 0.18$$

$$q_{bv*} = 40\tau_*^3 = 40(0.206)^3 = 0.35$$

$$X_e = \frac{36\nu_m^2}{(G-1)gd_s^3} = \frac{36 \times 10^{-10}\ \text{ft}^4\ \text{s}^2}{\text{s}^2(1.65) \times 32.2\text{ft} \times (0.00282)^3\ \text{ft}^3} = 0.003$$

$$q_{bv} = q_{bv*}\sqrt{(G-1)gd_s^3}\left(\sqrt{\frac{2}{3}+0.003} - \sqrt{0.003}\right)$$

$$= 0.35\sqrt{1.65 \times \frac{32.2\ \text{ft}}{\text{s}^2}(0.00282)^3\ \text{ft}^3 \times 0.763}$$

$$= 2.91 \times 10^{-4}\frac{\text{ft}^2}{\text{s}}$$

$$q_{bw} = \gamma_s\, q_{bv} = 0.048\ \text{lb/ft s}$$

Notice that $u_* = \sqrt{g\,h\,S_f} = 0.176$ ft/s and $\omega = 0.337$ ft/s, such that $u_*/\omega = 0.52$ and most of the sediment move as bedload.

Calculations by size fractions

The weight fraction Δp_i for each size fraction is first determined from the sediment size distribution. Calculations by size fractions for each method at different flow depths are summarized in the following table. The calculated results are compared with field measurements on Figure CS-9.1.1. The resulting transport rate $q_{bw} \approx 0.08$ lb/ft s in a 14 ft-wide channel corresponds to $Q_{bw} = 0.08 \times 14 \times 86,400/2,000$ tons/day $= 48$ tons/day, which is a relatively low transport rate.

d_s (mm)	Δp_i	DuBoys $q_{bvi}\,\Delta p_i$ $(10^{-6}\text{ ft}^2/\text{s})$	MPM $q_{bvi}\,\Delta p_i$ $(10^{-6}\text{ ft}^2/\text{s})$	Einstein–Brown $q_{bvi}\,\Delta p_i$ $(10^{-6}\text{ ft}^2/\text{s})$
$h = 0.2$ ft, $\tau_0 = \gamma_m\, hS = 0.02$ lb/ft^2				
0.074	0.0020	0.295	0.28	0.410
0.125	0.0082	0.682	1.10	1.100
0.246	0.0293	0.810	3.30	1.900
0.351	0.0865	0.520	8.30	3.900
0.495	0.1555	0	11.60	4.700
0.700	0.2075	0	9.60	2.700
0.990	0.2105	0	3.20	0.700
1.400	0.1495	0	0	0.050
1.980	0.0995	0	0	0.001
3.960	0.0515	0	0	0
Total		$q_{bv} = 2.31 \times 10^{-6}$ ft^2/s $q_{bw} = 0.0004$ lb/fts	37.5×10^{-6} ft^2/s 0.0062 lb/fts	15.6×10^{-6} ft^2/s 0.0026 lb/fts
$h = 0.6$ ft, $\tau_0 = 0.06$ lb/ft^2				
0.074	0.0020	6.7	1.6	11.2
0.125	0.0081	18.1	6.3	31.2
0.246	0.0293	37.2	21.7	52.0
0.351	0.0865	80.1	61.1	95.4
0.495	0.1555	103.7	103.0	106.0
0.700	0.2075	95.8	124.8	85.8
0.990	0.2105	62.8	109.1	55.2
1.400	0.1495	25.0	61.0	26.9
1.980	0.0995	6.0	26.7	8.4
3.960	0.0515	0	0	0.16
Total		$q_{bv} = 435.6 \times 10^{-6}$ ft^2/s $q_{bw} = 0.072$ lb/fts	515.5×10^{-6} ft^{2}s 0.085 lb/fts	472.3×10^{-6} ft^2/s 0.078 lb/fts

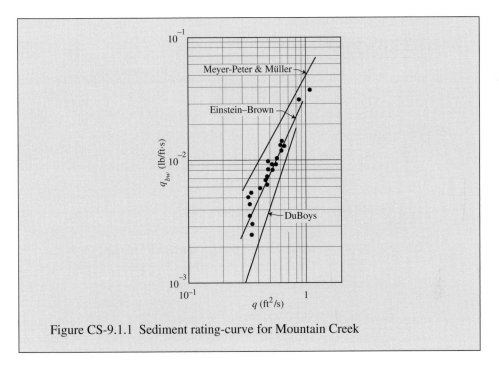

Figure CS-9.1.1 Sediment rating-curve for Mountain Creek

Case study 9.2 Little Granite Creek, Wyoming

This case study illustrates sediment transport calculation near incipient motion, or $u_*/\omega < 0.5$. Little Granite Creek is a steep cobble-bed stream with gravel transport during floods. The author is grateful to S. Ryan, K. Bunte, and M. Weinhold for the discussions on data collection and analyses at the site (Weinhold, 2002). The main characteristics have been presented in Problem 7.9, with a bed slope of 0.02, top width of 7.2 m, stage–discharge $h_m \sim 0.3 Q_{m^3/s}^{0.42}$ and bankfull discharge $Q = 6.22$ m^3/s. Based on d_{50} of the pavement layer shown in Figure 7.11, calculate the daily bedload in metric tons per day at bankfull discharge calculations based on the median grain size. Bedload calculation for $d_{50} = 110$ mm, $S_f = 0.02$, $h = 0.3\ 6.22^{0.42} = 0.65$ m, $d_* = 2,800$ and $\omega_{50} \approx 1.25$ m/s from Chapter 5. $\tau_0 = \gamma h S_f = 127$ Pa $= 2.66$ lb/ft^2, $v \cong 1 \times 10^{-6}$ m^2/s, $u_* = \sqrt{gh S_f} = 0.35$ m/s, and $u_*/\omega_{50} \approx 0.28$, thus bedload is dominant.

(a) DuBoys' equation:

$$q_{bv} = \frac{0.173}{110^{3/4}} 2.66\,(2.66 - 0.0125 - 0.019 \times 110)$$

$$= 0.0076\,\text{ft}^2/\text{s} = 0.0007\,\text{m}^2/\text{s}$$

$$Q_{bm} = q_{bv} W \rho_s = 0.0007 \frac{m^2}{s} \times 7.2\,m$$

$$\times \frac{2.65 \text{ metric ton}}{m^3} \times \frac{86,400\,s}{\text{day}} = \frac{1,157 \text{ metric tons}}{\text{day}}$$

(b) MPM with $\tau_{*c} = 0.047$:

$$\tau_* = \frac{\tau_0}{(\gamma_s - \gamma)\,d_s} = \frac{hS}{(G-1)\,d_s} = \frac{0.65\,m \times 0.02}{1.65 \times 0.11\,m} = 0.0716$$

$$q_{bv} = \sqrt{(G-1)\,g\,d_s^3}\,8\,(\tau_* - 0.047)^{3/2}$$

$$= \sqrt{1.65 \times 9.81\,m^2/s \times (0.11\,m)^3}\,8\,(0.0716 - 0.047)^{3/2}$$

$$= 0.0045\,m^2/s$$

$$Q_{bm} = 7,480 \text{ metric tons/day}$$

(c) Einstein–Brown:

$$d_* = d_s \left(\frac{(G-1)\,g}{v^2} \right)^{1/3} = 0.11 \left(\frac{1.65 \times 9.81}{\left(1 \times 10^{-6}\right)^2} \right)^{1/3} = 2,782$$

$$\omega = \frac{8v}{d_s} \left\{ \left(1 + \frac{d_*^3}{72}\right)^{0.5} - 1 \right\} = \frac{8 \times 1 \times 10^{-6}}{0.11} \frac{m}{s} \left\{ \left(1 + \frac{2,782^3}{72}\right)^{0.5} - 1 \right\}$$

$$= 1.26\,m/s$$

Notice here that $u_*/\omega = 0.35/1.26 = 0.28$, which is very low and close to incipient motion.

$$q_{bv*} = 2.15\,e^{-0.391/0.0716} = 0.0091$$

$$q_{bv} = q_{bv*}\omega d_s = 0.0091 \times 1.26\,m/s \times 0.11\,m = 0.00127\,m^2/s$$

$$Q_{bm} = 2,088 \text{ metric tons/day}$$

(d) Equation (9.4d) should not be used because $\tau_* = 0.0716 < 0.1$.

The calculated sediment loads from these methods are fairly consistent around 2,000 metric tons/day. However, field measurements from the data in Problem 7.9 are about 4–40 metric tons/day. Calculations far exceed measurements because the coarser fractions of this cobble-bed stream do not move, and equal mobility cannot be extrapolated to the coarser fractions of the bed material. It is clear that a much higher value of the critical Shields parameter ($\tau_{*c} > 0.047$) would have to be used to match the field measurements. Site calibration using the Bakke *et al.* (1999) procedure would be possible as shown by Weinhold (2002).

Exercise

9.1 Consider the angle of repose of a spherical particle on top of three spheres of equal diameter from Figure 7.1c. Estimate the critical Shields parameter from Equations (7.1b and 7.3b). Compare the results with the data in Figure 9.3b when $k_s = d_s$.

Problems

♦*Problem 9.1*

Calculate the unit bedload discharge for a channel given the slope $S_o = 0.01$, the flow depth $h = 20$ cm and the grain size $d_{50} = 15$ mm. From DuBoys' equation, calculate q_{bw} in lb/ft.s, and q_{bv} in ft²/s. (*Answer:* $q_{bw} = 0.17$ lb/ft.s; $q_{bv} = 1.03 \times 10^{-3}$ ft²/s.)

♦*Problem 9.2*

Use Meyer-Peter and Müller's method to calculate q_{bm} in kg/ms and q_{bv} in m²/s for the conditions given in Problem 9.1.

Problem 9.3

Use Einstein–Brown's method to calculate the bedload transport rate in a 100 m-wide coarse sand-bed channel with slope $S_o = 0.003$ when the applied shear stress τ_o equals τ_c. Determine the transport rate Q_{bv} in m³/s and in ft³/s. (*Answer:* $Q_{bv} = 3.27 \times 10^{-8}$ m³/s $= 1.15 \times 10^{-6}$ ft³/s.)

♦*Problem 9.4*

Use Einstein–Brown's bedload equation to calculate q_{bm} in kg/ms and q_{bw} in lb/ft.s for the conditions given in Problem 9.1. Estimate the bedload particle velocity and estimate the near-bed sediment concentration.

♦♦*Problem 9.5*

Use the methods detailed in this chapter to calculate the daily bedload in metric tons in a 20 m-wide medium gravel-bed canal with a slope $S_o = 0.001$ and at a flow depth of $h = 2$ m. Compare Q_{bm} the results in metric tons per day. (*Answer:* DuBoys $-Q_{bm} = 791$ tons/day; MPM $-Q_{bm} = 2,431$ tons/day; Einstein–Brown $-Q_{bm} = 884$ tons/day.)

◆*Problem 9.6*

Which bed sediment sampler would you recommend for the canal in Problem 9.5?

◆*Problem 9.7*

With reference to the Bergsche Maas bedform data given in Computer problem 8.1: (a) calculate the bedload sediment transport; and (b) estimate the time required for the bedload to fill the volume of a representative dune. Compare with the celerity from the method of Example 9.1.

◆◆*Problem 9.8*

With reference to Case study 9.2 and Figure 7.11: (a) use the d_{65} of the surface layer (pavement) and calculate the sediment transport rates in tons per day using DuBoys, MPM and Einstein–Brown at discharges 100, 200, and 400 cfs; and (b) repeat the calculation with sub-pavement material.

◆◆*Problem 9.9*

Consider the field measurement in the table below for the North Fork of the Toutle River at Hoffstadt Creek Bridge (Pitlick, 1992). Calculate τ_0, τ_c, τ_0/τ_c, σ_g, and h/d_{84}, Darcy–Weisbach f and Manning n. Compare resistance to flow measurements and Figure 6.6a, and calculate transport rates based on the methods from this chapter.

h (m)	V (m/s)	S	d_{50} (mm)	d_{84} (mm)	C_s (mg/l)	q_{bm} (kg/ms)
0.91	2.96	0.010	26	66	5,950	6.8
1.17	3.75	0.011	18	62	12,000	19.9
0.77	2.81	0.0057	12	53	6,760	13.2
0.83	3.06	0.0077	15	55	8,230	11.4
0.55	2.35	0.0068	8	36	3,020	6.2
0.70	1.96	0.0074	16	37	1,700	0.25
0.47	1.94	0.0078	8	33	690	3.6

♦Problem 9.10

Combine Equation (9.7) with q_{bv*} (Equation (9.3a)) and $q_{bv} = a\,v_a\,C_{av}$ to determine the thickness of the bed layer a when $u_* \approx \omega$.

♦♦Computer problem 9.1

Consider the channel reach analyzed in Computer problems 3.1 and 8.2. Select an appropriate bedload relationship to calculate the bed sediment discharge in metric tons/m•day for the uniform 1-mm sand in Computer problem 8.2. Plot the results along the 25-km reach and discuss the method, assumptions, and results.

♦Computer problem 9.2

Write a computer program to calculate the bedload transport rate by size fraction from the methods of DuBoys, Meyer-Peter and Müller, and Einstein–Brown, and repeat the calculations of the tabulation in Case study 9.1 at $h = 0.4$ ft and $\tau_0 = 0.04$ lb/ft^2. (*Answer:* DuBoys: $q_{bv} = 122 \times 10^{-6}$ ft^2/s, $q_{bw} = 0.0202$ lb/ft s; Meyer-Peter and Müller: $q_{bv} = 221 \times 10^{-6}$ ft^2/s, $q_{bw} = 0.036$ lb/ft s; Einstein–Brown: $q_{bv} = 140 \times 10^{-6}$ ft^2/s, $q_{bw} = 0.023$ lb/ft s.)

10

Suspended load

As the hydraulic forces exerted on sediment particles exceed the threshold condition for beginning of motion, coarse sediment particles move in contact with the bed surface as described in Chapter 9. Finer particles are brought into suspension when turbulent velocity fluctuations are sufficiently large to maintain the particles within the mass of fluid without frequent bed contact.

This chapter examines the concentration of sediment particles held in suspension (Section 10.1). The governing equations of turbulent diffusion are presented in Section 10.2 leading to turbulent mixing of washload in Section 10.3. Equilibrium vertical concentration profiles in Section 10.4 serve the analysis of suspended load in Section 10.5 and hyperconcentrations in Section 10.6. Field measurement techniques are covered in Section 10.7. Five examples illustrate the computation procedures.

10.1 Sediment concentration

The term sediment concentration deserves clarification to avoid misinterpretations. The units used in the measurement of sediment concentration vary with the range of concentrations and the standard measurement techniques utilized in different countries. The most commonly used unit for sediment concentration is mg/l which describes the ratio of the mass of sediment particles to the volume of the water–sediment mixture. Other units include kg/m^3 ($1mg/l = 1g/m^3$), the volumetric sediment concentration C_v, the concentration in parts per million C_{ppm}, and the concentration by weight C_w:

$$C_v = \frac{\text{sediment volume}}{\text{total volume}} = \frac{\forall_s}{\forall_t} = 1 - p_o \qquad (10.1a)$$

$$C_w = \frac{\text{sediment weight}}{\text{total weight}} = \frac{W_S}{W_T} = \frac{C_v G}{1 + (G - 1)C_v} \qquad (10.1b)\blacklozenge$$

Table 10.1. *Equivalent concentrations and mass densities*

C_v	C_w	C_{ppm}	$C_{mg/l}$	ρ_m (kg/m^3)	ρ_{md} (kg/m^3)
Suspension					
0.0001	0.00026	265	265	1,000.2	0.26
0.0005	0.0013	1,324	1,325	1,000.8	1.32
0.001	0.00264	2,645	2,650	1,001.6	2.65
0.0025	0.00659	6,598	6,625	1,004.1	6.65
0.005	0.01314	13,141	13,250	1,008.2	13.3
0.0075	0.01963	19,632	19,875	1,012.4	19.9
0.01	0.02607	26,069	26,500	1,016.5	26.5
0.025	0.06363	63,625	66,250	1,041.2	66.3
Hyperconcentration					
0.05	0.12240	122,401	132,500	1,083	133
0.075	0.17686	176,863	198,750	1,124	199
0.1	0.22747	227,467	265,000	1,165	265
0.25	0.46903	469,027	662,500	1,412	662
0.5	0.72603	726,027	1,325,000	1,825	1,325
0.75	0.88827	888,268	1,987,500	2,237	1,987
1.0	1.0	1,000,000	2,650,000	2,650	2,650

Note: Calculations based on mean density of water of 1 g/ml and specific gravity of sediment $G = 2.65$.

in which $G = \gamma_s/\gamma$

$$C_{ppm} = 10^6 C_w \qquad (10.1c)\blacklozenge$$

Note that the concentration in parts per million C_{ppm} is given by 1,000,000 times the weight of sediment over the weight of the water–sediment mixture. The corresponding concentration in mg/l and C_v is then given by the following formula.

$$C_{mg/l} = \frac{1\,mg/l\, G\, C_{ppm}}{G + (1-G)10^{-6}\,C_{ppm}} = \rho\, G\, C_v = 10^6\, mg/l\, G\, C_v \qquad (10.1d)$$

The conversion factors from C_{ppm} to $C_{mg/l}$ are given in Table 10.1. Notice that there is less than 10% difference between C_{ppm} and $C_{mg/l}$, at concentrations $C_{ppm} < 145,000$, and less than 1% difference when $C < 10,000$ ppm.

In the laboratory, the sediment concentration in $C_{mg/l}$ is measured as 1,000,000 times the ratio of the dry mass of sediment in grams to the volume of the water–sediment mixture in cubic centimeters (1 cm^3 = 1 ml). Two methods are commonly used: evaporation and filtration. The evaporation method is used where the sediment concentration of samples exceeds 2,000 mg/l–10,000 mg/l; the filtration method is preferred at lower concentrations. The lower limit applies when the sample consists mostly of fine material (silt and clay) and the upper limit when the sample is mostly sand. For samples having low sediment concentration, the evaporation method requires a correction if the dissolved solids content is high.

Another important consideration is that the concentration C varies with space (x,y,z) and time (t). Several average values are considered:

(1) The time-averaged concentration C_t is measured at a fixed location and integrated over the sampling time t_s, typically of the order of 10–60 seconds:

$$C_t(x_0, y_0, z_0, t) = \frac{1}{t_s} \int_{t_o}^{t_o+t_s} C(x, y, z, t)dt \qquad (10.2a)$$

Time-averaged concentrations are commonly called point measurements.

(2) The volume-averaged concentration $C_\forall$ integrated over a volume $\forall$:

$$C_\forall(x, y, z, t_o) = \frac{1}{\forall} \int \int \int_\forall C(x, y, z, t)d\forall \qquad (10.2b)$$

Volume-averaged concentrations are obtained from instantaneous measurements like bucket samples from the free surface.

(3) The flux-averaged concentration C_f, when multiplied by the total flow discharge Q, gives the exact advective mass flux passing through a given cross-section A, thus

$$C_f = \frac{1}{Q} \int_A C v_x \, dA \qquad (10.2c)$$

The flux-averaged concentration C_f is used when defining the concentration of sediments at a given stream cross-section. Measured sediment concentration profiles will vary around the flux-averaged concentration as discussed in Section 10.4.

The different types of concentration will be measured from different sampling devices as described in Section 10.7. It is essential to understand advection and diffusion processes before analyzing vertical concentration profiles.

10.2 Advection–diffusion equation

The equation governing the conservation of sediment mass can be applied to a small cubic control volume to derive the sediment continuity relationship. The rate of sediment changes per unit volume $\partial C/\partial t$ simply equals the difference of sediment fluxes across the faces of the control volume. The derivation previously detailed in Example 3.1 states that the sediment continuity relationship can be written as

$$\frac{\partial C}{\partial t} + \frac{\partial q_{tx}}{\partial x} + \frac{\partial q_{ty}}{\partial y} + \frac{\partial q_{tz}}{\partial z} = \dot{C} \qquad (10.3)\blacklozenge$$

in which C is the volume-averaged sediment concentration inside the infinitesimal control volume; $\hat{q}_{tx}, \hat{q}_{ty}, \hat{q}_{tz}$ are the sediment fluxes per unit area through the faces of the control volume; and $\dot{C}$ is the rate of sediment production per unit volume. This term $\dot{C}$ is zero for conservative substances and most sedimentation engineering applications. It is however kept in this equation to relate to possible internal mass changes such as chemical reactions, phase changes, adsorption, dissolution, flocculation, radioactive decay, etc. Note that the units of C can be C_{ppm}, C_v, or $C_{mg/l}$, as long as they are consistent with those of q_t and $\dot{C}$.

Expanding upon the derivation in Example 3.1, one recognizes several types of mass fluxes per unit area across the faces of the control volume: advective fluxes, diffusive fluxes, mixing fluxes, as well as dispersive fluxes. This can be written in a simple mathematical form as:

$$\hat{q}_{tx} = v_x C - (D + \varepsilon_x) \frac{\partial C}{\partial x} \tag{10.4a}$$

$$\hat{q}_{ty} = v_y C - (D + \varepsilon_y) \frac{\partial C}{\partial y} \tag{10.4b}$$

$$\hat{q}_{tz} = \underbrace{v_z C}_{\substack{advective \\ fluxes}} - \underbrace{(D + \varepsilon_z) \frac{\partial C}{\partial z}}_{\substack{diffusive\ and \\ mixing\ fluxes}} \tag{10.4c}$$

The advective fluxes describe the transport of sediments imparted by velocity currents. The rate of mass transport per unit area carried by advection is obtained from the product of sediment concentration and the velocity components v_x, v_y, and v_z, respectively (e.g. Example 3.1).

Two additional fluxes must be considered in sediment transport. Molecular diffusion describes the scattering of sediment particles by random molecular motion as described by Fick's law. In Figure 10.1, the example of molecular diffusion of cream in a still cup of coffee illustrates the slow molecular diffusion process. Turbulent mixing induces the motion of sediment particles due to turbulent fluid motion. The turbulent mixing process is far more effective than molecular diffusion and the example of stirring up the coffee with a spoon (Figure 10.1) illustrates this process. The rate of sediment transport for both fluxes is proportional to the concentration gradient. For diffusion in laminar flow, the proportionality constant D is called the molecular diffusion coefficient and has dimensions of L^2/T. The minus sign indicates that the mass flux is directed toward the direction of decreasing concentration (or negative concentration gradient). For mixing in turbulent flow, the turbulent mixing coefficients ε_y and ε_z describe the processes of turbulent diffusion. Mass transport in turbulent flow is also proportional to the concentration gradient. The

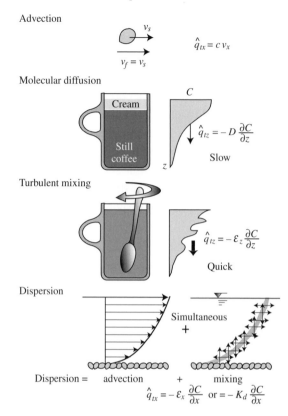

Figure 10.1. Sketch of advection, diffusion, mixing, and dispersion processes

molecular diffusion coefficient D is several orders of magnitude smaller than the turbulent mixing coefficients ε in turbulent flow. Dispersion is a process acting in the flow direction, which combines the interaction between advection and turbulent mixing, as sketched in Figure 10.1. This dispersion coefficient ε_x or K_d is usually very large.

The general relationship describing conservation of sediment mass for incompressible dilute suspensions subjected to diffusion, mixing, dispersion, and advection with point sediment sources is obtained from substituting Equation (10.4) into Equation (10.3):

$$\frac{\partial C}{\partial t} + \frac{\partial v_x C}{\partial x} + \frac{\partial v_y C}{\partial y} + \frac{\partial v_z C}{\partial z}$$

$$= \dot{C} + (D + \varepsilon_x)\frac{\partial^2 C}{\partial x^2} + (D + \varepsilon_y)\frac{\partial^2 C}{\partial y^2} + (D + \varepsilon_z)\frac{\partial^2 C}{\partial z^2} \tag{10.5a}$$

Owing to the conservation of fluid mass (Equation (3.6d)) for incompressible fluid at a low sediment concentration, the advection–diffusion relationship can be rewritten as

$$\underbrace{\frac{\partial C}{\partial t}}_{\substack{mass \\ change}} + \underbrace{v_x\frac{\partial C}{\partial x} + v_y\frac{\partial C}{\partial y} + v_z\frac{\partial C}{\partial z}}_{advective\ terms}$$

$$= \underbrace{\dot{C}}_{\substack{phase \\ change}} + \underbrace{(D+\varepsilon_x)\frac{\partial^2 C}{\partial x^2} + (D+\varepsilon_y)\frac{\partial^2 C}{\partial y^2} + (D+\varepsilon_z)\frac{\partial^2 C}{\partial z^2}}_{diffusive\ and\ mixing\ terms} \qquad (10.5b)$$

In laminar flow, the turbulent mixing and dispersive coefficients vanish ($\varepsilon_x = \varepsilon_y = \varepsilon_z = 0$). Conversely, in turbulent flows, the molecular diffusion coefficient D is negligible compared to the turbulent mixing and dispersion coefficients ($D \ll \varepsilon$).

As opposed to the Navier–Stokes equations discussed in Chapter 5, Equation (10.5a) can be easily solved owing to the linearity of the independent parameter C. This Equation (10.5a) is sometimes called advection–dispersion, or the diffusion–dispersion equation and has numerous applications in the field of sediment and contaminant transport in open channels.

10.3 Turbulent mixing of washload

This section focuses on sediment particles that are too fine to settle. The concept of washload refers to fine particles which are easily washed away by the flow – more specifically, particles in transport which are too fine to be found in significant amounts in the bed material. Washload often corresponds to silts and clays in sand-bed channels.

10.3.1 One-dimensional diffusion

Consider the case of one-dimensional molecular diffusion in still water as sketched in Figure 10.2. A mass m of fine sediment diffuses without settling in the y direction given the molecular diffusion coefficient D. The governing one-dimensional equation (Equation (10.5)) for molecular diffusion given $\partial C/\partial x = \partial C/\partial z = v_y = v_z = \partial^2 C/\partial x^2 = \partial^2 C/\partial z^2 = \varepsilon_y = \dot{C} = 0$, reduces to

$$\frac{\partial C}{\partial t} = \frac{D\partial^2 C}{\partial y^2} \qquad (10.6)$$

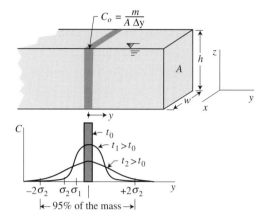

Figure 10.2. Sketch of one-dimensional diffusion

The solution to this equation defines the concentration $C(y,t)$ in space and time,

$$C(y,t) = \frac{m}{A\sqrt{4\pi Dt}} e^{-\frac{y^2}{4Dt}} \qquad (10.7)$$

in which the mass $m = \int_\forall C_{mg/l}\, d\forall$ is the volume integral of the concentration given the cross-sectional area A. The properties of this normal distribution are that the variance increases linearly with time $\sigma_D^2 = 2Dt$, and a practical estimate of the width of the sediment cloud (95% of the mass) is simply contained within $4\sigma_D = 4\sqrt{2Dt}$.

10.3.2 Mixing and dispersion coefficients, length and time scales

Consider a straight rectangular channel of width W, flow depth h and shear velocity u_*. The vertical mixing coefficient ε_v, the transversal mixing coefficient ε_t, and the longitudinal dispersion coefficient K_d are empirical functions of the product hu_* as suggested by Fischer *et al.* (1979):

$$\varepsilon_v \cong 0.067\, hu_* \qquad (10.8a)$$

$$\varepsilon_t \cong 0.15\, hu_* \text{ in straight channels} \qquad (10.8b)$$

$$\text{or } \varepsilon_t \cong 0.6\, hu_* \text{ in natural channels} \qquad (10.8c)$$

$$K_d \cong 250\, hu_* \text{ or } 0.011\frac{V^2 W^2}{hu_*} \qquad (10.8d)$$

The basic relationship for vertical mixing (Equation (10.8a)) between the coefficient ε_v and hu_* will be demonstrated in Section 10.4.

It is noticeable that vertical, transversal, and dispersion coefficients increase by orders of magnitude, respectively. Given the property of the variance for normal distributions, in the previous section, $\sigma^2 \approx 2Dt$, the vertical time scale t_v, the transversal time scale, t_t, and the longitudinal dispersion time scale t_d in a stream are defined after substituting t_t respectively with the average flow depth $h(h^2 = 2\varepsilon_v t_v)$, the channel width $W(W^2 = 2\varepsilon_t t_t)$, and the dispersion length $X_d(X_d^2 = 2K_d t_d)$.

$$t_v = \frac{X_v}{V} \cong \frac{h}{0.1u_*} \tag{10.9a}$$

$$t_t = \frac{X_t}{V} \cong \frac{W^2}{hu_*} \tag{10.9b}$$

$$t_d = \frac{X_d}{V} \cong \frac{X_d^2}{500hu_*}; \text{ where } t_d > t_t \tag{10.9c}$$

These time scales provide very rough approximations, but are very useful to determine which physical process is dominant in open channels, as sketched in Figure 10.3. Similarly, first-order approximations for length scales can be defined.

Traveling downstream at the mean flow velocity V, the corresponding length scales for complete vertical mixing X_v, complete transversal mixing X_t, and longitudinal dispersion X_d are respectively given by

$$X_v = \frac{hV}{0.1u_*} \tag{10.10a}$$

$$X_t = \frac{VW^2}{hu_*} \tag{10.10b}$$

and $X_d = \frac{500hu_*}{V}$, when lateral mixing is complete at a distance

$$X > VW^2/hu_* \tag{10.10c}$$

These first-order approximations of the distances X_v and X_t required for vertical and transversal mixing indicate that $\frac{t_t}{t_v} = \frac{X_t}{X_v} = 0.1\frac{W^2}{h^2}$. It is concluded that vertical mixing occurs before transversal mixing unless $W < 3h$. Consequently, studies of

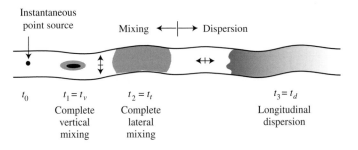

Figure 10.3. Sketch of mixing and dispersion processes

turbulent mixing in open channels generally assume that vertical mixing is complete, and depth-integrated 2-D models x–y will often be sufficient. Example 10.1 provides calculations of time and length scales.

It is also important to consider that longitudinal dispersion will only be possible when the concentration is uniform within a cross-section. This implies that dispersion can only be considered as the dominant process when $t_d > t_t = W^2/hu_*$. In a river flowing at a velocity V, this corresponds to saying that dispersion will start when lateral mixing is completed, or when $X > X_t = VW^2/hu_*$.

Example 10.1 Application of mixing time and length scales

Determine an approximate time scale and downstream distance required for complete vertical and horizontal mixing of washload in a gently meandering 600 ft-wide stream. The stream is 30 ft deep. The average flow velocity is $V = 2$ ft/s and the shear velocity u^* is 0.2 ft/s.

Step 1. Vertical mixing occurs at a time t_v approximated by Equation (10.9a)

$$t_v \cong \frac{h}{0.1u*} = \frac{30\,\text{ft} \times \text{s}}{0.1 \times 0.2\,\text{ft}} = 1,500\,\text{s} = 25\,\text{minutes}$$

The corresponding downstream distance is $X_v = Vt_v = \frac{2.0\,\text{ft}}{\text{s}} \times 1,500\,\text{s} = 3,000\,\text{ft}$.

Notice that complete vertical mixing occurs at a downstream distance roughly equal to 100 times the flow depth.

Step 2. Transversal mixing should be complete at a time t_t approximated by Equation (10.9b).

$$t_t \cong \frac{W^2}{hu_*} = \frac{(600)^2\,\text{ft}^2\,\text{s}}{30\,\text{ft} \times 0.2\,\text{ft}} = 60,000\,\text{s} \cong 17\,\text{hours}$$

The corresponding downstream distance X_t is

$$X_t = Vt_t = \frac{2.0\,\text{ft}}{\text{s}} \times 60,000\,\text{s} = 120,000\,\text{ft} \cong 23\,\text{miles}$$

Vertical mixing is completed long before transversal mixing.

10.3.3 Lateral mixing from steady point sources

Consider the steady supply of fine sediment, or contaminant, as a centerline point source of concentration C_o at a mass rate $\dot{m}$ into a stream of width W, depth h, and average flow velocity V, as sketched on Figure 10.4a. Assuming that advection is dominant in the downstream x direction while mixing occurs in the transversal y

direction $\varepsilon_y = \varepsilon_t$, the governing equation [Equation (10.5)] reduces to

$$\frac{\partial C}{\partial t} + \frac{V \partial C}{\partial x} = \varepsilon_t \frac{\partial^2 C}{\partial y^2} \tag{10.11}$$

The solution, after a coordinate transformation $x' = x - Vt$, becomes similar to Equation (10.7) in which D is replaced by ε_t, and x by x'. Advection and mixing are separate and additive processes. In other words, advective mixing is the same as mixing in a stagnant fluid when viewed in a coordinate system moving at speed V. Assuming that complete vertical mixing occurs long before transversal mixing, the approximate solution when $t > t_v$ in an infinitely wide channel is obtained from Equation (10.7) after considering $t = x/V$, $A = hVt$, and $\dot{m} = m/t$.

$$C(x,y) = \left[\frac{\dot{m}}{h\sqrt{4\pi \varepsilon_t x V}} \right] e^{\frac{-y^2 V}{4\varepsilon_t x}} \tag{10.12}$$

In channels of finite width W, the relative concentration C/C_o with $\bar{C} = \dot{m}/hVW$ for mid-stream point source is plotted on Figure 10.4a. The term in brackets of Equation (10.12) represents the maximum concentration along a cross-section centerline at a distance x from the point source. A reasonable length X_t for complete transversal mixing is $X_t \cong 0.15 \, V \, W^2/\varepsilon_t$ for centerline injection. In the case of side injection, the solution can be found by analogy with a double channel width. Thus, the distance X_t for complete transversal mixing is $X_t \cong 0.6 \, V \, W^2/\varepsilon_t$ for side injection of fine sediment such as washload from bank erosion. The case of mixing of two streams

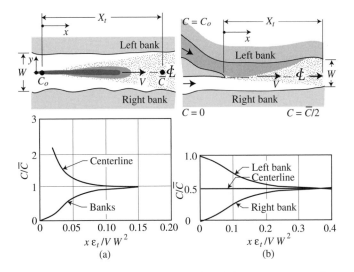

Figure 10.4. a) Plume for steady point source b) mixing at stream confluence (after Fischer *et al.*, 1979)

of equal discharge with $c_o = \bar{c}$ is shown on Figure 10.4b. A reasonable length for complete transversal mixing is $X_t \approx 0.4\ V\ W^2/\varepsilon_t$. It is interesting to note that the ratio of $X_t/W \approx 10\,W/h$. This means that deep rivers (low W/h) will mix rather quickly while shallow rivers (large W/h) will require very long distances for complete lateral mixing.

Example 10.2 presents calculations of turbulent mixing of a steady point source of washload.

Example 10.2 Application to resuspension from a dredge

As sketched in Figure E-10.2, sidecasting dredge causes the steady resuspension of 3 million gallons per day of clay at a 200,000 ppm concentration near the centerline of a 30 ft-deep and 1,300 ft-wide river. If the mean flow velocity is 2 ft/s and the shear velocity is 0.2 ft/s, determine the width of the plume and the maximum concentration 10,000 ft downstream of the dredge.

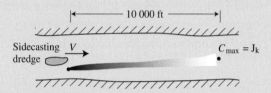

Figure E-10.2 Steady point source

1 million gallons/day $= 1\text{mgd} = 0.04382\ \text{m}^3/\text{s}$ (From Table 2.2).

Step 1. The mass flux

$$\dot{m} = QC = 3 \times 0.04382 \frac{\text{m}^3}{\text{s}} \times 35.32 \frac{\text{ft}^3}{\text{m}^3} \times 2 \times 10^5 \text{ppm} = 9.3 \times 10^5 \text{ppm ft}^3/\text{s}$$

Step 2. The transverse mixing coefficient $\varepsilon_t \cong 0.6\,h u_*$

$$\varepsilon_t = 0.6 \times 30\,\text{ft} \times 0.2 \frac{\text{ft}}{\text{s}} = 3.6 \frac{\text{ft}^2}{\text{s}}$$

Step 3. The plume width is approximately $4\,\sigma_t$ from Section 10.3.1

$$4\sigma_t \cong 4\sqrt{2\varepsilon_t x/V} = 4\sqrt{2 \times \frac{3.6\,\text{ft}^2}{\text{s}} \times \frac{10,000\,\text{ft}}{2\,\text{ft}}\,\text{s}}$$

$$4\sigma_t \cong 760\,\text{ft}$$

Step 4. The maximum centerline concentration from Equation (10.12), at $x = 10{,}000$ ft and $y = 0$, is

$$C_{\max} = \frac{\dot{m}}{h\sqrt{4\pi\,\varepsilon_t xV}} = \frac{9.30 \times 10^5\,\text{ft}^3\,\text{ppm}}{\text{s}30\,\text{ft}\sqrt{4\pi \times 3.6\,\text{ft}^2/\text{s} \times 10{,}000\,\text{ft} \times 2\,\text{ft/s}}}$$

$$C_{\max} = 33\ \text{ppm}$$

10.3.4 Longitudinal dispersion of an instantaneous point source

As sketched in Figure 10.1, dispersion is caused primarily by the non-uniformity of the velocity profile in the cross-section of shear flow. Dispersion is due to the combined and simultaneous action of advection and turbulent mixing. The dispersion coefficient K_d describes the diffusive property of the velocity distribution and is given here without derivation (Fischer *et al.* 1979):

$$K_d = \frac{-1}{h\varepsilon} \int\limits_0^h (v_x - V) \int\limits_0^z \int\limits_0^z (v_x - V)\,dz\,dz\,dz \qquad (10.13)$$

where v_x is the local velocity, V is the depth-averaged flow velocity, h is the flow depth and ε is the turbulent mixing coefficient. Practical estimates of K_d are given by Equation (10.8d). The dispersion coefficient is usually very large (order of 100 ft^2/s or 10 m^2/s). Washload transport in the streamwise direction is proportional to the concentration gradient and the dispersion coefficient.

Dispersion only becomes the dominant mixing process after the sediment concentration becomes uniform in a cross-section as sketched in Figure 10.4. This requires complete vertical mixing, $t > t_v$, and complete transverse mixing, $t > t_t$. The critical requirement is usually that lateral mixing must be completed $t > t_t > t_v$. This is possible at distances $x > 0.4VW^2/\varepsilon_t$ and at time $t > 0.4W^2/\varepsilon_t$. At this inception point, the corresponding concentration obtained from Equations (10.7, 10.8b, and 10.9c) will be approximately $C_{di} \approx m/hW^2$. This means that longitudinal dispersion will become of practical significance only in relatively small channels. In large channels, the concentration becomes very small.

For a practical application, consider the instantaneous point source of fine sediment. The cloud of sediment will approach a Gaussian distribution in the downstream direction only after an initial period of complete vertical and lateral mixing, $t > t_{di} = 0.4W^2/\varepsilon_t$. After this initial period, a cloud of total length approximately equal to $4\sigma_d = 4\sqrt{2K_d t}$ will propagate in the downstream direction at a peak concentration given by:

$$C_{max} = \frac{m}{Wh\sqrt{4\pi K_d t}} \tag{10.14}$$

where $t > 0.4 W^2/\varepsilon_t$. Example 10.3 outlines the application of this method to the longitudinal dispersion of an instantaneous point source.

Example 10.3 Application to longitudinal dispersion of an instantaneous point source

As sketched in Figure E-10.3, localized mass wasting of an overhanging stream bank causes a $M_s = 6{,}000\,\mathrm{kg}$ block of saturated very fine silt to dissolve into a 20 m-wide and 1 m-deep stream. Assuming rapid erosion of the block under a 1.5 m/s flow velocity and a stream slope of 100 cm/km, determine the peak concentration and the length of the dispersed sediment cloud observed at a bridge located at $X_b = 10\,\mathrm{km}$ downstream of the sediment source. How long will the pulse of increased concentration last under the bridge?

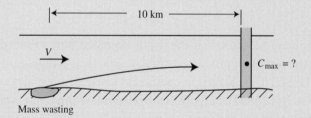

Figure E-10.3 Instantaneous point source

Step 1. The shear velocity is

$$u_* = \sqrt{ghS} = \sqrt{\frac{9.8\,\mathrm{m}}{\mathrm{s}^2} \times 1\,\mathrm{m} \times 10^{-3}} = 0.1\,\mathrm{m/s}$$

Step 2. The dispersion coefficient K_d and the transverse mixing coefficient ε_t are, respectively,

$$K_d \cong 250\,h u_* = 250 \times \frac{1\,\mathrm{m} \times 0.1\,\mathrm{m}}{\mathrm{s}} = \frac{25\,\mathrm{m}^2}{\mathrm{s}}$$

$$\varepsilon_t = 0.6\,h u_* = 0.6 \times 1\,\mathrm{m} \times \frac{0.1\,\mathrm{m}}{\mathrm{s}} = \frac{0.06\,\mathrm{m}^2}{\mathrm{s}}$$

Step 3. The initial period ends when

$$t_{di} = \frac{0.4W^2}{\varepsilon_t} = \frac{0.4 \times (20\,\mathrm{m})^2\mathrm{s}}{0.06\,\mathrm{m}^2} = 2{,}667\,\mathrm{s} = 45\,\mathrm{min}$$

at a downstream distance X_{di}

$$X_{di} = V\,t_{di} = \frac{1.5\,\mathrm{m}}{\mathrm{s}} \times 2{,}667\,\mathrm{s} = 4{,}000\,\mathrm{m} < 10\,\mathrm{km}$$

Step 4. At the bridge, the length L of the dispersed cloud is

$$L = 4\sqrt{\frac{2K_dX_b}{V}} = 4\sqrt{2 \times \frac{25\,\mathrm{m}^2}{\mathrm{s}} \times \frac{10{,}000\,\mathrm{m\,s}}{1.5\,\mathrm{m}}} = 2.3\,\mathrm{km}$$

and will be centered at the bridge at a time $t_b = \frac{X_b}{V} = \frac{10{,}000\,\mathrm{m\,s}}{1.5\,\mathrm{m}} = 6{,}666\,\mathrm{s} = 1.85$ hours after the injection upstream.

Step 5. The maximum concentration of the dispersed cloud under the bridge at time t_b is given by:

$$C_{max} = \frac{M_s}{Wh\sqrt{4\pi K_d t_b}} = \frac{6{,}000\,\mathrm{kg}}{20\,\mathrm{m} \times 1\,\mathrm{m}\sqrt{4\pi \times \frac{25\,\mathrm{m}^2}{\mathrm{s}} \times 6{,}666\,\mathrm{s}}}$$

$$= \frac{0.207\,\mathrm{kg}}{\mathrm{m}^3} = 207\,\mathrm{mg/l}$$

Step 6. The increase in sediment concentration under the bridge will last $L/V = 2{,}300\,\mathrm{m}/1.5\,\mathrm{m/s} = 1{,}500$ or about 25 minutes.

10.4 Suspended sediment concentration profiles

This section focuses on particles that are large enough to settle. Consider steady uniform turbulent flow in a wide rectangular channel without any phase change ($\dot{C} = 0$). The settling of sediment particles due to the density difference between the particles and the surrounding fluid induces a downward particle settling flux. All the terms of the diffusion–dispersion equation (Equation 10.5) vanish except those describing the sediment fluxes in the vertical z direction:

$$v_z \frac{\partial C}{\partial z} = \varepsilon_z \frac{\partial^2 C}{\partial z^2} \tag{10.15}$$

This equation can be integrated with respect to z. As sketched in Figure 10.5, equilibrium is obtained when the upward turbulent flux is balanced by the downward

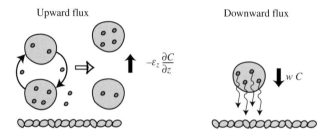

Figure 10.5. Sketch of vertical sediment fluxes

settling flux. Although the net vertical velocity is zero, the downward fall velocity of sediment particles, $v_z = -\omega$, in Equation (10.15) gives

$$\omega C + \varepsilon_z \frac{\partial C}{\partial z} = 0 \tag{10.16}$$

The resulting concentration profile for constant values of ω and ε_z is

$$C = C_0 e^{-\omega z/\varepsilon_z} \tag{10.17}$$

This relationship shows that as $\omega z < \varepsilon_z$, the concentration profile becomes gradually uniform, whereas most of the sediment is concentrated near the bed when $\omega z > \varepsilon_z$. Since $\varepsilon_z \approx 0.06hu_*$, this relationship points to the fact that concentration profiles depend on u_*/ω.

The most general case where the turbulent mixing coefficient of sediment ε_z varies with depth is examined by analogy with the momentum exchange coefficient ε_m defined in Equation (6.7).

$$\varepsilon_z = \beta_s \varepsilon_m = \beta_s \frac{\tau \, dz}{\rho_m d v_x} \tag{10.18}$$

in which β_s is the ratio of the turbulent mixing coefficient of sediment to the momentum exchange coefficient. This coefficient β_s has been found to remain sufficiently close to unity for most practical applications, but is kept here for the derivation of the concentration profile.

The vertical sediment concentration can be determined after substituting the relationships for shear stress (Equation 8.6) and for turbulent velocity profiles (Equation 6.11) into Equation (10.18) to give

$$\varepsilon_z = \beta_s \kappa u_* \frac{z}{h} (h - z) \tag{10.19}$$

The resulting mixing coefficient ε_z varies with z as shown in Figure 10.6. Notice that the maximum value of ε_z with $\beta_s = 1$ and $\kappa = 0.4$ equals $\varepsilon_z = 0.1u_* h$ at mid-depth

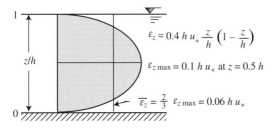

Figure 10.6. Vertical mixing coefficient

$z = 0.5\,h$ (result quite similar to Equation (10.8a) for the average of the parabolic distribution of ε_z). The expression for ε_z in Equation (10.19) is substituted into Equation (10.16), and solved after separating the variables C and z.

$$\frac{C}{C_a} = \left(\frac{h-z}{z}\,\frac{a}{h-a}\right)^{\text{Ro}=\frac{\omega}{\beta_s \kappa u_*}} \qquad (10.20a) \blacklozenge\blacklozenge$$

in which C_a represents the reference sediment concentration at a reference elevation "a" above the bed elevation. The relative concentration C/C_a depends on the elevation z above the reference elevation as derived by Rouse (1937). The exponent Ro is referred to as the Rouse number and reflects the ratio of the sediment properties to the hydraulic characteristics of the flow. Also, with $\beta_s = 1$ and $\kappa = 0.4$, $u_*/\omega = 2.5/\text{Ro}$.

Simplifications are possible for reciprocal points like $z = h - a$. For instance, the concentration at 80% of the flow depth $C_{0.8}$ and the concentration at 20% of the flow depth $C_{0.2}$ can be defined from Equation (10.20a) as

$$\frac{C_{0.8}}{C_{0.2}} = \left(\frac{1}{16}\right)^{\text{Ro}} \qquad (10.20b)$$

Conversely, one can thus use two-point concentration measurement to define the Rouse number. For this instance above, $\text{Ro} = 0.83\log C_{0.2}/C_{0.8} = 0.52\log C_{0.1}/C_{0.9}$. As examples, when $u_*/\omega = 25$, or $\text{Ro} = 0.1$, Equation (10.20b) gives $C_{0.8}/C_{0.2} = 0.75$, and when $u_*/\omega = 100$, or $\text{Ro} = 0.025$, then $C_{0.8} = 0.93\,C_{0.2}$. The suspended sediment concentration profile becomes uniform under $u_* > 100\,\omega$. In practice, this corresponds to settling velocities $\omega < 0.001$ m/s, which corresponds to silts and clays. This is why the methods covered previously in Section 10.3 are usually applicable to silts and clays.

Figure 10.7 illustrates concentration profiles where $a/h = 0.05$. The concentration of sediment particles becomes increasingly large near the bed as the sediment size increases. It is interesting to note that $u_*/\omega = 2.5/\text{Ro}$. Accordingly, sediment is predominantly transported near the bed (Chapter 9) when $u_*/\omega < 1$, or $\text{Ro} > 2.5$.

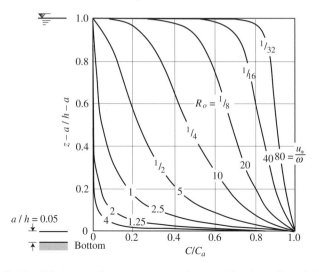

Figure 10.7. Equilibrium sediment concentration profiles for $a/h = 0.05$

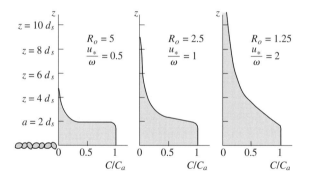

Figure 10.8. Near-bed concentration profiles for $R_o > 1.25$, or $u_* < 2\omega$

In the case where bedload is dominant, the sediment concentration profile near the bed can be calculated from a simplification of Equation (10.20a) where $h - z \approx h - a$, or

$$\frac{C}{C_a} \approx \left(\frac{a}{z}\right)^{Ro} \quad \text{near the bed} \tag{10.20c}$$

An example of near-bed concentration profiles is shown in Figure 10.8 based on $a = 2d_s$ for $Ro = 1.25$, 2.5, and 5. This example shows that the thickness of the layer where sediment transport takes place is larger than $2d_s$ and varies with the Rouse number, or u_*/ω. The concept of a bed layer of variable thickness seems interesting at first. One must consider the difficulties of setting exact values for: (1) bed elevation; (2) representative grain diameter; and (3) near-bed concentration.

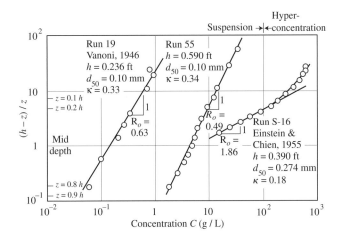

Figure 10.9. Suspended sediment concentration profiles (modified after Woo *et al.*, 1988)

It is also very difficult to measure sediment concentration in a layer that will be typically a few millimeters in size. After all, it must be recognized that exact values of C_a are very difficult to measure. It remains that when Ro > 2.5, most of the sediment transport will take place in a thin layer very close to the bed.

As the Rouse number decreases, a greater fraction of the sediment will be transported in suspension. Typically, the suspension becomes significant as Ro < 2.5, or as $u_* > \omega$. For steady uniform flow, the concentration of suspended sediment C varies with flow depth h and distance above bed z. A log–log plot of C versus $(h-z)/z$ shows a straight line from which the exponent Ro can be graphically defined, as shown on Figure 10.9.

This Figure 10.9 shows the usual procedure to determine Ro from point concentration measurements. Straight lines can normally be fitted to the field concentration measurements as long as the sediment concentration is less than about 100,000 mg/l, which corresponds to about $C_v < 0.05$. As the concentration increases beyond that point, hyperconcentrated sediment flows have more uniform sediment concentration profiles, as shown in run S-16. The maximum near-bed concentrations of granular material were around $C_v \approx 0.5$, or about 1×10^6 mg/l. The second interesting feature of Rouse plots, like Figure 10.9, is that $h-z=z$ at mid-depth. The sediment concentration at mid-depth C_{mid} from Equation (10.20a) becomes approximately

$$C_{\mathrm{mid}} \approx C_a \, (2d_s/h)^{\mathrm{Ro}} \qquad\qquad (10.20\mathrm{d})$$

This shows that one can obtain an extremely large variability in sediment concentration in deep sand-bed rivers when the Rouse number is fairly large (Ro > 0.5). Case study 10.1 on the Mississippi River is quite instructive in this regard.

Case study 10.1 Mississippi River, United States

This case study illustrates sediment transport calculations when $u_*/\omega > 2$. The suspended sediment concentration profiles of the Mississippi River near Tarbert Landing were examined by Akalin (2002). Bed sediment samples were collected with a drag bucket and point suspended sediment samples were collected with a P-63 sampler. The gradation curves of bed and suspended material excluding silts and clays are shown in Figure CS-10.1.1. At a flow depth of 66 ft (20 m) and slope of 3.7 cm/km, the shear velocity is $u_* = \sqrt{ghS} = 0.085$ m/s. The corresponding flow properties are shown in Table CS-10.1.1. At values of Ro < 2.5, or $u_* > \omega$, it becomes very difficult to determine sediment transport as a function of bedload. Field measurements of sediment concentration thus become very useful.

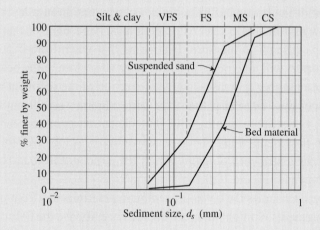

Figure CS-10.1.1 Gradation curves of bed and suspended material (sand fraction only) (after Akalin, 2002)

Table CS-10.1.1. *Main properties of the Mississippi River at Tarbert Landing*

Sand size (mm)	d_*	ω (mm/s)	u_*/ω	Ro calculated	τ_*
0.1	2.5	8.2	10.3	0.24	4.46
0.17 (d_{50})	4.3	21	4.0	0.61	2.62
0.2	5	26	3.2	0.77	2.23
0.4	10	57	1.5	1.67	1.11
0.8	20	96	0.88	2.84	0.56

The suspended sediment concentration profile is shown in Figure CS-10.1.2. The overall measured Rouse number is 0.66 and compares well with the calculated Rouse number of 0.61 for the median grain diameter of the suspension $d_s = 0.17$ mm in Table CS-10.1.1. It is worthwhile estimating the concentration near the bed from $C_a \approx C_{mid\text{-}depth}(h/2d_s)^{Ro} = 50\,\text{mg/l}\,(20/2 \times 0.00017)^{0.66}$ 70,000 mg/l.

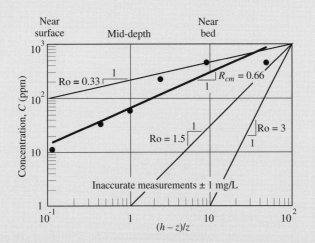

Figure CS-10.1.2 Suspended sediment concentration profile (after Akalin, 2002)

Calculations by size fractions for sand-bed rivers can become quite difficult when Ro > 0.5. The reason is that the near-bed concentration becomes excessively large as the Rouse number increases. To illustrate this, assume that the measured mid-depth concentration is as low as possible, e.g. 1 mg/l. Notice that sediment concentration measurement errors are of the order of a few mg/l. For coarse sand ($d_s = 0.7$ mm) with a calculated Rouse number $Ro_c = 2$, the near-bed concentration would become $C_a \approx 1\,\text{mg/l}\,(20/2 \times 0.0007)^2 = 204\,\text{kg/l}$, which is physically impossible. In reality, measurement errors can cause discrepancies between calculated and measured values of the Rouse number when Ro > 0.5, or $u_*/\omega < 5$. For example, measured mid-depth concentration of 1 mg/l and a near-bed concentration of 70,000 mg/l would result in a measured Rouse number of $Ro_m = \log\,(C_a/C_{mid})/\log\,(h/2d_s) = \log\,(70,000)/\log\,(20/2 \times 0.0007) = 1.2$, compared with a calculated Rouse number of 2. It can be concluded from this analysis that the measured Rouse number will be smaller than the calculated Rouse number in deep sand-bed rivers.

Figure CS-10.1.3 shows the results for the Mississippi River. The figure shows good agreement between the calculated and measured values of the Rouse

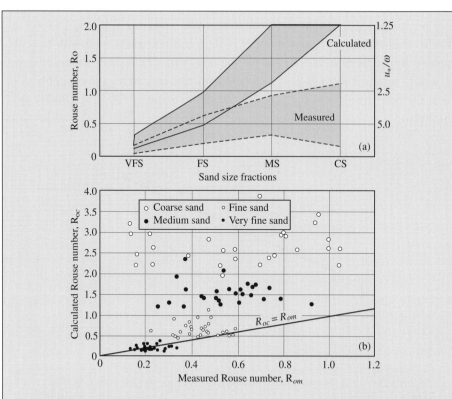

Figure CS-10.1.3 Rouse numbers a) calculated and measured Rouse numbers
b) comparison of the measured and calculated Rouse numbers (after Akalin,
2002)

number for very fine sands (Ro ≈ 0.2). The apparent discrepancy for coarser
grain size is due to the fact that the real sediment concentrations for medium
(M), coarse (C), and very coarse (VC) sands in this example should be much
lower than 1 mg/l. As shown in Figure CS-10.1.2 as $h/z > 100$, large values of
the Rouse number will yield concentrations in most of the water column below
the accuracy in concentration measurement of a few mg/l.

10.5 Suspended load

The unit suspended sediment discharge q_s in natural streams and canals is computed
from the depth-integrated advective flux of sediment Cv_x above the bed layer $z > a$:

$$q_s = \int_a^h Cv_x \, dz \qquad (10.21)$$

The corresponding total suspended sediment discharge Q_s is obtained from integration of the unit suspended sediment discharge over the entire width of the channel, or

$$Q_s = \int_{width} q_s \, d\,W \tag{10.22}$$

The suspended load L_s defines the amount of sediment passing a cross-section in suspension over a certain period of time, thus

$$L_s = \int_{time} Q_s \, dt \tag{10.23}$$

By analogy with bedload, described in Chapter 9, the units can be either by weight, mass, or volume. It is common to use the suspended load in metric tons (1,000 kg).

The comparison of suspended load to bedload delineates which mode of sediment transport is dominant. The suspended unit sediment discharge q_s can be calculated from Equation (10.21) after substituting C from Equation (10.20a) and v_x from Equation (6.16):

$$q_s = \int_{z=2d_s}^{h} C_a \frac{u_*}{\kappa} \left[\left(\frac{h-z}{z} \right) \left(\frac{a}{h-a} \right) \right]^{\frac{\omega}{\beta_s \kappa u_*}} \ln \frac{30\,z}{k_s'} \, d\,z \tag{10.24}\blacklozenge$$

Similarly, the total unit sediment discharge can be obtained by integrating from $z = k'_s/30$ to the free surface $z = h$, thus:

$$q_t = \int_{z=\frac{k'_s}{30}}^{h} C_a \frac{u_*}{\kappa} \left[\left(\frac{h-z}{z} \right) \left(\frac{a}{h-a} \right) \right]^{\frac{\omega}{\beta_s \kappa u_*}} \ln \frac{30\,z}{k_s'} \, d\,z \tag{10.25}$$

The ratio of suspended to total unit sediment discharges indicates whether most of the sediment transport occurs in suspension or in the bed layer.

After substituting $\beta_s = 1$, $\kappa = 0.4$, $k'_s = d_s$, the ratio q_s/q_t from Equations (10.24) and (10.25) becomes independent of C_a and u_*, but varies as a function of the ratio of shear to fall velocities u_*/ω and relative submergence h/d_s. It is found that sediment transport can be subdivided into three zones describing which mode of transport is

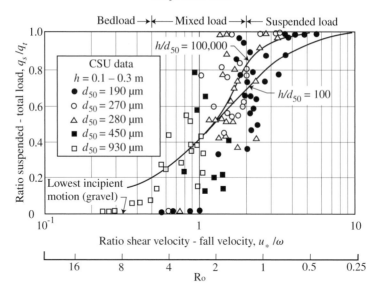

Figure 10.10. Ratio of suspended to total load versus ratio of shear to fall velocities

dominant: (1) bedload; (2) mixed load; or (3) suspended load. Figure 10.10 shows the ratio of suspended to total load as a function of u_*/ω and h/d_s. The CSU Laboratory data for the mixed load are from Guy, Simons, and Richardson (1966). It is interesting to note that for turbulent flow over rough boundaries, incipient motion corresponds to $u_*/\omega \cong 0.2$. The lines shown are those from the Einstein Integrals of Guo and Julien (2004) as obtained by Shah-Fairbank (2009) for $h/d_s = 100$ and 100,000. Bedload is dominant at values of u_*/ω less than about 0.5, and the methods detailed in Chapter 9 should be used to determine the sediment transport rate. A zone called mixed load is found where $0.5 < u_*/\omega < 2$ in which both the bedload and the suspended load contribute to the transport. It is instructive to note that when $u_*/\omega < 2$, the ratio of the suspended load to the bedload is approximately equal to $(u_*/\omega)^2$. The bedload and suspended load are approximately equal when $u_* = \omega$. In the case of mixed load, $u_*/\omega < 2$, the total load will be less than 5 times the bedload. Methods to calculate the bed material load are also presented in Chapter 11. Suspended load is dominant when $u_*/\omega > 2$, and gravitational effects on the particles are negligible compared to turbulent mixing as u_*/ω becomes very large. Bedload equations and the Einstein procedure from Equation (10.25) and Appendix A should be used in the range of $0.2 < u_*/\omega < 2$ but should not be used when $u_*/\omega > 2$. In this case, field measurements of the suspended load and the Modified Einstein procedure are preferable. These observations are summarized in Table 10.2, and the results are also shown in the Shields diagram of Figure 10.11.

Table 10.2. *Modes of sediment transport given shear and fall velocities*

Ro	u_*/ω	Mode of sediment transport
> 12.5	< 0.2	no motion
$\cong$ 12.5	$\cong$ 0.2	incipient motion
12.5–5	0.2–0.5	bedload – Chapter 9
5–1.25	0.5–2	mixed load – Chapter 11
0.5–1.25	2–5	suspension – Chapter 10
< 0.5	> 5	suspension
0.1	25	$C_{0.8} = 0.75C_{0.2}$
0.025	100	$C_{0.8} = 0.93C_{0.2}$

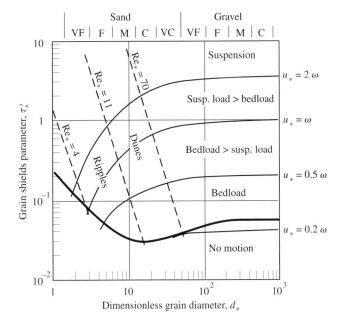

Figure 10.11. Bedload and suspended load

10.6 Hyperconcentrations

Hyperconcentrations refer to heavily sediment-laden flows in which the presence of fine sediments materially affects fluid properties and bed material transport. In general, the volumetric sediment concentration C_v of hyperconcentrations ranges from 5–60%. The mass density of hyperconcentrations ρ_m is calculated from $\rho_m = \rho + (\rho_s - \rho)C_v$. The dry specific mass of a mixture is the mass of solids per unit

Table 10.3. *Physical properties of mudflows and debris flows*

C_v	C_w	C_{ppm}	$C_{mg/l}$ mg/l	$\rho_m{}^a$ kg/m^3	ρ_{md} kg/m^3	p_o	e	$w\%$	B	λ^b
0.05	0.122	122,401	132,500	1,082	132	0.95	19	717	20	0.76
0.10	0.227	227,468	265,000	1,165	265	0.9	9	340	10	1.20
0.20	0.4	398,496	530,000	1,330	530	0.8	4	150	5	2.20
0.30	0.53	531,773	795,000	1,495	795	0.7	2.3	88	3.3	3.70
0.40	0.64	638,554	1,060,000	1,660	1,060	0.6	1.5	56	2.5	6.48
0.50	0.73	726,027	1,325,000	1,825	1,325	0.5	1.0	38	2.0	14.0
0.60	0.8	798,994	1,590,000	1,990	1,590	0.4	0.67	25	1.67	121
0.70	0.86	860,789	1,855,000	2,155	1,855	0.3	0.42	16	1.42	–

[a] Calculations assuming $G = \rho_s/\rho = 2.65$
[b] Linear concentration based on $C_v* = 0.615$

total volume. It is identified as ρ_{md} and calculated as $\rho_s C_v$. The porosity p_0 is the ratio of the volume of water per unit total volume, or $p_0 = 1 - C_v$. The void ratio e corresponds to the volume of water per unit volume of solids, or $e = (1 - C_v)/C_v$. The bulking factor B is the ratio of the total volume to the volume of solids, $B = 1/C_v = 1 + e$. A linear sediment concentration λ has been defined by Bagnold (1956) as a function of the volumetric sediment concentration C_v and the maximum volumetric sediment concentration $C_{v*} \simeq 0.615$. The relationship for λ is $\lambda = \left((0.615/C_v)^{1/3} - 1\right)^{-1}$.

The fluid properties can also be defined from measurements of weight of solids W_s, weight of fluid W, and total weight W_t. The sediment concentration by weight C_w is defined as the ratio of W_s/W_t. The water content of soils as "w" is defined from the ratio of the weight of water W to the weight of solids W_s, or $w = W/W_s$. Conversions to the volumetric concentration are obtained as $C_v = 1/(1 + wG)$, or $w = (1 - C_v)/GC_v$. Typical values of these physical properties are listed in Table 10.3.

Atterberg limits define four states for a soil depending on the value of the water content. The solid state refers to water contents less than the shrinkage limit w_s. The shrinkage limit is determined as the water content after just enough water is added to fill all the voids of a dry pat of soil. A semi-solid state is defined when the water content is less than the plastic limit w_p. The plastic limit is determined by measuring the water content of a soil when threads of soil 1/8 inch in diameter begin to crumble. A plastic state is found when the water content is less than the liquid limit w_l. The liquid limit is determined from standard geotechnical procedures. Finally, the liquid state is found when the water content exceeds the liquid limit w_l. The plasticity index I_p is the difference between the liquid and plastic limits, or $I_p = w_l - w_p$. The liquidity index IL is defined as a function of w_l and w_p as $IL = (w - w_p)/I_p$. Although the limits between the various states have been set arbitrarily, the concept of Atterberg limits is particularly suited to the analysis of landslides and debris flows because it defines water contents at which a specific soil behavior will be observed. Typical values of the Atterberg limits for clay minerals are presented in Table 10.4. It is observed that the liquid limit of clay minerals are about $C_v = 0.4$ for kaolinite, $C_v = 0.25$ for illite and can be as low as $C_v < 0.05$ for montmorillonite. The plastic limit can also be about $C_v = 0.5$ for kaolinite, $C_v = 0.4$ for illite and $C_v = 0.28$ for montmorillonite. At a given concentration, one can thus expect much larger values of yield strength and viscosity for soils containing montmorillonite than kaolinite.

10.6.1 Rheology of hyperconcentrations

Rheology describes the forces required for the deformation of matter. More specifically, the graphical measure of the shear stress applied at a given rate of deformation

Table 10.4. *Volumetric concentrations of the Atterberg limits for clays*

Mineral	Exchangeable ion	Liquid limit C_v	Plastic limit C_v
Montmorillonite	Na	0.05	0.41
	K	0.054	0.28
	Ca	0.069	0.32
	Mg	0.084	0.38
	Fe	0.115	0.33
Illite	Na	0.24	0.42
	K	0.24	0.39
	Ca	0.27	0.46
	Mg	0.28	0.45
	Fe	0.25	0.44
Kaolinite	Na	0.42	0.54
	K	0.43	0.56
	Ca	0.50	0.56
	Mg	0.41	0.55
	Fe	0.39	0.50
Attapulgite	H	0.12	0.20

Source: Lambe and Whitman, 1969.

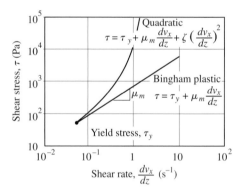

Figure 10.12. Rheogram for non-Newtonian fluids

dv_x/dy of a fluid defines a rheogram (Figure 10.12). At low rates of deformation, the shear stress in clear water increases linearly with the rate of deformation and the fluid is said to be Newtonian. The dynamic viscosity of a mixture μ_m in the laminar flow regime is then defined as the slope of the rheogram. Under large shear stresses, the boundary layer flow becomes turbulent (except in the laminar sublayer), and the shear stress increases with the second power of the rate of deformation, as discussed in Chapter 6.

The Bingham rheological model (Bingham, 1922) is to some extent a useful simplified rheological model. Beyond a finite shear stress, called yield stress τ_y, the rate of deformation dv_x/dz is linearly proportional to the excess shear stress. The constitutive equation is

$$\tau = \tau_y + \mu_m \frac{dv_x}{dz} \tag{10.26}$$

in which μ_m is the dynamic viscosity of the mixture. The Bingham plastic model is well-suited to homogeneous suspensions of fine particles, particularly muds, under low rates of deformation. Experimental laboratory results by Qian and Wan (1986) and others confirm that under rates of deformation observed in the field ($\partial v_x/\partial z < 10\,\mathrm{s}^{-1}$), fluids with large concentrations of silts and clays behave like Bingham plastic fluids.

The analysis of coarse sediment mixtures is somewhat more complex and involves an additional shear stress due to particle impact. Bagnold (1954) pioneered laboratory investigations on the impact of sediment particles. He defined the dispersive shear stress τ_d induced by the collision between sediment particles as:

$$\tau_d = c_{Bd}\rho_s\left(\left(\frac{0.615}{C_v}\right)^{1/3} - 1\right)^{-2} d_s^2 \left(\frac{dv_x}{dz}\right)^2 \tag{10.27}$$

The dispersive shear stress is shown to increase with three parameters, namely: (1) the second power of the particle size; (2) the volumetric sediment concentration; and (3) the second power of the rate of deformation. It is important to recognize that the dispersive stress is proportional to the product of these three parameters, therefore high values of all three parameters are required to induce a significant dispersive shear stress.

A quadratic rheological model has been proposed by O'Brien and Julien (1985) which combines the following stress components of hyperconcentrated sediment mixtures: (1) cohesion between particles; (2) internal friction between fluid and sediment particles; (3) turbulence; and (4) inertial impact between particles. The resulting quadratic model is:

$$\tau = \tau_y + \mu_m \frac{dv_x}{dz} + \zeta\left(\frac{dv_x}{dz}\right)^2 \tag{10.28} \blacklozenge$$

where μ_m is the dynamic viscosity of the mixture, and ζ is the turbulent–dispersive parameter. The last term of the quadratic model combines the effects of turbulence with the dispersive stress induced by inertial impact of sediment particles. Combining the conventional expression for the turbulent stress in sediment-laden flows with Bagnold's dispersive stress gives:

$$\zeta = \rho_m l_m^2 + c_{Bd}\,\rho_s \lambda^2 d_s^2 \tag{10.29}$$

where ρ_m and l_m are respectively the mass density and mixing length of the mix-ture; d_s is the particle diameter; λ is Bagnold's linear concentration; and c_{Bd} is an empirical parameter defined by Bagnold ($c_{Bd} \cong 0.01$).

10.6.2 Parameter evaluation

The rheological properties of hyperconcentrations are determined from laboratory analyses of rheograms obtained from viscometric measurements. At least three kinds of devices are commercially available for the measurement of rheograms: (1) the capillary viscometer; (2) the concentric cylindrical viscometer; and (3) the cone and plate viscometer. Concentric cylindrical viscometers seem to be best suited for a wide range of shear rates. Field observations, however, indicate that shear rates rarely exceed 100/s. It is therefore recommended to measure the rheological properties of hyperconcentrations under similarly low rates of shear. Rheological properties of hyperconcentrations are generally formulated as a function of sediment concentration as shown on Figure 10.13. The recommended empirical formulas are the exponential relationships for yield stress and viscosity at large concentrations of fines, and the Bagnold equation to calculate the dispersive shear stress of coarse particles.

The measurements in Figure 10.14 show that both the yield stress and viscosity increase exponentially with the sediment concentration.

$$\log \tau_y \approx \log a + b\, C_v \tag{10.30a}$$

$$\log \mu_m \approx -3 + c\, C_v \tag{10.30b}$$

where τ_y is the yield stress in Pa, μ_m is the dynamic viscosity in Pa.s and C_v is the volumetric sediment concentration. The constants $a, b,$ and c are coefficients determined from the measurements. For instance, typical values of the coefficients for different types of muds, clays, and lahars are presented in Table 10.5. Values of b and c are equal to 10, meaning that a 10% increase in C_v gives a tenfold increase in τ_y and μ_m.

A generic relationship for typical soils as a function of the total sediment con-centration can be obtained for $a = 0.005$, $b = 7.5$ and $c = 8$. Accordingly, the yield stress of a typical soil at a volumetric sediment concentration of 70% can be esti-mated from Equation (10.30a) as $\tau_y = a \times 10^{bC_v} \simeq 0.005 \times 10^{(7.5 \times 0.7)} = 889\,\mathrm{Pa} = 8,890\,\mathrm{dynes/cm^2}$. Similarly, the dynamic viscosity of the same mixture is calculated from Equation (10.30b) as $\mu_m \simeq 0.001 \times 10^{cC_v} \cong 0.001 \times 10^{(8 \times 0.7)} = 398\,\mathrm{Pa.s} =$

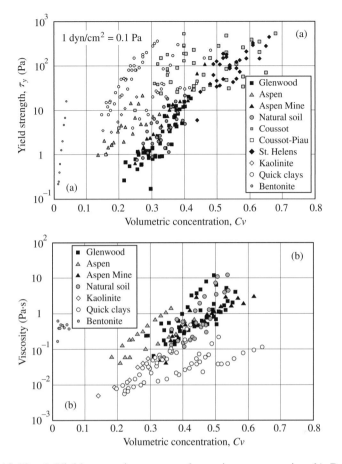

Figure 10.13. a) Yield strength versus volumetric concentration b) Dynamic viscosity versus volumetric concentration

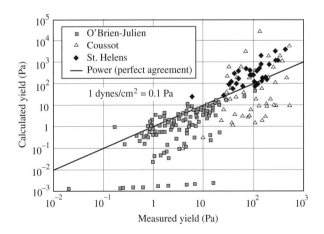

Figure 10.14. Calculated and measured yield strength of hyperconcentrations

Table 10.5. *Coefficients a, b, c of the yield strength and viscosity relationships*

Material	Liquid limit C_v	Yield strength in Pa $\tau_y = a10^{bC_v}$		Viscosity in Pa.s $\mu_m = 0.001 \times 10^{cC_v}$
		a	b	c
Bentonite (montmorillonite)	0.05–0.2	0.002	100	100
Sensitive clays	0.35–0.6	0.3	10	5
Kaolinite	0.4–0.5	0.05	9	8
Typical soils	0.65–0.8	0.005	7.5	8
Granular material	–	–	2	3

3,980 poise. At such high sediment concentration this soil is probably not moving because its yield strength exceeds the critical shear stress of large boulders and the viscosity is 400,000 times larger than the viscosity of clear water. Figure 10.14 shows the comparison of Equation (10.30) with laboratory and field measurements.

10.6.3 Fall velocity and particle buoyancy

As presented in Equation (5.22a), and Happel and Brenner (1965), the fall velocity of particles of diameter d_s and specific weight γ_s in a Bingham plastic fluid of specific weight γ_m is given by:

$$\omega^2 = \frac{4}{3} \frac{gd_s}{C_D} \frac{\gamma_s - \gamma_m}{\gamma_m} \tag{10.31}$$

in which the drag coefficient C_D depends on the dynamic viscosity μ_m and the yield stress τ_y of the Bingham fluid. After defining the Bingham Reynolds number ($\text{Re}_B = \frac{\rho_m d_s \omega}{\mu_m}$) and the Hedstrom number $\text{He} = \frac{\rho_m d_s^2 \tau_y}{\mu_m^2}$, the drag coefficient can be rewritten as

$$C_D = \frac{24}{\text{Re}_B} + \frac{2\pi \, \text{He}}{\text{Re}_B^2} + 1.5 \tag{10.32}\blacklozenge$$

The settling velocity ω can be obtained from solving Equations (10.31) and (10.32) as a function of particle diameter d_s, yield strength τ_y, mixture viscosity μ_m and mass density ρ_m as:

$$\omega = \frac{8\mu_m}{\rho_m d_s} \left(\left(1 + \frac{\rho_m g \, d_s^3 (\rho_s - \rho_m)}{72 \mu_m^2} - \frac{\pi}{48} \frac{\rho_m d_s^2 \tau_y}{\mu_m^2}\right)^{0.5} - 1 \right) \tag{10.33}\blacklozenge$$

This relationship reduces to Equation (5.23d) when $\tau_y = 0$, or He $= 0$, $\rho_m = \rho$ and $\mu_m = \mu$. Interestingly, the fall velocity reduces to zero for particles smaller than the buoyant particle diameter d_{sb} given by:

$$d_{sb} = \frac{3\pi}{2} \frac{\tau_y}{(\gamma_s - \gamma_m)} \tag{10.34}\blacklozenge$$

The particles $d_s < d_{sb}$ remain neutrally buoyant in the mixture and do not settle (Qian and Wan, 1986). At values of $\tau_y = 100$ Pa and $C_v = 0.6(\rho_m \approx 2,000 \text{ kg/m}^3)$, cobbles as large as 70 mm in diameter would not settle in the sediment mixture.

10.6.4 Dimensionless rheological model and classification

The relative magnitude of the terms in the quadratic equation (Equation (10.28)) defines conditions under which simplified rheological models are applicable. The dimensionless rheological model was obtained by Julien and Lan (1991) after rewriting Equations (10.28) and (10.29) as

$$\Pi_\tau = 1 + (1 + \Pi_{td})\,\Pi_{dv} \tag{10.35}$$

in which the dimensionless parameters are defined as:

1. Dimensionless excess shear stress Π_τ

$$\Pi_\tau = \frac{\tau - \tau_y}{\mu_m \dfrac{dv_x}{dz}} \tag{10.36a}$$

2. Dimensionless dispersive–viscous ratio Π_{dv}

$$\Pi_{dv} = c_{Bd}\frac{\rho_s d_s^2}{\mu_m}\left(\left(\frac{0.615}{C_v}\right)^{1/3} - 1\right)^{-2}\left(\frac{dv_x}{dz}\right) \tag{10.36b}$$

3. Dimensionless turbulent–dispersive ratio Π_{td}

$$\Pi_{td} = \frac{\rho_m l_m^2}{c_{Bd} \cdot \rho_s d_s^2}\left[\left(\frac{C_v^*}{C_v}\right)^{1/3} - 1\right]^2 \tag{10.36c}$$

where $C_v^* \cong 0.615$, $c_{Bd} \cong 0.01$ and for field applications $l_m \simeq 0.4\,\text{h}$.

The usefulness of the dimensionless rheological model is demonstrated in Figure 10.15 where Π_τ is plotted versus Π_{dv}. When Equation (10.35) is fitted to the experimental data sets of Govier *et al.*, Savage and McKeown, and Bagnold, it is interesting to notice that the value $c_{Bd} = 0.01$ suggested by Bagnold is comparable to the value $(1 + \Pi_{\tau d}) = 0.87$ obtained from the slope of the line on Figure 10.15.

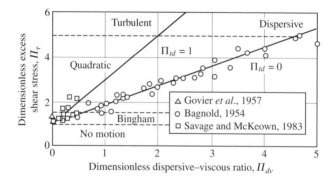

Figure 10.15. Classification of rheological models (modified after Julien and Lan, 1991)

Simplifications of the quadratic rheological model are possible under the following conditions: (1) the Bingham model (Equation (10.26)) is applicable when $\Pi_\tau \to 1$, and viscosity will be dominant when $\tau < 5\tau_y$; (2) the flow is turbulent when $\Pi_\tau > 5$ and $\Pi_{\tau d} > 1$; and (3) the dispersive shear stress is dominant when $\Pi_\tau > 5$ and $\Pi_{dv} > 4$. This analysis suggests that hyperconcentrations could be classified as: (1) mudflows when the Bingham rheological model is applicable; (2) hyperconcentrated flows or mud flood, when the turbulent shear stress is dominant; and (3) debris flows when the dispersive stress controls. Flood mitigation design must take into consideration the rheological behavior of the three types of hyperconcentrated sediment flows (Julien and Leon, 2000).

10.6.5 Classification and flood mitigation

Mud floods are very fluid hyperconcentrated flows in steep mountain channels. The grain size is very small compared to flow depth ($d_{50} \ll 0.02$ h). Flow velocities are very high and the turbulent flow is often supercritical. The water content is lower than the liquid limit and the concentration of clay is low. The turbulent shear stress is dominant. Conveyance design for mud floods should include consideration of sediment bulking, surging (roll waves), supercritical flow, debris plugging, sediment abrasion, super elevation, and potential for sediment scour and deposition. It is preferable to maintain the channel cross-section as straight and uniform as possible. Straight, steep channels will result in high velocities and high Froude numbers and will prevent the formation of cross waves and local deposition behind channel irregularities. Channel lining with concrete, or grouted riprap can be effective, but very expensive. Two important design considerations with lining channels on steep alluvial fan slopes are abrasion of the lining and excess pore water pressure.

Mudflows have a fluid viscosity that is several orders of magnitude higher than that of water. The content of clay is high, and the clay type is often montmorillonite in the United States. The yield strength is very high ($\tau_y > 100\,\mathrm{Pa}$) and some size fractions do not settle according to Equation (10.34). The water content is fairly close to the plastic limit and the Bingham model is applicable. The hydraulic properties of mudflows are typically low flow velocities and large flow depths, thus low values of the Froude number. These hydraulic properties sustain motion of mudflows on flat slopes. Flood mitigation design must include consideration of flow avulsion, debris and mud plugging of channel and conveyance facilities, and cleanup/maintenance. Effective mitigation measures for mudflows include storage, deflection, spreading, and frontal wave dissipation. Mudflow detention basins can be very effective where the mudflow volume is relatively small and can be estimated for the design flood event.

Debris flows involve the motion of large rocks and debris characterized by destructive frontal impact surging and flow cessation on steep slopes. The mixture must contain very large concentrations of very coarse material. Dispersive stress arising from the collision of large particles (Equation (10.27)) which controls the exchange of flow momentum and energy dissipation. The fluid matrix is essentially non-cohesive. The interstitial fluid does not significantly inhibit particle contact, permitting frequent collisions and impact between the solid clasts. Debris flows originate on steep slopes and attain high velocities. The impact forces generated by fast-moving coarse material can be exceedingly destructive. Structures such as sabo dams, debris rakes, and fences are designed to separate out the debris material. The purpose of sabo dams is to arrest the frontal wave of debris, store as much solid material as possible, and drain the debris flow of the fluid matrix.

10.6.6 Velocity of hyperconcentrations

In Figure 10.16, measured values of resistance to flow in terms of the Darcy–Weisbach friction factor f from $V/u_* = \sqrt{8/f}$ are shown as a function of the relative submergence h/d_{50}. It is interesting to observe that despite the complexity of mudflows and debris flows, the ratio of $V/u*$ is rarely larger than 30. There is a slight increase in $V/u*$ with h/d_{50}, as described by the logarithmic relationship. A straight line with slope 1/6 also shows that the Manning–Strickler approach is also equally applicable. The dispersive stress relationship (Equation (10.27)) is only comparable to the measurements when $h/d_s < 50$.

For mud floods and debris flows, the mean flow velocity can be estimated from the logarithmic equation

$$V = 5.75 \log \left(\frac{h}{d_{50}} \right) (ghS)^{1/2} \qquad (10.37a)\blacklozenge$$

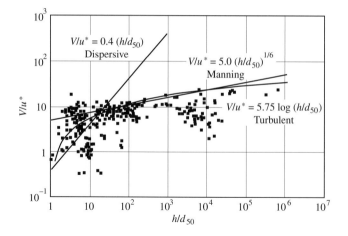

Figure 10.16. Resistance diagram for hyperconcentrated flows

Alternatively, the Manning approach with $n = 0.064\, d_{50}^{1/6}$ where d_s is the median grain diameter in meters also provides reasonable estimates. This corresponds to

$$V = 5\left(\frac{h}{d_{50}}\right)^{1/6}(ghS)^{1/2} \qquad\qquad (10.37b)$$

As a calculation example, the mean flow velocity in a steep mountain channel can be estimated from $S = 0.05$, the flow depth is 3 m and the median grain diameter is 30 mm. First the shear velocity is calculated with gravitational acceleration $g = 9.81$ m/s^2 as $u^* = \sqrt{ghS} = (9.81 \times 3 \times 0.05)^{1/2} = 1.21$ m/s. The logarithmic approach gives the following mean flow velocity of $V \cong u^*5.75 \times \log h/d_s = 1.21 \times 5.75 \times \log(3/0.03) = 13.9$ m/s. Alternatively, the Manning equation yields $V = 9.81^{1/2}h^{2/3}S^{1/2}/(0.2d_{50}^{1/6}) = 9.81^{1/2}\, 3^{2/3}\, 0.05^{1/2}/0.2 \times 0.03^{1/6} = 13$ m/s.

In the case of mudflows, the total shear stress is comparable to the yield strength and the fluid is highly viscous, such that the Bingham model can be used as a first approximation. In this case, the mean flow velocity can be approximated by $V = h\,(\tau - \tau_y)/2\mu_m$. For instance, consider a 4 m layer of mud at $C_v = 0.6$ containing a significant proportion of bentonite on a slope $S = 0.02$ at $\tau_y = 600$ Pa and $\mu_m = 1000$ Pa.s. Because the yield strength ($\tau_y = 600$ Pa) is comparable to the applied shear ($\tau = \rho_m \times g \times h \times S \sim 2{,}000$ kg/m^3 $\times 9.81$ m/s^2 $\times 4$ m $\times 0.02 = 1{,}570$ Pa), the mean flow velocity can be estimated from $V \sim 4$ m $(1{,}570 - 600)$ Pa/2$\times 1{,}000$ Pa.s $= 1.94$ m/s. Notice that this velocity of mudflows is only about two times the shear velocity. Example 10.4 provides detailed calculations of some hyperconcentrated flow characteristics. Example 10.5 provides a similar example for mudflows.

Example 10.4 Debris flow and mud flood

A mixture of uniform medium sand $d_{50} = 0.5$ mm flows on a very steep slope $S_o = 0.25$. If the flow depth is 40 cm and the volumetric concentration is $C_v = 0.4$, estimate the following: (1) the mass density of the mixture; (2) the yield stress; (3) the dynamic viscosity of the mixture; (4) the non-settling particle diameter; (5) the Hedstrom number of 5 mm gravel; (6) the fall velocity of 5 mm gravel; (7) which rheological model applies when $dv_x/dz = 50$/s; and (8) the mean flow velocity.

Step 1. The mass density from Equation (2.13)

$$\rho_m = \rho(1 + (G-1)C_v) = \frac{1,000\,\text{kg}}{\text{m}^3}(1 + (2.65-1)0.4) = \frac{1,660\,\text{kg}}{\text{m}^3}$$

Step 2. The yield stress from Equation (10.30a) and Table 10.5 for a typical soil:

$$\tau_y \cong 0.005 \times 10^{7.5 \times 0.4} = 5\,\text{Pa}$$

Step 3. The dynamic viscosity from Equation (10.30b), Table 10.5 and μ at 20°C

$$\mu_m \cong 1 \times 10^{-3} \times 10^{8 \times 0.4} = 1.6\,\text{Pa.s or } 1.6\,\text{Ns/m}^2$$

Step 4. The non-settling grain diameter from Equation (10.34)

$$d_{sb} = \frac{3\pi\,\tau_y}{2(\rho_s - \rho_m)g} = \frac{3\pi\,5\text{N} \times \text{m}^3 \times \text{s}^2}{2\,\text{m}^2(2,650 - 1,660)\,\text{kg}\,9.81\,\text{m}} = 2.4\,\text{mm}$$

Step 5. The Hedstrom number

$$\text{He} = \frac{\rho_m d_s^2 \tau_y}{\mu_m^2} = \frac{1,660\,\text{kg}}{\text{m}^3} \times \frac{25 \times 10^{-6}\text{m}^2}{(1.6)^2\,\text{Pa}^2\text{s}^2} 5\,\text{Pa} = 0.08$$

Step 6. (a) $\omega = 0$ for 0.5 mm sand because $d_s < d_{sb} = 2.4$ mm; (b) the fall velocity from Equation (10.33) for 5-mm gravel is

$$d_*^3 = \frac{\rho_m g d_s^3}{\mu_m^2}(\rho_s - \rho_m)$$

$$= \frac{1,660\,\text{kg}}{\text{m}^3} \times \frac{9.81\,\text{m}}{\text{s}^2} \times \frac{125 \times 10^{-9}\,\text{m}^3(2,650 - 1,660)\,\text{kg}}{(1.6)^2\text{Pa}^2\,.\text{s}^2\,\text{m}^3} = 0.787$$

and

$$\omega = \frac{8\,\mu_m}{\rho_m\,d_s}\left(\left(1+\frac{d_*^3}{72}-\frac{\pi\,\mathrm{He}}{48}\right)^{0.5}-1\right)$$

$$= \frac{8\times1.6\,\mathrm{Pa.s\,m^3}}{1{,}660\,\mathrm{kg}\,0.005\,\mathrm{m}}\left(\left(1+\frac{0.787}{72}-\frac{0.08\pi}{48}\right)^{0.5}-1\right) = 4.4\,\mathrm{mm/s}$$

which is equivalent to the settling velocity of very fine sand.

Step 7. The applied boundary shear τ_o is:

$$\tau_o = \rho_m\,g\,h\,S_o = \frac{1{,}660\,\mathrm{kg}}{\mathrm{m^3}}\times\frac{9.8\,\mathrm{m}}{\mathrm{s^2}}\times0.4\,\mathrm{m}\times0.25 = 1{,}628\,\mathrm{Pa}$$

Notice that the yield stress τ_y from Step 2 is small compared to the total shear stress τ_o. The dimensionless parameters from Equation (10.35) at $\frac{dv_x}{dz} = 50/s$ give

$$\Pi_\tau = \frac{\tau-\tau_y}{\mu_m\frac{dv_x}{dz}} = \frac{(1{,}628-5)\mathrm{N}\times\mathrm{m^2}\times\mathrm{s}}{\mathrm{m^2}\times1.6\,\mathrm{N}\times\mathrm{s}\times50} = 20$$

From Figure 10.15, both the viscous and yield stresses are negligible, and the dominant stress is either turbulent or dispersive.

Considering $l_m = 0.4h$, $c_{Bd} = 0.01$, and assuming the impact of particles $d_{50} = 5\,\mathrm{mm}$, in Equation (10.36c)

$$\Pi_{td} = \frac{\rho_m(0.4)^2}{c_{Bd}\rho_s}\left(\left(\frac{0.615}{C_v}\right)^{1/3}-1\right)^2\frac{h^2}{d_{50}^2}$$

$$= \frac{1{,}660\,\mathrm{kg}}{\mathrm{m^3}}\frac{(0.4)^2(0.4)^2\mathrm{m^2}\times\mathrm{m^3}}{0.01\times2{,}650\,\mathrm{kg}\,(0.005)^2\mathrm{m^2}}\left(\left(\frac{0.615}{0.4}\right)^{1/3}-1\right)^2 = 1{,}525$$

The dispersive stress is clearly negligible, and a turbulent model is applicable. Another way to check this is that the dispersive stress can only be significant when $h/d_s < 20$. In this case turbulence is the main factor.

Step 8. From Figure 10.16 the mean flow velocity of this mud flood with a boundary roughness $d_{50} = 0.5\,\mathrm{mm}$ is

$$V = \sqrt{ghS}\,5.75\,\log h/d_s$$

$$V = \sqrt{\frac{9.81\,\mathrm{m}}{\mathrm{s^2}}\times0.4\,\mathrm{m}\times0.25}\,5.75\,\log\left(\frac{0.4\,\mathrm{m}}{0.0005\,\mathrm{m}}\right) = 16.5\,\mathrm{m/s}$$

Example 10.5 Mudflow

A 3 m-thick layer of mud flows down an alluvial fan $S_0 = 0.02$. The analysis of a sample reveals that the total volumetric concentration $C_v = 0.7$ and the sample is comprised of 5% clay, 75% silt, and 20% sand with $d_{50} = 0.05$ mm. Estimate the following: (1) the mass density of the mixture; (2) the yield stress; (3) the dynamic viscosity of the mixture; (4) the non-settling particle diameter; (5) the Hedstrom number; (6) the fall velocity of a 1m boulder; and (7) the mean flow velocity.

Step 1. The mass density from Equation (2.13)

$$\rho_m = \rho(1 + (G-1)C_v) = \frac{1,000\,\text{kg}}{\text{m}^3}(1 + (2.65 - 1)0.7) = \frac{2,155\,\text{kg}}{\text{m}^3}$$

Step 2. The yield stress from Equation (10.30a) and Table 10.5

$$\tau_y \cong 0.005 \times 10^{7.5 \times 0.7} = 889\,\text{Pa}$$

The applied stress is

$$\tau = \rho_m g h S = \frac{2,155\,\text{kg}}{\text{m}^3} \times \frac{9.81\,\text{m}}{\text{s}^2} \times 3\,\text{m} \times 0.02 = 1,268\,\text{Pa}$$

The applied shear is $\tau < 5\tau_y = 1,268\,\text{Pa}$, and the yield and viscous stresses are dominant.

Step 3. From Equation (10.30b) and Table 10.5

$$\mu_m \cong 0.001 \times 10^{8 \times 0.7} = 398\,\text{Pa.s}$$

Step 4. From Equation (10.34), the non-settling particle diameter is

$$d_{sb} = \frac{3\pi}{2}\frac{\tau_y}{(\rho_s - \rho_m)g} = \frac{3\pi \times 889\,\text{Pa} \times \text{m}^3\,\text{s}^2}{2(2,650 - 2,155)\,\text{kg} \times 9.81\,\text{m}} = 860\,\text{mm}$$

It is interesting to note that medium boulders $d_s < 0.85\,\text{m}$, would not settle in this mud.

Step 5. The Hedstrom number for a 1 m boulder

$$\text{He} = \frac{\rho_m d_s^2 \tau_y}{\mu_m^2} = \frac{2,155\,\text{kg}}{\text{m}^3} \times \frac{1\,\text{m}^2 \times 889\,\text{Pa}}{398^2\,\text{Pa}^2\,\text{s}^2} = 12.1$$

Step 6. The settling velocity of a 1 m boulder from Equation (10.33) is

$$\omega = \frac{8\mu_m}{\rho_m d_s}\left[\left(1+\frac{\rho_m g\, d_s^3(\rho_s-\rho_m)}{72\mu_m^2}-\frac{\pi}{48}\mathrm{He}\right)^{0.5}-1\right]$$

$$\omega = \frac{8\times398\,\mathrm{Pa.s\,m^3}}{2,155\,\mathrm{kg\,1\,m}}\left[\left(1+\frac{2,155\,\mathrm{kg}}{\mathrm{m^3 s^2}}\times\frac{9.81\,\mathrm{m}}{\mathrm{m^3}}\times\frac{1\,\mathrm{m^3}(2,650-2,155)\mathrm{kg}}{72\times398^2\,\mathrm{Pa^2 s^2}}\right.\right.$$

$$\left.\left.-\frac{\pi}{48}\times12.1\right)^{0.5}-1\right]=0.09\,\mathrm{m/s}$$

Notice that the hyperconcentrated sediment mixture reduced the settling velocity of a 1 m-diameter boulder to the settling velocity of an equivalent sand particle in clear water.

Step 7. The yield stress is a significant fraction of the total shear stress. The Bingham model is applicable. The mean flow velocity will be approximately

$$V\cong\frac{h(\tau-\tau_y)}{2\mu_m}=\frac{3\,\mathrm{m}\,(1,268-889)\,\mathrm{Pa}}{2\times398\,\mathrm{Pa.s}}=1.42\,\mathrm{m/s}$$

This velocity is only twice the shear velocity

$$u_*=\sqrt{ghS}=\sqrt{\frac{9.81\,\mathrm{m}\times3\,\mathrm{m}\times0.02}{\mathrm{s^2}}}=0.76\,\mathrm{m/s}$$

Mudflows are thus observed when τ_y is large compared to τ and viscosity is very high.

10.7 Field measurements of suspended sediment

The quantity of sediments held in suspension in a stream can be measured from representative samples of the water–sediment mixture. Samples are divided into three types according to the desired type of concentration measurement discussed in Section 10.1. Additional information can be found in Edwards and Glysson (1988) and Shen and Julien (1993).

10.7.1 Instantaneous samplers

Instantaneous samplers trap a volume of the suspension flowing through a cylindrical tube by simultaneously closing off both ends. Bucket sampling from the free surface also gives an instantaneous sample. The sample is then filtered and dried to provide a measure of volume-averaged sediment concentration, $C_\forall$.

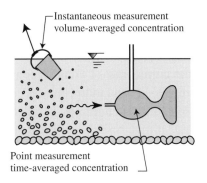

Figure 10.17. Instantaneous and point sampling

Instruments of a container type that can be opened and closed instantaneously have been designed. Studies of their effectiveness indicated that they are not suitable for general field use. They are best used in reservoir studies and at river locations where the particle size in suspension is sufficiently small to prevent settling at low turbulence levels.

10.7.2 Point samplers

Point samplers are designed to collect through time a sample at a given point in the stream vertical as sketched in Figure 10.17. The dried sample measures the time-averaged sediment concentration, C_t. The body of the sampler contains air which is compressed by the inflowing liquid so that its pressure balances the external hydrostatic head. A remotely operated rotary valve opens and closes the sampler. During the sampling period, the valve is opened and the air escapes the sampler at a nozzle intake velocity nearly equal to the local stream velocity.

Point samplers such as the P-61 (100 lb), the P-63 (200 lb) and the P-50 (300 lb) are commonly used to provide information on sediment concentration and particle size distribution along a vertical. The P-61 uses a pint milk bottle container while the P-63 is designed to use either the pint or the quart sample container. The P-50 is very similar to the P-63 and has been developed for use on major streams such as the lower Mississippi River. The capacity of the container is a quart.

10.7.3 Depth-integrating samplers

Integrating samplers move vertically at a constant speed with an upstream-pointed nozzle. Good samplers are designed to maintain isokinetic conditions. Accordingly, the sampling intake velocity equals the natural flow velocity at every point. The sample is then dried to measure the flux-averaged concentration, C_f.

Depth-integrating samplers accumulate water and sediment as they are lowered or raised along the vertical at a uniform rate. The air in the sample container is compressed by the inflowing liquid. If the lowering speed is such that the rate of air compression exceeds the normal rate of liquid inflow, the actual rate of inflow will exceed the local stream velocity and inflow may occur through the air exhaust. If the transit speed is too low, the container will be filled before the total depth is reached. In practice, the transit time must be adjusted so that the container is not completely filled. The descending and ascending velocities need not be equal, but the velocity must remain constant during each phase.

Specifically, the lowering rate should not exceed 0.4 of the mean flow velocity to avoid excessive angles between the nozzle and the approaching flow. Laboratory tests provide useful information on the recommended lowering rates R_T as a function of the flow velocity v and the flow depth for a 1 pint container in a depth-integrating sampler with three intake nozzle diameters (Figure 10.18).

The DH-48 sampler (4.5 lb) is used in shallow flows when the unit discharge does not exceed 1 m²/s. The DH-59 (24 lb) is used in streams with low velocities and depths beyond the wading range. The D-49 (62 lb) with cable suspension is designed for use in streams beyond the range of hand-operated equipment.

10.7.4 Sediment discharge measurements

Sediment transport in natural streams is very important and the accuracy of sediment discharge measurements not only depends on the field methods and equipment utilized for data collection, but also upon representative measurements of the sediment distribution in the flow. In natural streams, the sediment concentration at a given cross-section varies in both the vertical and the transversal directions and also changes with time. Therefore, the concentration of sediment in suspension may increase or decrease both in space and time although the flow remains steady in a straight uniform reach.

The standard procedure used for the measurement of the rate of sediment transport at a given instant is essentially based on the definition of the advective flux-averaged concentration Equation (10.2c); hence

$$Q_s = C_f Q = \int_A C v_x \, dA \qquad (10.38)$$

in which A is the cross-sectional area; C_f is the flux-averaged concentration and v_x is the velocity of sediment particles.

In small streams it is sometimes possible to measure the total stream discharge by using one of the following methods: (1) volumetric method; (2) dilution method;

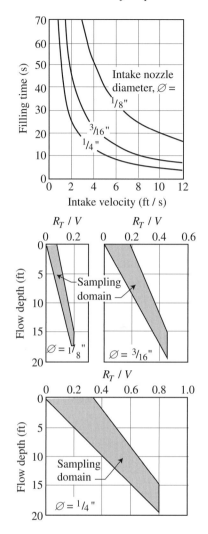

Figure 10.18. Depth-integrating sampling characteristics

or (3) weir equation. In the first method, if the total flow of a small stream can be directed into a given basin or container of fixed volume, the ratio of the volume to the filling time determines the time-averaged discharge entering the container.

With the second method, a dilute suspension at concentration C_1 of a conservative substance is steadily injected in a stream at a constant rate Q_1 into a small turbulent stream of unknown discharge. Note that C_1 represents the difference in concentration between the solution and the natural concentration of the stream. With reference to Section 10.3, a sample of water is taken at a certain distance, downstream of the source point where the vertical and transversal mixing are completed.

With a measure of the uniform concentration C_2 of the sample, the discharge Q_2 of the stream can be simply obtained by $Q_2 = C_1Q_1/C_2$.

With the third method, man-made structures with a fixed geometry, such as weirs, spillways, and pipes are first located. The discharge, through these elements, is then evaluated from the geometry, stage, and velocity measurements. For example, the weir equation can be used with simple width and stage measurements.

The flux-averaged concentration in small streams can often be measured at locations where the turbulence intensity is very high such as plunge pools, downstream of contractions, or in turbulence flumes. Particles from all size fractions are held in suspension and a bulk sample provides an accurate evaluation of the flux-averaged sediment concentration. Turbulence flumes consisting of series of baffles anchored to a concrete slab can be installed. The turbulence induced by the baffles is sufficient to transport in suspension almost the entire load in the stream. In streams where mostly fine sediments (silts and clays) are held in suspension, the exponent of the Rouse equation is generally very small and uniform concentration of sediment can be assumed for those fractions (e.g. Table 10.2 when $u_*/\omega > 25$). In such cases, a sample taken at any location along the vertical is sufficient to describe the flux-averaged concentration for those size fractions.

Exercises

◆10.1 Derive concentration by weight C_w (Equation 10.1b) from concentration by volume C_v (Equation 10.1a) given the density of sediment particles $G = \gamma_s/\gamma$. Also write C_v as a function of C_w.

◆10.2 (a) Derive Equation (10.12) from Equation (10.7) for the steady point source;
(b) Derive Equation (10.19) from Equation (10.18); and
(c) Evaluate the maximum value of ε_z on a vertical.

◆◆10.3 Define the Rouse number from concentration measurements at 25% and 75% of the flow depth from Equation (10.20a).

◆10.4 Derive the settling velocity of hyperconcentrations in Equation (10.33) from Equations (10.31) and (10.32).

Problems
Problem 10.1

Assume $a = 2d_s$ and plot the dimensionless concentration profiles C/C_a for medium silt, fine sand, and coarse sand in a 3 m-deep stream sloping at $S_o = 0.002$.

(*Answer: C/C_a* > 0.95 for medium silt; $C/C_a \simeq 0.36$ at mid-depth for fine sand; and $C/C_a < 0.05$ for most of the coarse sand profile.)

◆◆Problem 10.2

Given the sediment concentration profile from Problem 6.1: (a) plot the concentration profile log *C* versus log *(h − z)/z*; (b) estimate the particle diameter from the Rouse number; and (c) determine the unit sediment discharge from the given data.

◆Problem 10.3

Calculate the daily sediment load in a near-rectangular 50 m-wide stream with an average flow depth $h = 2$ m and slope $S_o = 0.0002$ when 25% of the sediment load is fine silt, 25% is very fine silt and 50% is clay, and the mid-depth concentration is $C = 50,000$ mg/l. (*Answer: $L_s \equiv 750,000$ tons/day.*)

◆Problem 10.4

A physical model of the stream in Problem 10.3 is to be constructed in the hydraulics laboratory at a scale of 1:100 horizontal and 1:20 vertical. Calculate the ratio of transversal to vertical mixing time scales: (a) for the model; and (b) for the prototype. Also calculate the mixing coefficients and length scales for vertical and lateral mixing for the prototype. At what downstream distance would longitudinal dispersion become the dominant process? Are vertical and lateral mixing complete at the same location in the model and prototype?

◆Problem 10.5

Calculate the length required for complete transversal mixing in the St-Lawrence river at a discharge of 500,000 cfs. Assume an average river width of 2 miles, a slope of about 0.4 ft/mile and Manning $n = 0.02$. (*Answer*: More than 30,000 miles)

◆◆Problem 10.6

The table in Problem 6.8 gives the velocity distribution and the suspended sand concentration for the fraction passing the 0.105 mm sieve and retained on the 0.074 mm sieve on the Missouri river. Determine the following: (a) plot the velocity profile *V* versus log *z* and concentration log *C* versus log *(h − z)/z*; and (b) compute from the graphs and given data the following: $u_* =$ shear velocity, V = mean velocity, κ = von Kármán constant, f = Darcy–Weisbach friction factor, and Ro = Rouse number; (c) compute the unit sediment discharge for this size fraction from field measurements; (d) calculate the flux-averaged concentration; and (e) estimate the near-bed concentration.

◆Problem 10.7

Bank erosion of fine silts occurs on a short reach of a 100 m-wide meandering river at a discharge of 750 m³/s. The riverbed slope is 50 cm/km and the flow depth is 5 m. The mass wasted is about 10 metric tons per hour. Determine the distance required for complete mixing in the river, the maximum concentration at that point, and the average sediment concentration.

(*Answer:* $X_t \approx 19$ km; $C_{max} = 4.3$ mg/l; $\bar{C} = 3.7$ mg/l.)

◆◆Problem 10.8

Consider the mudflow characteristics at different locations along Rudd Creek, Utah in the table below. Determine the following: (a) plot the yield strength, viscosity, and velocity measurement; (b) compare the yield strength and viscosity with Table 10.5 on Figures 10.13 and 10.16; (c) if the flow velocity increases linearly with depth, determine $\partial v_x/\partial z$; (d) calculate the dimensionless parameters from Equation (10.36), assuming $C_v^* = 0.8$; (e) what is the relative magnitude of the different shear stresses? (f) what is the diameter of particles that are non-settling? and (g) what is the settling velocity of a 1 m boulder?

Depth (m)	Bed slope	Grain size (m)	τ_y (Pa)	μ_m (Pa.s)	C_v	ρ_m (kg/m³)	Observed velocity (m/s)
2	0.09	0.1	1,250	200	0.75	2,237	3
0.4	0.09	0.01	200	25	0.65	2,072	4.3
0.5	0.09	0.01	180	20	0.64	2,056	2.8
0.3	0.09	0.004	30	5	0.52	1,858	3
0.05	0.09	0.001	0	0.1	0.1	1,165	1.5

◆Problem 10.9

Consider the lahar flow data shown below. Calculate the shear velocity and plot the results assuming $d_s = 1$ mm on Figure 10.16.

Slope (mm/m)	Velocity (m/s)	Depth (m)
167	9	11.5
79	7.8	14.9
113	13.9	12.2
32	8.7	11.9
18	6.4	7.4

♦Problem 10.10

Consider the lahar flow data at Mt. St. Helens below. At a median grain size of 5 mm, calculate $u_* = \sqrt{g h S_f}$ and plot the velocity on Figure 10.16.

Depth (m)	Discharge (m³/s)	Width (m)	Velocity (m/s)	Slope
15.2	26,800	106	16.6	0.092
10.6	25,900	99	24.6	0.065
14.5	28,200	148	13.1	0.041
14.9	21,700	117	12.4	0.042
14.8	19,900	123	10.9	0.043
13.9	21,000	106	14.2	0.036
10.7	19,200	85	21.1	0.031
9.4	16,600	116	15.2	0.026
9.3	6,250	72	9.3	0.027
6.0	7,320	100	12.2	0.03

Problem 10.11

Consider the data from the Rio Grande at Bernalillo on June 4, 1953 for Culbertson *et al.* (1972). The total flow depth is 3.5 ft, the channel top width is 270 ft, energy grade line S_f is 0.00083 ft/ft, mean particle size is approximately

z (ft)	v (ft/s)	C (ppm)
0.3	3.72	2,359
0.8	4.09	1,809
1.3	4.36	1,540
1.7	4.36	1,240
2.2	4.25	1,151

0.69 mm, and the temperature is 62°F. The velocity and concentration profiles are provided in the table above. Answer the same questions as Problem 10.6, and determine the flux-average concentration.

♦♦Problem 10.12

Consider the sediment concentration profile by size fraction of the Enoree River, S. Carolina in the table below. Plot the Rouse profiles for each size fraction in mm and find the Rouse numbers. Also determine the total sediment concentration at each point, plot the Rouse profile and determine the Rouse number and the concentration at $a = 2d_s \approx 1$ mm about the bed. The flow depth is 5 ft, the slope is 0.00084 and the water temperature is 6.2°C.

Elevation z (ft)	Velocity v (ft/s)	Concentration by size fraction (ppm)					
		0.074–0.124 (mm)	0.124–0.175	0.175–0.246	0.246–0.351	0.351–0.495	0.495–0.701
0.1		2.4	8.1	33	86	285	422
0.15	1.87	2.4	8.5	45	67	131	82
0.45	2.30	2.3	3.8	17	44	95	89
0.75	2.15	1.4	5.5	17	33	65	59
1.35	2.90	1.5	4.8	13	26	38	30
1.95	2.85	2.2	5.3	14	18	27	20
2.55	3.25	7.1	10.3	15	14	10	10
3.15	3.63	2.8	3.9	9	10	10	7

♦Problem 10.13

Examine the possibility of removing the sill retaining about 37,000 m³ of contaminated sediment during a 100-year flood that would last 12 hours at a discharge close to 355 m³/s. The channel is 25 m wide, 3.8 m deep, and $S = 0.005$. The mean flow velocity is 3.7 m/s. The riverbed material is $d_{10} = 6$ mm, $d_{50} = 90$ mm. The alluvial wedge sketched in Figure P-10.13 is finer with 80% of the material finer than 2 mm, 16% between 2 and 16 mm and 4% coarser than 16 mm.

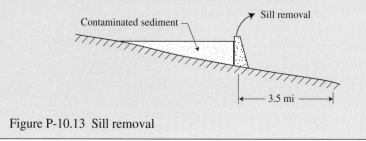

Figure P-10.13 Sill removal

Determine the following: (1) shear velocity; (2) Darcy–Weisbach f; (3) Manning n; (4) which size fractions will move as bedload/suspension; (5) length and time for vertical mixing; (6) length and time for lateral mixing; (7) initial time for dispersion and compare with flood duration; (8) is dispersion important?; (9) bedload transport rates in m³/s for particles between 2 and 16 mm; (10) methods from Chapter 11 that can estimate the bed material load (*answer*: $q_{bw} \simeq 150\,\text{N/ms}$); and (11) if 30,000 m³ are removed during the 12 h flood, determine the maximum sediment concentration of contaminated sediment (*answer*: $C_{max} \simeq 5{,}000\,\text{mg}/l$).

11

Total load

This chapter describes methods to estimate the total sediment load of a river. In Section 11.1, several sediment transport formulas commonly used in engineering practice are presented with detailed calculation examples. Section 11.2 discusses sediment-rating curves; Section 11.3 covers short- and long-term sediment load. Section 11.4 focuses on sediment sources and sediment yield from upland areas. This chapter includes three case studies of capacity- and supply-limited sediment transport.

Every sediment particle which passes a given stream cross-section must satisfy the following two conditions: (1) it must have been eroded somewhere in the watershed above the cross-section; and (2) it must be transported by the flow from the place of erosion to the cross-section. To his statement, Einstein (1964) added that each of these two conditions may limit the rate of sediment transport depending on the relative magnitude of the two controls: (1) the transporting capacity of the stream; and (2) the availability of material in the watershed. The amount of material transported in a stream therefore depends on two groups of variables: (1) those governing the sediment transport capacity of the stream such as channel geometry, width, depth, shape, wetted perimeter, alignment, slope, vegetation, roughness, velocity distribution, tractive force, turbulence, and uniformity of discharge; and (2) those reflecting the quality and quantity of material made available for transport by the stream including watershed topography, geology, magnitude-intensity-duration of rainfall and snowmelt, weathering, vegetation, cultivation, grazing and land use, soil type, particle size, shape, specific gravity, resistance to wear, settling velocity, mineralogy, cohesion, surface erosion, bank cutting, and sediment supply from tributaries.

The sediment transport capacity of a stream under unlimited sediment supply can be determined as a function of the hydraulic variables and the shape of the stream cross-section. As sketched in Figure 11.1, the total sediment load in a stream can be divided three different ways:

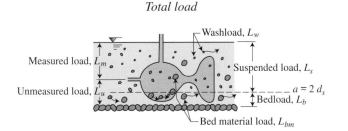

Figure 11.1. Sketch of ways to determine the total load

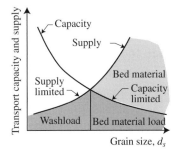

Figure 11.2. Sediment transport capacity and supply

(1) *By the type of movement.* The total sediment load L_T can be divided into the bedload L_b treated in Chapter 9, and the suspended load L_s (covered in Chapter 10).

$$L_T = L_b + L_s \qquad (11.1a)$$

(2) *By the method of measurement.* In this case, the total sediment load L_T consists of the measured load L_m and the unmeasured load L_u. Because one can only use point samples from the water surface to a distance of approximately 10 centimeters above the bed surface, the measured sediment load L_m is only part of the suspended load L_s. The unmeasured sediment load L_u consists of the entire bedload L_b plus the fraction of the suspended load L_s, transported below the lowest sampling elevation.

$$L_T = L_m + L_u \qquad (11.1b)$$

Helley–Smith samplers can measure part of the unmeasured load for the size fractions coarser than the mesh size of the sampling bag.

(3) *By the source of sediment.* In this case, the total sediment load L_T is equal to the fine sediment fraction coming from upstream, also called washload L_w plus the coarser grain sizes from the bed material load L_{bm}. The d_{10} of the bed material is therefore commonly used as the breakpoint between washload ($d_s < d_{10}$) and bed material load ($d_s > d_{10}$). As sketched in Figure 11.2, washload sediment transport is limited by the upstream supply of fine particles. The bed material load is determined by the capacity

of the flow to transport the sediment sizes found in the bed.

$$L_T = L_w + L_{bm} \tag{11.1c}$$

It must be recognized that it is difficult to determine the total sediment load of a stream, from any Equations (11.1 a–c). Because the washload depends on sediment supply rather than transport capacity, it is impossible to determine the total sediment load from the sediment transport capacity based on flow characteristics alone.

11.1 Sediment transport capacity

Numerous sediment transport formulas have been proposed in the past fifty years and subsequent modifications of original formulations have been prescribed. Although significant progress has been made, none of the existing sediment transport formulas can determine the total load. In engineering practice, several formulas are compared with field observations to select the most appropriate equation at a given field site.

For given streamflow conditions, any sediment transport equation can only predict the sediment transport capacity of a given bed sediment mixture. Existing sediment transport formulas can be classified into several categories owing to their basic approaches: (1) formulation based on advection; (2) formulation based on energy concepts in which the rate of work done for transporting sediment particles in turbulent flow is related to the rate of energy expenditure; and (3) graphical methods and empirical equations based on regression analysis.

The following methods reflect more recent developments in sediment transport calculations. Einstein's method is still viewed as a landmark, despite the complexity of the advection-based procedure and some recent developments include the series expansion of the Einstein integrals by Guo and Julien, and the improvement of the Modified Einstein Procedure. The method of Simons, Li, and Fullerton was derived from Einstein's method and offers simplicity in the calculations for steep sand-bed channels. Four methods based on energy and stream power concepts are presented. The methods of Bagnold and Engelund–Hansen offer simplicity while those of Ackers–White and Yang gained popularity in computer models. Finally the methods of Shen–Hung, Brownlie and Karim–Kennedy are essentially the result of regression analysis using comprehensive data sets.

11.1.1 Einstein's approach

The total bed sediment discharge per unit width q_t can be calculated from the sum of the unit bed sediment discharge q_b and the unit suspended sediment discharge q_s

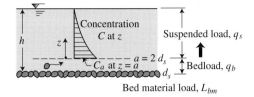

Figure 11.3. Sketch of the Einstein approach

from Equation (10.21). As sketched in Figure 11.3, the Einstein approach estimates the suspended load from the bedload.

$$\text{Thus, } q_t = q_b + \int_a^h C\,v_x\,dz \tag{11.2}$$

Given the velocity profile for a hydraulically rough boundary from Equation (6.16) with $k'_s = d_s$ and the Rouse concentration profiles from Equation (10.20a), the total unit bed sediment discharge is written as:

$$q_t = q_b + \int_a^h C_a \frac{u_*}{\kappa} \left(\frac{h-z}{z} \frac{a}{h-a} \right)^{\frac{\omega}{\beta_s \kappa u_*}} \ln\left(\frac{30z}{d_s} \right) dz \tag{11.3}\blacklozenge$$

The reference concentration $C_a = q_b/a\,v_a$ is obtained from the unit bed sediment discharge q_b transported in the bed layer of thickness $a = 2d_s$, given the velocity v_a at the top of the bed layer, $v_a = (u_*/\kappa) \ln (30a/d_s) = 4.09\,u_*/\kappa$, Einstein used $v_a = 11.6\,u_*$. Rewriting Equation (11.3) in dimensionless form with $z^* = z/h$, $E = 2d_s/h$ and $\text{Ro} = \omega/\beta_s \kappa\,u_*$ gives:

$$q_t = q_b \left[1 + I_1 \ln \frac{30h}{d_s} + I_2 \right] \tag{11.4a}$$

where

$$I_1 = 0.216 \frac{E^{\text{Ro}-1}}{(1-E)^{\text{Ro}}} \underbrace{\int_E^1 \left[\frac{1-z^*}{z^*} \right]^{\text{Ro}} dz^*}_{J_1(\text{Ro})} \tag{11.4b}$$

$$I_2 = 0.216 \frac{E^{\text{Ro}-1}}{(1-E)^{\text{Ro}}} \underbrace{\int_E^1 \left[\frac{1-z^*}{z^*} \right]^{\text{Ro}} \ln z^*\,dz^*}_{J_2(\text{Ro})} \tag{11.4c}$$

The two integrals I_1 and I_2 have been solved with the use of nomographs prepared by Einstein. The reader is referred to Appendix A for the details of the original method.

11.1.2 Guo and Julien's method

Recent developments include the series expansion of the Einstein integrals in Equation (11.4) by Guo and Julien (2004). For $u_* > 2\omega$, the function of $E = 2d_s/h$ and Rouse number $\text{Ro} = \omega/\beta_S \kappa u_*$ is rapidly convergent as soon as the series expansion term $k > 1 + \text{Ro}$. There are four steps for this method:

Step 1. Estimate F_1 (Ro) using a maximum of 10 terms

$$F_1(\text{Ro}) = \int_0^E \left(\frac{1-z^*}{z^*}\right)^{\text{Ro}} dz^* = \frac{(1-E)^{\text{Ro}}}{E^{\text{Ro}-1}} - \text{Ro} \sum_{k=1}^{10} \frac{(-1)^k}{k} - \text{Ro}\left(\frac{E}{1-E}\right)^{k-\text{Ro}}$$

(11.5a)

Step 2. Estimate J_1 (Ro) from Equation (11.4b)

$$J_1(\text{Ro}) = \int_E^1 \left(\frac{1-z^*}{z^*}\right)^{\text{Ro}} dz^* = \frac{\pi \, \text{Ro}}{\sin(\pi \, \text{Ro})} - F_1(\text{Ro})$$

(11.5b)

Step 3. Estimate F_2 (Ro) with a maximum of 10 terms

$$F_2(\text{Ro}) = \int_0^E \left(\frac{1-z^*}{z^*}\right)^{\text{Ro}} \ln z^* \, dz^* = F_1 \text{Ro}\left(\ln E + \frac{1}{\text{Ro}-1}\right)$$

$$+ \text{Ro} \sum_{k=1}^{10} \frac{(-1)^k F_1(\text{Ro}-k)}{(\text{Ro}-k)(\text{Ro}-k-1)}$$

(11.5c)

Step 4. Estimate J_2 (Ro) from Equation (11.4c)

$$J_2(\text{Ro}) = \int_E^1 \left(\frac{1-z^*}{z^*}\right)^{\text{Ro}} \ln z^* \, dz^* = \frac{\text{Ro}\pi}{\sin \text{Ro}\pi}$$

$$\times \left\{\pi \cot(\text{Ro}\pi) - 1 - \frac{1}{\text{Ro}} + \frac{\pi^2}{6} \frac{\text{Ro}}{(1+\text{Ro})^{0.7162}}\right\} - F_2(\text{Ro}) \quad (11.5d)$$

This formulation is very accurate for any non-integer value of Ro.

For integer values of the Rouse number, $\text{Ro} = n \pm 10^{-3}$, the formulations for J_1(Ro) and J_2 (Ro) become:

$$J_1(\text{Ro} = n) = \sum_{k=0}^{n-2\geq 0} \frac{(-1)^k n!}{(n-k)!k!} \frac{E^{k-n+1}-1}{n-k-1} + (-1)^n (n \ln E - E + 1) \quad (11.6a)$$

and

$$J_2(\mathrm{Ro}=n) = \sum_{k=0}^{n-2\geq0} \frac{(-1)^k\, n!}{(n-k)!k!} \cdot \left\{ \frac{E^{1+k-n}\ln E}{n-k-1} + \frac{E^{1+k-n}-1}{(n-k-1)^2} \right\}$$

$$+ (-1)^n \left\{ \frac{n}{2}\ln^2 E - E\ln E + E - 1 \right\} \tag{11.6b}$$

For instance, for $\mathrm{Ro} = n = 1$, $J_1(1) = E - 1 - \ln E$, and for $\mathrm{Ro} = n = 3$, $J_1(3) = -3\ln E + \frac{1}{2E^2} - \frac{3}{E} + \frac{3}{2} + E$. An application example is presented in Case study 11.1.

In a very simplified form, the sediment transport when $u_* < 2\omega$ can also be approximated by

$$q_\tau = q_b \left[1 + \left(\frac{u_*}{\omega}\right)^2 \right] \tag{11.6c}\blacklozenge$$

This approach implies $q_s/q_b = (u_*/\omega)^2$ as shown in Figure 10.10. It is recommended when $u_* < 2\omega$, or $\mathrm{Ro} > 1.25$.

11.1.3 Simons, Li, and Fullerton's method

Simons, Li, and Fullerton (1981) developed easy-to-apply power relationships that estimate sediment transport based on the flow depth h and velocity V. The power relationships were developed from a computer solution of the Meyer-Peter and Muller bedload transport equation and Einstein's integration of the suspended bed sediment discharge

$$q_s = c_{S1}\, h^{c_{S2}}\, V^{c_{S3}} \tag{11.7}$$

The results of the total bed sediment discharge are presented in Table 11.1. The high values of c_{S3} $(3.3 < c_{S3} < 3.9)$ show that sediment transport rates depend highly on velocity. The influence of depth is comparatively less important $(-0.34 < c_{S2} < 0.7)$. For flow conditions within the range outlined in Table 11.2, the regression equations should be accurate within ten percent. The equations were obtained for steep sand- and gravel-bed channels under supercritical flow. They do not apply to cohesive material. The equations assume that all sediment sizes are transported by the flow without armoring. Case study 11.1 provides sample calculations of sediment transport using these power relationships.

Table 11.1. *Power equations for total bed sediment discharge in sand- and fine gravel-bed streams*

$q_s = c_{S1} h^{cS2} V^{cS3}$	d_{50} (mm)							
	0.1	0.25	0.5	1.0	2.0	3.0	4.0	5.0
Gr = 1.0								
c_{S1}	3.30×10^{-5}	1.42×10^{-5}	7.6×10^{-6}	5.62×10^{-6}	5.64×10^{-6}	6.32×10^{-6}	7.10×10^{-6}	7.78×10^{-6}
c_{S2}	0.715	0.495	0.28	0.06	−0.14	−0.24	−0.30	−0.34
c_{S3}	3.30	3.61	3.82	3.93	3.95	3.92	3.89	3.87
Gr = 2.0								
c_{S1}		1.59×10^{-5}	9.8×10^{-6}	6.94×10^{-6}	6.32×10^{-6}	6.62×10^{-6}	6.94×10^{-6}	
c_{S2}		0.51	0.33	0.12	−0.09	−0.196	−0.27	
c_{S3}		3.55	3.73	3.86	3.91	3.91	3.90	
Gr = 3.0								
c_{S1}			1.21×10^{-5}	9.14×10^{-6}	7.44×10^{-6}			
c_{S2}			0.36	0.18	−0.02			
c_{S3}			3.66	3.76	3.86			
Gr = 4.0								
c_{S1}				1.05×10^{-5}				
c_{S2}				0.21				
c_{S3}				3.71				

Definitions: q_s, unit sediment transport rate in ft^2/s (unbulked); V, velocity in ft/s; h, depth in feet; Gr, gradation coefficient.

Table 11.2. *Range of parameters for the Simons–Li–Fullerton method*

Parameter	Value range
Froude number	1 – 4
Velocity	6.5 – 26 (ft/s)
Manning n	0.015 – 0.025
Bed slope	0.005 – 0.040
Unit discharge	10 – 200 (ft^2/s)
Particle size	$d_{50} \geq 0.062$ mm
	$d_{90} \leq 15$ mm

Case study 11.1 Big Sand Creek, United States

This case study illustrates sediment transport calculations when $u_*/\omega < 2$. A test reach of the Big Sand Creek near Greenwood, Mississippi has been used for bed sediment discharge calculations by size fractions using the methods of Einstein, Guo and Julien, and Simons–Li–Fullerton. The sand-bed channel has a bed slope $S_o = 0.00105$. An average of four bed sediment samples is shown in the following table, showing that 95.8% of the bed material is between $d_s = 0.589$ mm and $d_s = 0.147$ mm, which is divided into four size fractions for the calculations:

Grain size distribution mm	Average grain size			Settling velocity	
	mm	ft	Δp_i in %	cm/sec	ft/s
$d_s > 0.60$	—	—	2.4	—	—
$0.60 > d_s > 0.42$	.50	0.00162	17.8	6.2	0.205
$0.42 > d_s > 0.30$	0.36	0.00115	40.2	4.5	0.148
$0.30 > d_s > 0.21$	0.25	0.00081	32.0	3.2	0.106
$0.21 > d_s > 0.15$	0.18	0.00058	5.8	2.0	0.067
$0.15 > d_s$	—	—	1.8	—	—

$d_{16} = 0.24$ mm $\sigma_g = 1.35$
$d_{35} = 0.29$ mm $Gr = 1.35$
$d_{50} = 0.34$ mm
$d_{65} = 0.37$ mm
$d_{84} = 0.44$ mm

Einstein's method

Calculations by size fractions using the Einstein method are detailed in Appendix A, given the cross-sectional geometry information from Figure A.7.

Guo and Julien's method

Calculations in English units are based on $d_{50} = 0.34 \text{ mm} = 1.1 \times 10^{-3}$ ft, or $\omega = 0.164$ ft/s. At a slope $S = 0.00105$, the shear velocity is $u_* = \sqrt{ghS} = 0.184 h^{1/2}$ as a function of flow depth h in feet. The Rouse number is $\text{Ro} = \omega/0.4u_*$ and the Shields parameter is $\tau_* = 0.57 h$.

The unit bed load by volume in ft²/s is $q_{bv} = 15\tau_*^{1.5}\omega d_s = 1.18 \times 10^{-3}h^{1.5}$. The unit bed material load by volume is obtained from Eq. (11.4) with $E = 2d_s/h$

$$q_{tv} = q_{bv}\left\{1 + \underbrace{\frac{0.216 E^{\text{Ro}-1}}{(1-E)^{\text{Ro}}}[J_1(\text{Ro})\ln(60/E) + J_2(\text{Ro})]}_{q_{Sv}/q_{bv}}\right\}$$

and the total sediment load in tons per day is

$$Q_{tw} = q_{tv} \times W \times \gamma_s = q_{tv}\left(\text{in ft}^2/\text{s}\right) \times W\,(\text{in ft}) \times 2.65$$

$$\times \frac{62.4\,\text{lb}}{\text{ft}^3} \times \frac{\text{ton}}{2,000\,\text{lb}} \times \frac{86,400\,\text{s}}{d}$$

$$= 8.43 h^{1.5} W (1 + q_{Sv}/q_{bv})$$

For instance, at a flow depth $h = 4.14$ ft, $u_* = 0.374$ ft/s, $\text{Ro} = 1.096$, $\tau_* = 2.36$, $q_{bv} = 0.01$ ft²/s, $E = 5.3 \times 10^{-4}$, $J_1 = 9.863$, $J_2 = -45.19$, and $q_{tv} = 8.28 q_{bv}$. At a channel width $W = 234$ ft, the daily bed material load $Q_{tw} = 137,000$ tons per day.

It is instructive to notice from this example that the ratio of the suspended load q_{Sv} to the bedload q_{bv} is very roughly proportional to the square of u_*/ω. This could be inferred from Figure 10.10 and is a reasonable approximation when $u_* < 2\omega$. The Einstein method is less accurate when $u_* > 2\omega$ because the suspended load q_{Sv} is much larger than the bedload q_{bv}.

W (ft)	h (ft)	E	Ro	u_*/ω	J_1	J_2	q_{Sv}/q_{bv}	Q_s (tons/day)
103	1.36	1.6×10^{-3}	1.91	1.31	375	−2,029	1.18	3,010
136	1.76	1.25×10^{-3}	1.67	1.48	126	−675	1.67	7,150
170	2.50	8.8×10^{-4}	1.41	1.77	39	−203	2.79	21,500
194	3.30	6.7×10^{-4}	1.226	2.04	17.2	−84	4.64	55,300
234	4.14	5.3×10^{-4}	1.096	2.28	9.86	−45.2	7.28	137,000

Simons, Li, and Fullerton's method

This method does not involve calculations by size fractions. However, the coefficients of the bed sediment transport equation can be interpolated from Table 11.1 given the gradation coefficient Gr and the median grain size d_{50}. In the case in point, $d_{50} = 0.34$ mm and $Gr = 1.35$, one obtains

$$q_s(\text{in ft}^2/\text{s}) = 1.25 \times 10^{-5} h^{0.43}(\text{in ft}) V^{3.65}(\text{in ft/s})$$

or

$$Q_s(\text{tons/d}) = 0.0893 \, W(\text{ft}) \, h^{0.43}(\text{ft}) \, V^{3.65} \, (\text{ft/s})$$

				Sediment load (tons/day)		
W ft	h ft	V ft/s	Q ft³/s	Q_s Einstein	Q_s Guo & Julien	Q_s Simons et al.
103	1.36	2.92	409	670	3,010	525
136	1.76	4.44	1,063	3,940	7,150	3,580
170	2.50	6.63	2,818	30,500	21,500	22,400
194	3.30	8.40	5,377	113,000	55,300	68,400
234	4.14	9.92	9,610	324,000	137,000	167,000

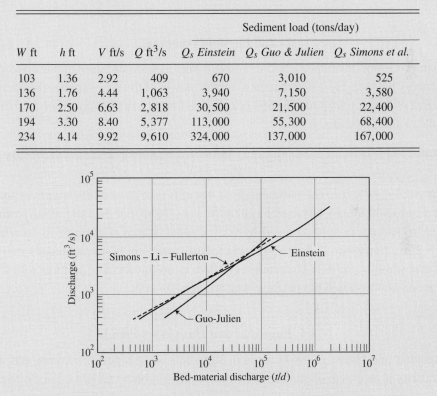
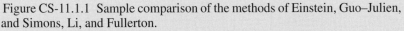

Figure CS-11.1.1 Sample comparison of the methods of Einstein, Guo–Julien, and Simons, Li, and Fullerton.

The results of bed sediment discharge calculations for Big Sand Creek using the methods of Einstein, Guo–Julien, and Simons–Li–Fullerton are shown on Figure CS-11.1.1. In all cases, the bed sediment discharge increases very rapidly with water discharge.

The following methods predict sediment transport as a function of mean flow velocity V. This requires knowledge of resistance to flow, usually in the presence of bedforms. At a given stage, bedforms increase resistance to flow as described in Section 8.3, and thus reduce V and sediment transport. The following four methods (Sections 11.1.4–8) also involve the concept of stream power. Bagnold used the product of shear stress and mean flow velocity while Engelund–Hansen and Yang used the product of velocity and slope. Finally, it should be noted that the stream power approach does not always consider the concept of threshold of motion. For instance the rate of sediment transport does not reduce to zero in the methods of Bagnold and Engelund–Hansen, and this even at very low velocities or large grain sizes.

11.1.4 Bagnold's method

Bagnold (1966) developed a bed sediment transport formula based on the concepts of energy balance. He stated that the available power of the flow supplies the energy for bed sediment transport. The resulting bed sediment transport equation combines bedload and suspended load:

$$q_t = q_b + q_s = \frac{\tau_o V}{(G-1)}\left(e_B + 0.01\frac{V}{\omega}\right) \tag{11.8}$$

where $0.2 < e_B < 0.3$. The sediment discharge q_t is expressed as dry weight per unit time and width in any consistent system of units. Note that the ratio of suspended load to bedload is approximated by $0.01\ V/e_B\omega$, which is close to u_*/ω when $V = 25u_*$. Equation (11.8) is applicable to fully turbulent flows and results should be best when $u_*/\omega < 2$. Unfortunately, there is no threshold of motion with this equation and it should not perform well when $u_*/\omega < 0.2$.

11.1.5 Engelund and Hansen's method

Engelund and Hansen (1967) applied Bagnold's stream power concept and the similarity principle to obtain the sediment concentration by weight C_w as follows:

$$C_w = 0.05\left(\frac{G}{G-1}\right)\frac{V\,S_f}{[(G-1)gd_s]^{1/2}}\frac{R_h\,S_f}{(G-1)d_s} \tag{11.9}$$

where d_s is the grain size, S_f is the friction slope, R_h is the hydraulic radius, V is the depth-averaged velocity, g is the gravitational acceleration, and G is the specific gravity of sediment. This method does not have any threshold value for incipient motion and will calculate concentrations for size fractions larger than incipient motion.

11.1.6 Ackers and White's method

Ackers and White (1973) postulated that only a part of the shear stress on the channel bed is effective in causing motion of coarse sediment, while in the case of fine sediment, suspended load movement predominates for which the total shear stress is effective in causing the sediment motion. On this premise, the sediment mobility was described by the parameter

$$c_{AW5} = \frac{u_*^{c_{AW1}}}{\sqrt{(G-1)g\,d_s}} \left(\frac{V}{\sqrt{32}\,\log(10h/d_s)} \right)^{1-c_{AW1}} \tag{11.10}$$

in which $c_{AW1} = 0$ for coarse sediment and unity for fine sediment. The total sediment concentration by weight is given by

$$C_w = c_{AW2}G \frac{d_s}{h} \left(\frac{V}{u_*} \right)^{c_{AW1}} \left(\frac{c_{AW5}}{c_{AW3}} - 1 \right)^{c_{AW4}} \tag{11.11}$$

in which c_{AW1}, c_{AW2}, c_{AW3} and c_{AW4} depend on the dimensionless particle diameter $d_* = \left[\frac{(G-1)g}{v^2} \right]^{1/3} d_s$. The relationships for c_{AW1}, c_{AW2}, c_{AW3} and c_{AW4} obtained using flume data for particle sizes ranging from 0.04 mm to 4.0 mm are:

(1) for

$$1.0 < d_* \le 60.0,$$

$$c_{AW1} = 1.0 - 0.56 \log d_*$$

$$\log c_{AW2} = 2.86 \log d_* - (\log d_*)^2 - 3.53$$

$$c_{AW3} = \frac{0.23}{d_*^{1/2}} + 0.14$$

$$c_{AW4} = \frac{9.66}{d_*} + 1.34$$

(2) for

$$d_* > 60.0,$$

$$c_{AW1} = 0, \ c_{AW2} = 0.025, \ c_{AW3} = 0.17, \ c_{AW4} = 1.50.$$

It can be seen that incipient motion occurs where $c_{AW3} = c_{AW5}$. Such a condition for incipient motion agrees well with Shields' value for coarse sediment. Ackers and White's method tends to largely overestimate the concentration and sediment transport of fine and very fine sands.

11.1.7 Yang's method

Yang (1973) suggested that the total sediment concentration is related to potential energy dissipation per unit weight of water, i.e. the unit stream power which he expressed as the product of the velocity and slope. The dimensionless regression relationships for the total sediment concentration C_t in ppm by weight are:

(1) for sand:

$$\log C_{ppm} = 5.435 - 0.286 \log \frac{\omega d_s}{\nu} - 0.457 \log \frac{u_*}{\omega}$$

$$+ \left(1.799 - 0.409 \log \frac{\omega d_s}{\nu} - 0.314 \log \frac{u_*}{\omega} \right) \log \left(\frac{VS}{\omega} - \frac{V_c S}{\omega} \right) \quad (11.12\text{a})$$

(2) for gravel,

$$\log C_{ppm} = 6.681 - 0.633 \log \frac{\omega d_s}{\nu} - 4.816 \log \frac{u_*}{\omega}$$

$$+ \left(2.784 - 0.305 \log \frac{\omega d_s}{\nu} - 0.282 \log \frac{u_*}{\omega} \right) \log \left(\frac{VS}{\omega} - \frac{V_c S}{\omega} \right) \quad (11.12\text{b})$$

in which the dimensionless critical velocity, V_c/ω, at incipient motion, can be expressed as:

$$\frac{V_c}{\omega} = \frac{2.5}{\left[\log \left(\frac{u_* d_s}{\nu} \right) - 0.06 \right]} + 0.66; \text{ for } 1.2 < \frac{u_* d_s}{\nu} < 70 \quad (11.13\text{a})$$

and

$$\frac{V_c}{\omega} = 2.05; \text{ for } \frac{u_* d_s}{\nu} \geq 70 \quad (11.13\text{b})$$

These empirical equations are dimensionless, including the total sediment concentration C_{ppm} in parts per million by weight, V_c is the average flow velocity at incipient motion, VS is the unit stream power, and VS/ω is the dimensionless unit stream power. Flume and field data ranged from 0.137 to 1.71 mm for sand sizes, and 0.037 to 49.9 feet for water depth. However, the majority of the data covered medium to coarse sands at flow depths rarely exceeding 3 feet. Yang's method tends to overestimate transport for very coarse sands and there is a significant discontinuity between sand and gravel equations at $d_s = 2$ mm.

11.1.8 Shen and Hung's Method

Shen and Hung (1972) recommended a regression formula based on available data for engineering analysis of sediment transport. They selected the sediment concentration as the dependent variable and the fall velocity ω in ft/s of the median diameter of bed material, flow velocity V in ft/s, and energy slope as independent variables. The concentration of bed sediment by weight in ppm is given as a power series of the flow parameter, based on 587 data points:

$$\log C_{ppm} = [-107{,}404.459 + 324{,}214.747\, Sh$$

$$-326{,}309.589\, Sh^2 + 109{,}503.872\, Sh^3] \tag{11.14a}$$

where

$$Sh = \left[\frac{V\, S^{0.57159}}{\omega^{0.31988}}\right]^{0.00750189} \tag{11.14b}$$

The fall velocity of sediment particle was corrected to the actual measured water temperature but not to include the effect of significant concentrations of fine sediment on bed material transport. It is most important not to round off the coefficients and exponents of Equation (11.14). This equation performs quite well with flume data, but tends to underpredict the total load of large rivers like the Rio Grande, the Mississippi, the Atchafalaya, the Red River, and some large canals in Pakistan.

11.1.9 Brownlie's method

Brownlie (1981) obtained the following equation for the concentration C_{ppm}

$$C_{ppm} = 7{,}115\, c_B \left(\frac{V - V_c}{\sqrt{(G-1)g d_s}}\right)^{1.978} S_f^{0.6601} \left(\frac{R_h}{d_s}\right)^{-0.3301} \tag{11.15a}$$

in which the value of V_c is given in terms of the Shields dimensionless critical shear stress τ_{*c}, the friction slope S_f, and the geometric standard deviation of the bed material σ_g, by the equation

$$\frac{V_c}{\sqrt{(G-1)g d_s}} = 4.596\, \tau_{*c}^{0.529} S_f^{-0.1405} \sigma_g^{-0.1606} \tag{11.15b}$$

The coefficient c_B is unity for laboratory data and 1.268 for field data.

11.1.10 Karim and Kennedy's method

Karim and Kennedy (1983) carried out a regression analysis of the bed sediment data from laboratory flumes and natural streams:

$$\log \frac{q_t}{\gamma_s \sqrt{(G-1)g\, d_s^3}} = -2.28 + 2.97\, c_{k1} + 0.30\, c_{k2}\, c_{k3} + 1.06\, c_{k1}\, c_{k3} \tag{11.16a}$$

where

$$c_{k1} = \log \frac{V}{\sqrt{(G-1)g\,d_s}}; \quad c_{k2} = \log(h/d_s); \quad c_{k3} = \log\left(\frac{u_* - u_{*c}}{\sqrt{(G-1)g\,d_s}}\right) \quad (11.16b)$$

The equations of Brownlie and Karim–Kennedy require further testing. Case study 11.2 compares bed sediment transport calculations on the Colorado river by the methods of Bagnold, Engelund–Hansen, Ackers–White, Yang, Shen–Hung, Karim–Kennedy, and Brownlie. These methods are expected to be most effective when $1 < u_*/\omega < 5$.

Case study 11.2 Colorado River, United States

This case study illustrates sediment transport calculations when $1 < u_*/\omega < 5$. The Colorado River at Taylor's Ferry carries significant volumes of sand. The bed material size has a geometric mean of 0.32 mm and standard deviation $\sigma_g = 1.44$, $d_{35} = 0.287$ mm, $d_{50} = 0.33$ mm, $d_{65} = 0.378$ mm. The detailed sieve analysis is given in the table below, and the rating curve shown in Figure CS-11.2.1.

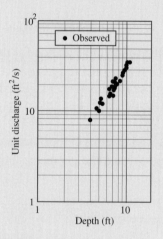

Figure CS-11.2.1 Rating curve

Sieve opening mm	% Finer
0.062	0.22
0.125	1.33
0.25	21.4
0.5	88.7
1	98
2	99
4	99.5

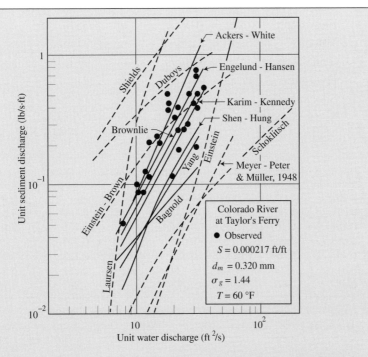

Figure CS-11.2.2 Sediment-rating curve of the Colorado River (after Vanoni *et al.*, 1960)

The stage–discharge relationship for the Colorado River is shown on Figure CS-11.2.1 at flow depths ranging from 4 to 12 feet and unit discharges from 8 to 35 ft²/s. Consider the channel slope 0.000217, channel width 350 ft, and the water temperature at 60°F, or $\nu = 1.21 \times 10^{-5}$ ft²/s. Calculate the unit sediment discharge by size fraction using the methods detailed in Sections 11.1.4 to 11.1.10. Plot the results on the given sediment-rating curve shown in Figure CS-11.2.2.

For the sake of comparison of several bed sediment transport equations, flow depths (a) $h = 10$ ft; and (b) 4 ft have been selected respectively. At each flow depth, several sediment transport equations are compared in terms of potential concentration in ppm for each size fraction, and then the fraction weighted sediment discharge in lb/ft·s is given in the second tabulation.

Note that from Equation (10.1), $C_{ppm} = 10^6\, C_w$ and $C_{mg/l} = \frac{G\,C_{ppm}}{G+(1-G)C_w}$. The unit sediment discharge q_s in N/m·s is obtained from

$$q_s(\text{in N/m/s}) = 10^{-3}g\,(\text{in m/s}^2)\,C\,(\text{in mg/l})\,q\,(\text{in m}^2/\text{s})$$

$$q_s(\text{in lb/ft} \times \text{s}) = q_s(\text{in N/ms}) \times \frac{\text{lb}}{4.45\,\text{N}} \times \frac{\text{m}}{3.28\,\text{ft}} = 0.0685\,q_s\,(\text{in N/ms})$$

(a) Flow depth $h = 10\,\text{ft}$ or $3.05\,\text{m}$
     Velocity $V = 3.2\,\text{ft/s}$ or $0.97\,\text{m/s}$
     Unit discharge $q = 32\,\text{ft}^2/\text{s}$ or $2.97\,\text{m}^2/\text{s}$
     Hydraulic radius $R_h = 9.45\,\text{ft}$ or $2.9\,\text{m}$
     Shear stress $\tau_o = 0.128\,\text{lb/ft}^2$ or $6.14\ \text{N/m}^2$
     Shear velocity $u_* = 0.257\,\text{ft/s}$ or $0.078\,\text{m/s}$

Δp_i	d_s (mm)	d_*	ω (m/s)	u_*/ω	τ_c (N/m^2)	τ_*
0.002	0.042	0.97	0.001	56.7	—	9.6
0.011	0.083	1.95	0.005	14.5	0.166	4.8
0.201	0.167	3.89	0.019	4.1	0.166	2.4
0.673	0.333	7.79	0.049	1.65	0.203	1.2
0.093	0.667	15.5	0.085	0.92	0.328	0.6
0.01	1.33	31.1	0.132	0.59	0.761	0.3
0.01	2.66	62.3	0.193	0.41	1.86	0.15

d_s	Bagnold	Engelund–Hansen	Ackers–White	Yang	Shen–Hung	Karim–Kennedy	Brownlie
Potential sediment concentration C (ppm)							
0.042	960	2,003	a	3,529	1,439	7,584	1,133
0.083	270	1,013	2,017	567	586	1,289	491
0.167	101	504	584	162	222	454	313
0.333	60	252	140	93	97	181	191
0.667	48	126	69	91	54	79	107
1.33	43	63.3	31	113	34	35	48
2.67	40	31.5	5.9	1.8	22	14	14
Fraction weighted sediment transport $\Delta p_i q_s$ (N/ms)							
0.042	0.06	0.13	a	0.23	0.09	0.49	0.07
0.083	0.09	0.32	9	0.18	0.19	0.42	0.16
0.167	0.59	2.94	3.42	0.95	1.3	2.66	1.84
0.333	1.18	4.95	2.75	1.83	1.92	3.55	3.76
0.667	0.13	0.34	0.19	0.25	0.15	0.22	0.29
1.33	0.01	0.02	0.01	0.03	0.01	0.01	0.01
2.67	0.01	0	0	0	0.01	0	0
Total							
N/ms	2.07	8.7	15.37?^{a}	3.48	3.66	7.35	6.13
lb/fts	0.142	0.6	1.05	0.234	0.25	0.5	0.42

^{a} Extremely high.

(b)
Flow depth	$h = 4$ ft	or	1.22 m			
Velocity	$V = 2$ ft/s	or	0.61 m/s			
Unit discharge	$q = 8$ ft^2/s	or	0.74 m^2/s			
Hydraulic radius	$R_h = 3.91$ ft	or	1.2 m			
Shear stress	$\tau_o = 0.053$ lb/ft^2	or	2.538 N/m^2			
Shear velocity	$u_* = 0.1652$ ft/s	or	0.05 m/s			

Δp_i	d_s (mm)	d_*	ω	u_*/ω (m/s)	τ_c (N/m^2)	τ_*
0.002	0.042	0.97	0.001	36.4	—	3.8
0.011	0.083	1.94	0.005	9.3	0.166	1.9
0.201	0.167	3.89	0.019	2.7	0.166	0.96
0.673	0.333	7.78	0.047	1.06	0.203	0.48
0.093	0.667	15.57	0.085	0.59	0.328	0.24
0.01	1.333	31.15	0.132	0.38	0.761	0.12
0.01	2.667	62.3	0.193	0.26	1.86	0.06

d_s	Bagnold	Engelund–Hansen	Ackers–White	Yang	Shen–Hung	Karim–Kennedy	Brownlie
Potential sediment concentration C (ppm)							
0.042	612	797	a	1,570	543	1,522	598
0.083	181	403	2,017	250	184	255	198
0.167	75	200	142	69	56	116	127
0.333	50	100	57	37	21	58	75
0.667	42	50	30	34	10	31	38
1.33	39	25	5.2	42	5.7	14	11.1
2.67	37	12	0	1.2	3.4	3.3	0.03
Fraction weighted sediment transport $\Delta p_i q_s$ (N/ms)							
0.042	0.01	0.01	a	0.03	0.01	0.02	0.01
0.083	0.01	0.02	0.16	0.02	0.01	0.02	0.02
0.167	0.11	0.14	0.21	0.1	0.08	0.17	0.19
0.333	0.24	0.17	0.28	0.18	0.1	0.29	0.37
0.667	0.03	0.01	0.02	0.02	0.01	0.02	0.03
1.33	0	0	0	0	0	0	0
2.67	0	0	0	0	0	0	0
Total							
N/ms	0.41	0.36	0.67?[a]	0.36	0.21	0.52	0.61
lb/fts	0.028	0.024	0.046	0.024	0.015	0.036	0.042

[a] extremely high.

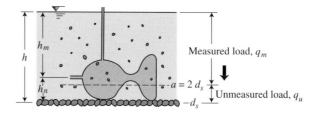

Figure 11.4. Sketch of the modified Einstein approach

11.1.11 Modified Einstein Procedure

As sketched in Figure 11.4, the Modified Einstein Procedure estimates the unmeasured load from the measured load. It can proceed from depth-integrated samples or point samples. The Modified Einstein Procedure (MEP) has been developed by Colby and Hembree (1955) to determine the total sediment load based on field measurements obtained from a depth-integrated suspended sediment sampler and a sample of the bed material. The method was developed on data from the Niobrara River, Nebraska. The original procedure assumed that the Rouse number (Ro) varied empirically with the settling velocity (ω) to a power of 0.7. In addition, the procedure arbitrarily divides the Einstein bedload transport rate by 2. Several re-modifications have been proposed (Colby and Hubbell 1961; Lara 1966; Burkham and Dawdy 1980; Shen and Hung 1983) that aid in the total load calculation based on the MEP. Holmquist-Johnson and Raff (2006) developed the Bureau of Reclamations Automated Modified Einstein Procedure (BORAMEP), which is based on Lara's modification.

11.1.11a SEMEP procedure for a depth-integrated sampler

The most recent development of the MEP is based on the series expansion of the Einstein integrals determined by Guo and Julien (2004), refer to Section 11.1.2. The series expansion of the modified Einstein procedure (SEMEP), was developed by Shah-Fairbank (2009) and tested on several laboratory and sand-bed river data from the Niobrara to the Mississippi River.

Sediment is found in suspension when the shear velocity is greater than the fall velocity. In order to quantify suspended sediment the logarithmic velocity law (Equation 6.16) and the concentration profile (Equation 10.20a) are inserted into Equation 11.1.

$$q_t = q_b + q_s = q_b + \int_a^h C v \, dz \qquad (11.17)$$

$$C_a = \frac{q_b}{v_a a} = \frac{q_b}{11.6\,u_* a} \tag{11.18}$$

$$q_t = q_b + 0.216 q_b \frac{E^{Ro-1}}{(1-E)^{Ro}} \left\{ \ln\left(\frac{30h}{d_s}\right) J_1 + J_2 \right\} \tag{11.19}$$

$$J_1 = \int_E^1 \left(\frac{1-z*}{z*}\right)^{Ro} dz* \tag{11.20a}$$

$$J_2 = \int_E^1 \ln z * \left(\frac{1-z*}{z*}\right)^{Ro} dz* \tag{11.20b}$$

Where, q_t is the unit total load, q_b is the unit bedload, C is the concentration, v is the velocity, h is the flow depth, a is the reference depth of $2d_s$, d_s is the particle size, C_a is the reference concentration at a, v_a is the reference velocity at a, u_* is the shear velocity, E is $2d_s/h$ and J_1 and J_2 are the Einstein integrals, and z^* is the ratio of z/h.

The value of q_b is determined directly from q_m measured between the sampler nozzle height at $z = d_n$ and the free surface at $z = h$.

$$q_m = \int_{d_n}^h Cv\, dz \tag{11.21}$$

$$q_m = 0.216 q_b \frac{E^{Ro-1}}{(1-E)^{Ro}} \left\{ \ln\left(\frac{60}{E}\right) J_{1A} + J_{2A} \right\} \tag{11.22}$$

$$J_{1A} = \int_A^1 \left(\frac{1-z*}{z*}\right)^{Ro} dz* \tag{11.23}$$

$$J_{2A} = \int_A^1 \ln z * \left(\frac{1-z*}{z*}\right)^{Ro} dz* \tag{11.24}$$

where A is d_n/h, and J_{1A} and J_{2A} are the modified Einstein integrals.

Once q_b is determined the unit total load q_t can be calculated based on Equation (11.19). The following are the advantages to SEMEP:

(1) based on median grain diameter (d_{50}) in suspension, no bins are required;
(2) bedload calculated based on measured load, no need to arbitrarily divide the Einstein bedload equation by 2;
(3) calculate Ro directly from settling equation, no Ro fitting based on power function;
(4) calculate total load even when there are not enough overlapping bins between suspended and bed material; and
(5) calculated total load cannot be less than measured load.

Example 11.1 Modified Einstein analysis for depth-integrated data

Shah-Fairbank (2009) analyzed data from three different USGS publications (Colby and Hembree 1955; Kircher 1981; Williams and Rosgen 1989) to test SEMEP. The data for the Platte River (Kircher 1981) and 93 other US streams (Williams and Rosgen 1989) are considered to be total load data sets because they contain both depth-integrated and Helley–Smith measurements. In addition, data from the Niobrara River collected by Colby and Hembree (1955) were also tested. The total load measurement on the Niobrara River occurs at a constricted section where it is assumed that a depth-integrated sampler can measure the total load.

Step 1: Determine the median particle size in suspension;
Step 2: Calculate the Ro based on the median particle size in suspension;
Step 3: Use the series expansion of the Einstein integrals to calculate J_{1A} and J_{2A};
Step 4: Determine the unit bedload discharge from the measured load based on Equation (11.22)
Step 5: Use the series expansion of the Einstein integrals to calculate J_1 and J_2; and Calculate the unit total load based on Equation (11.19).

Based on the analysis using the data from several USGS publications a range of applicability is developed. When the value of u_*/ω is greater than 5 (Ro less than 0.5) and measured total load is greater than 1000 tons/day SEMEP performs with a high level of accuracy. This is validated by the following statistical analysis using three parameters ($MAPE$, R^2 and CC) defined as:

$$MAPE = \frac{\displaystyle\sum_{i=1}^{n} \frac{abs\,(X_i - Y_i)}{X_i}}{n} \quad \text{Mean absolute percent error} \quad \text{(E-11.1.1)}$$

where X_i is the measured sediment discharge, Y_i is the calculated sediment discharge and n is the sample size.

$$R^2 = \left(\frac{\displaystyle\sum_{i=1}^{n} (X_i - \bar{X})(Y_i - \bar{Y})}{\sqrt{\displaystyle\sum_{i=1}^{n} (X_i - \bar{X})^2 \sum_{i=1}^{n} (Y_i - \bar{Y})^2}} \right)^2 \qquad \text{Coefficient of determination}$$

$$\text{(E-11.1.2)}$$

where $\bar{X}$ is the average of the measured and $\bar{Y}$ the average of the calculated values

$$CC = \frac{2S_{xy}}{S_x^2 + S_y^2 + (\bar{X} - \bar{Y})^2} \qquad \text{Concordance correlation coefficient} \qquad \text{(E-11.1.3)}$$

where S_{xy} is the covariance, S_x and S_y are the variance of the measured and calculated values respectively.

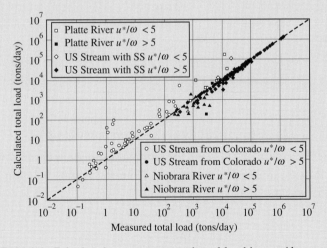

Figure E-11.1.1 Calculated versus measured total load in tons/day

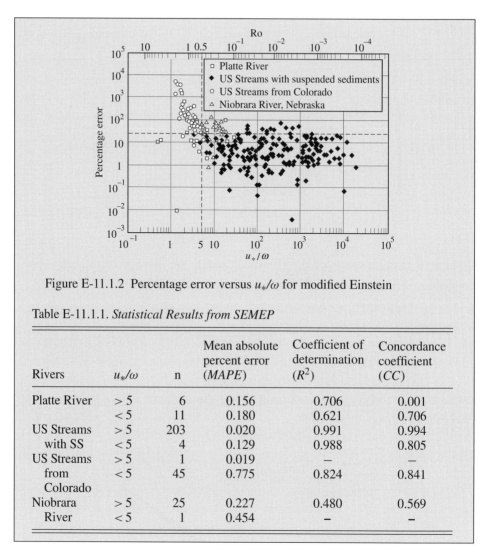

Figure E-11.1.2 Percentage error versus u_*/ω for modified Einstein

Table E-11.1.1. *Statistical Results from SEMEP*

Rivers	u_*/ω	n	Mean absolute percent error (*MAPE*)	Coefficient of determination (R^2)	Concordance coefficient (*CC*)
Platte River	> 5	6	0.156	0.706	0.001
	< 5	11	0.180	0.621	0.706
US Streams	> 5	203	0.020	0.991	0.994
with SS	< 5	4	0.129	0.988	0.805
US Streams	> 5	1	0.019	–	–
from	< 5	45	0.775	0.824	0.841
Colorado					
Niobrara	> 5	25	0.227	0.480	0.569
River	< 5	1	0.454	–	–

11.1.11b SEMEP procedure with point samples

Equation (11.25) is derived from the logarithmic fit to the measured velocity profile and the Rouse number Ro is obtained from the power function fitted to the concentration measurements.

$$v = \frac{u_*}{\kappa} \ln\left(\frac{z}{z_o}\right) \tag{11.25}$$

$$C = C_a \left(\frac{h-z}{z}\frac{a}{h-a}\right)^{Ro} \tag{11.26}$$

where, v is the velocity, u_* is the shear velocity, κ is the von Kármán constant of 0.4, z is the flow depth, z_o is the depth of flow where v is zero, C is the concentration, C_a is the reference concentration, h is the flow depth, a is the reference depth and Ro is the Rouse number. The values of u_*, z_o (depth of zero velocity), C_a and Ro are constants determined from the regression analysis.

The measured load and total load can be determined as follows:

$$q_m = \frac{C_a h u_*}{\kappa} \left(\frac{E}{1-E} \right)^{\text{Ro}} \left\{ \ln \left(\frac{h}{z_o} \right) J_{1A} + J_{2A} \right\} \tag{11.27}$$

$$q_t = C_a v_a a + \frac{C_a h u_*}{\kappa} \left(\frac{E}{1-E} \right)^{\text{Ro}} \left\{ \ln \left(\frac{h}{z_o} \right) J_{1E} + J_{2E} \right\} \tag{11.28}$$

Example 11.2 shows comparisons between the calculations and the measurements of the total sediment load calculated from SEMEP with point measurements of flow velocity and sediment concentration.

Example 11.2 Modified Einstein analysis for point data

Using a point sampler allows for the total load to be calculated directly by fitting a concentration and velocity profile to the measured data points. As a result, the following parameters are determined directly: u_*, y_o, C_a and Ro. This example compares total load between the measurements and calculated results based on regression analysis and SEMEP. The data sets summarized in Table E-11.2 contain point velocity and concentration measurements.

Table E-11.2. *Summary of point data*

Data	Source	h	d_n	h/d_s
Laboratory Data	Coleman (1986)	0.170 to 0.172 m	0.006 m	1,600
Enoree River, SC	Anderson (1942)	3 to 5.15 ft	0.06 to 0.103 ft	3,200 to 6,300
Rio Grande at Bernalillo, NM	Nordin and Dempster (1963)	2.35 to 2.56 ft	0.27 to 0.37 ft	11,500 to 12,500
Mississippi River, MS	Akalin (2002)	21 to 110 ft	0.4 to 2.2 ft	15,000 to 530,000

The original MEP only calculates total load based on sediment samples from a depth-integrated sampler. The following procedure shows how a point sampler can be used to quantify load. The first step is to quantify the total and measured loads. Figure E-11.2.1a shows a schematic of the trapezoidal method used to determine the actual sediment discharge.

The logarithmic velocity and concentration profiles are then fitted to the data sets. Figure E-11.2.1b shows a schematic of the trend lines generated for the velocity and concentration data set. Figure E-11.2.2 shows the values of the measured to total loads q_m/q_t as a function of u_*/ω and the percentage of flow depth sampled h_m/h. All the data are plotted in Figure E-11.2.3 to show how the accuracy of SEMEP varies with the ratio of measured depth to median grain diameter in suspension.

The results show that when h_m/d_s are greater than 1,000 over 70% of the data points have an error less than 25%. In addition, there is a high error with data from the Coleman Laboratory data and Enoree River. This is because the value of u_*/ω is smaller than 5 for those samples.

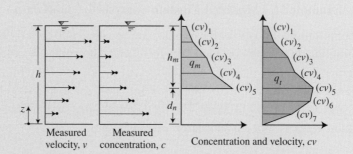

Figure E-11.2.1a Calculation of unit measured and total load based on point data

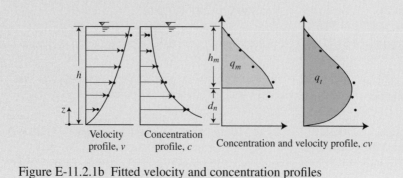

Figure E-11.2.1b Fitted velocity and concentration profiles

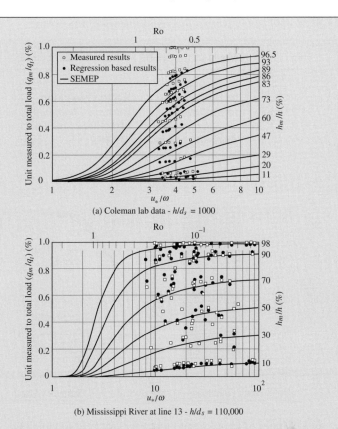

(a) Coleman lab data - $h/d_s = 1000$

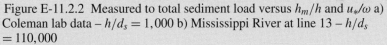

(b) Mississippi River at line 13 - $h/d_s = 110,000$

Figure E-11.2.2 Measured to total sediment load versus h_m/h and u_*/ω a) Coleman lab data – $h/d_s = 1,000$ b) Mississippi River at line 13 – $h/d_s = 110,000$

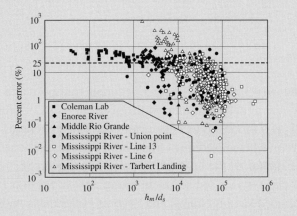

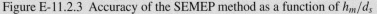

Figure E-11.2.3 Accuracy of the SEMEP method as a function of h_m/d_s

Table 11.3. *Modes of transport and recommended procedure*

u_*/ω	Ro	Mode of transport	Calculation procedure
< 0.2	> 12.5	no motion	–
0.2 to 0.5	5 to 12.5	bedload	Bedload equation
0.5 to 2	1.25 to 5	mixed load	Einstein Chap. 9 or Equation (11.6c)
2 to 5	0.5 to 1.25	suspended load	SEMEP or Section 11.1
> 5	< 0.5	suspended load	SEMEP

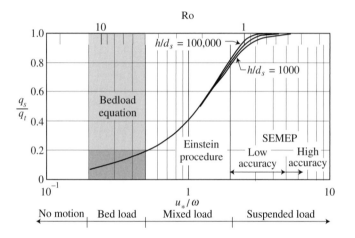

Figure 11.5. Modes of transport and procedure for load calculation

   Based on this analysis the mode of sediment transport and the recommended procedure to calculate the total sediment load are summarized in Table 11.3 and Figure 11.5.

11.2 Sediment-rating curves

Sediment-rating curves display the rate of sediment transport, as a function of flow discharge. The rate of sediment transport is given in terms of sediment discharge or alternatively as a flux-averaged concentration. The analysis of sediment-rating curves depends on whether sediment transport is limited by the sediment transport capacity of the stream, or the upstream supply of sediment. Section 11.2.1 covers the analysis of capacity-limited sediment-rating curves; Section 11.2.2 deals with the effects of graded sand mixtures; and Section 11.2.3 deals with supply-limited sediment-rating curves.

11.2.1 Capacity-limited sediment-rating curves

In the case where sediment transport is controlled by the transporting capacity of bed sediment, the analysis of sediment-rating curves can be considered in two ways: (1) comparative analysis of sediment transport capacity curves by size fractions; and (2) comparative analysis of sediment transport formulas with field measurements.

The analysis of sediment transport capacity curves by size fractions involves plotting the sediment transport capacity from various formulas versus the particle size. An example is shown in Figure 11.6 for four sediment transport formulas, at two flow depths 2 ft and 10 ft respectively, all other conditions being identical. The transport capacity is shown to increase largely with flow depths and also varies inversely with sand grain size. In this example from Williams and Julien (1989) a comparison of these four sediment transport formulas highlights that: (1) there is more variability in the predictions from different equations at high flows, e.g. high sediment transport rates; (2) the transport rates calculated using the methods of Ackers and White (1973) and of Toffaleti (1968) are very high for fine sands and very low for coarse sands; and (3) Yang's (1973) method shows a slight increase in sediment transport capacity with grain size for coarse sands.

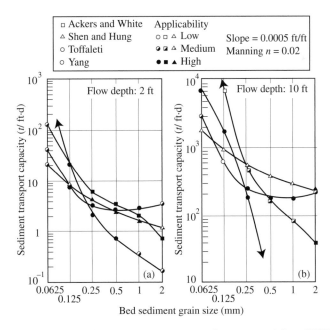

Figure 11.6. Examples of sand transport capacity curves (after Williams and Julien, 1989)

11.2.2 Graded sand mixtures

In general, the transport capacity calculated by size fraction exceeds the transport capacity calculations based on the median grain size. For instance, consider a sand size distribution 33% at 0.125 mm, 33% at 0.25 mm, and 33% at 0.5 mm. At a flow depth of 10 ft transport capacity calculations by size fractions using the Ackers–White equation on Figure 11.6 gives 0.33 ($\cong$ 7000 tons/ft $\cdot$ d) + 0.33 ($\cong$ 500 tons/ft. d) + 0.33 ($\cdot$ 180 tons/ft $\cdot$ d) = 2560 tons/ft $\cdot$ d which far exceeds calculations based on the median grain size 1 $\cdot$ 500 tons/ft $\cdot$ d = 500 tons/ft $\cdot$ d.

In sand-bed channels, the method of Wu *et al.* (2004) analyzed the relationship between sediment transport calculation by size fractions and calculations based on the median grain diameter d_{50}. The results depend on the gradation coefficient of the bed material $\sigma_g = \sqrt{d_{84}/d_{16}}$. The data in Figure 11.7a shows that the measured transport rate Q_{si} for a grain size d_i varies inversely with the grain size according

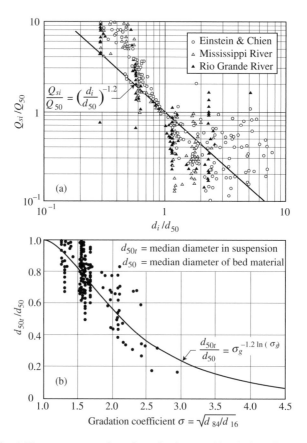

Figure 11.7. a) Transport capacity of sand mixtures b) relative diameter d_{50t}/d_{50} versus gradation coefficient σ_g (after Wu *et al.*, 2004)

to $Q_{si}/Q_{s50} \approx (d_i/d_{50})^{-b}$ where the value of $b \approx 1.2$. Wu *et al.* (2004) then defined the ratio of the transport rate by size fraction Q_{sf} to the transport rate Q_{s50} based on the median grain diameter d_{50} as

$$Q_{sf} = Q_{s50} \, e^{0.5\left(1.2\ln\sigma_g\right)^2} \qquad\qquad (11.29\text{a})\blacklozenge$$

Similarly, the d_{50t} of the sediment in suspension could be determined as a function of σ_g and d_{50t} of the bed material as

$$d_{50t} = d_{50}\sigma_g^{-1.2 \, \ln\sigma_g} \qquad\qquad (11.29\text{b})$$

where d_{50t} is the median diameter of sediment in suspension and d_{50} is the median diameter of the bed material. Comparisons with laboratory and field measurements are shown in Figure 11.7b. In practice, it is found that this correction is very significant when $\sigma_g > 2$ and negligible when $\sigma_g < 1.5$.

The bed sediment-rating curve in Figure CS-11.2.1 for the Colorado River illustrates the variability in calculated sediment discharge from various formulas at a given discharge. It has long been considered that at a given discharge, the sediment transport rate calculated from several formulas can vary by about two orders of magnitude, as reported by Vanoni *et al.* (1960), ASCE (1977, 2007). Nowadays, the state-of-the-art in sediment transport predictions has improved considerably. As shown in Case study 11.2, the formulas presented in this book compare very favorably with field observations in Figure CS-11.2.1. This is generally true for rivers with capacity-limited sediment transport.

Relatively little scatter in field measurements is observed when the sediment load is controlled by the sediment transport capacity. This is due to the fact that the bed material load directly depends on discharge. In such cases, variations in sediment-rating curves are due to variability in water temperature, stream slope, bed sediment size, particle size distribution, and measurement errors. For capacity-limited sediment transport, the sediment-rating curve often fits a power law of the form:

$$q_s = \bar{a}q^{\bar{b}} \qquad\qquad (11.30\text{a})$$

Fitting straight lines on logarithmic graphs within the range of observed discharge often gives an exponent $\bar{b}$ ranging between 1 and 2. The flux-averaged sediment concentration C_f is given by the ratio of sediment discharge to water discharge.

From Equation (11.30a) this concentration is

$$C_f = \frac{q_s}{q} = \bar{a}q^{\bar{b}-1} \qquad\qquad (11.30b)$$

It is reasonable to expect a concentration increase during floods, which corresponds to the exponent $\bar{b} > 1$. The determination of the coefficients $\bar{a}$ and $\bar{b}$ by regression analysis should preferably be based on Equation (11.30b), as opposed to Equation (11.30a), to avoid spurious correlation.

11.2.3 Supply-limited sediment-rating curves

The case of supply-limited sediment-rating curves is characterized by low concentrations and high variability. Sediment transport is limited by the supply of sediment, usually washload, which varies with the location and intensity of rainstorms on the watershed (forest versus agricultural fields), seasonal variation in temperature, weathering, vegetation, and type of precipitation (rain or snow). The source of sediment includes upland erosion, streambank erosion, point sources, and snowmelt. A comparison with field sediment discharge measurements is essential in the analysis of sediment-rating curves. Under supply-limited conditions, the transport rate calculated from formulas can be several orders of magnitude larger than field measurements. The sediment supply does not depend solely on discharge and this causes a large variability of the flow discharge at a given rate of sediment supply. The example of the La Grande River in Figure 11.8a is typical of supply-limited sediment transport with low concentration and large variability.

When the sediment concentration data are widely dispersed around the sediment curve obtained by regression analysis, better results can be obtained by subdividing the discharges into small intervals and taking the average value (without logarithmic transformation) of sediment concentration for each interval. The sediment-rating curve is then hand plotted from these mean values of the concentration. This procedure circumvents the bias introduced by linear regression analysis of log-transformed variables. This procedure also avoids the problems of mathematically fitting straight lines through curvilinear sediment-rating relationships.

Loop-rating curves describe the fact that at a given flow discharge, there can be a higher sediment concentration during the rising limb than falling limb of the hydrograph. An example shown in Figure 11.8a, b, and c illustrates a loop-rating curve, or hysteresis effects in sediment-rating curves. Hysteresis effects between discharge and concentration, seasonal variation, inaccuracies in flow and sediment measurements, and variability in the washload may explain the scatter of points on the sediment transport graph. Better results are sometimes achieved, provided sufficient data is available, by setting individual sediment-rating curves for each month.

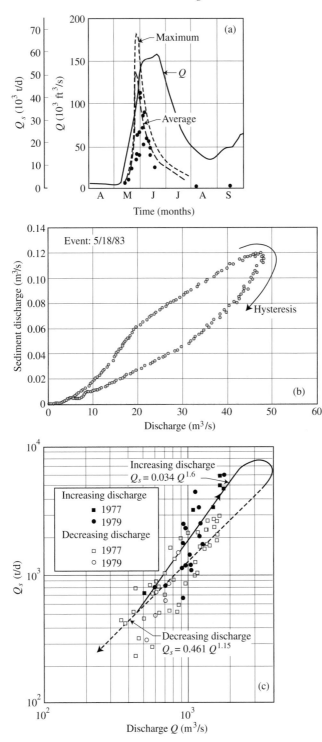

Figure 11.8. a) Suspended load in La Grande River (after Frenette and Julien, 1987) b) Example of loop-rating curve at Goodwin Creek c) Sediment-rating curves of the Bell River (after Frenette and Julien, 1987)

Different sediment-rating curves for the rising and falling limbs of hydrographs can sometimes be identified (Figure 11.8c). Sediment supply from streambank erosion can sometimes be separated from upland sediment sources. The identification of sediment sources is possible when sediments from different sources have different mineralogy, clay content, percentage of organic matter, color, concentration, and water chemistry. At times, the name of the river alone provides an indication of the type of sediment transport: e.g. Muddy Creek, Red River, Green River, Colorado River, Black River, White River, Chalk Creek, Clear Creek, Caine River, and Platte River. Finally, in cold regions, snowmelt erosion rates can be far different from rainfall erosion rates.

11.3 Short- and long-term sediment load

The short-term analysis of sediment load in Section 11.3.1 provides information, generally on a daily basis, on the magnitude and variability of sediment transport during rainstorm or snowmelt events. The long-term analysis, on the other hand, gives an estimate of the expected amount of sediment yielded by a stream. On an annual basis, it gives the mean annual sediment load of a stream, covered in Section 11.3.2. The long-term sediment load is required for reservoir sedimentation, sediment budget, and specific degradation studies.

11.3.1 Daily sediment load

The daily sediment load can be computed with a relatively high degree of accuracy when the discharge and sediment concentration do not change rapidly within one day. This can be a challenge on very small streams, but is a reasonably good approximation for most rivers. The total sediment discharge in tons per day is the product of the flux-averaged total sediment concentration, the daily mean water discharge, and a unit conversion factor. The daily sediment load is obtained by one of the following formulas:

$$Q_s(\text{metric tons/day}) = 0.0864\, C_{mg/l} Q(\text{in m}^3/\text{s}) \qquad (11.31a)\blacklozenge$$

or

$$Q_s(\text{metric tons/day}) = 2.446 \times 10^{-3} C_{mg/l} Q(\text{in ft}^3/\text{s}) \qquad (11.31b)$$

During periods of rapidly changing concentration and water discharge, the concentration and gage records are subdivided into hourly time increments. Incomplete sediment records in which daily discharge measurements are sparse can be analyzed by first obtaining the sediment-rating curve from the measurements using the

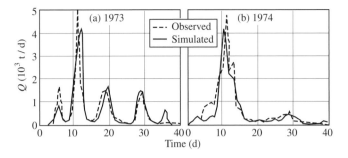

Figure 11.9. Daily sediment load simulation of the York River (after Frenette and Julien, 1987)

method of Section 11.2.2. Alternatively, non-linear regression of log-transformed concentration versus discharge measurements is also possible. The reconstitution of missing sediment concentration is then possible from discharge measurements and the sediment-rating curve. A typical graph of daily sediment discharge for the York River is shown in Figure 11.9. The results are usually good as long as there is minimal variability in the sediment-rating curve. The case of capacity-limited sediment transport usually provides the best results. In the case of supply-limited transport, a stochastic component can be added to the deterministic mean sediment-rating curve (e.g. Frenette and Julien, 1987).

11.3.2 Annual sediment load

There are two basic approaches for the determination of the long-term average sediment load of a river: (1) the summation approach; and (2) the flow duration curve approach. First, the summation over a long period of time of the measured and reconstituted daily sediment discharges from Section 11.3.1 can be accomplished using computers. Mass curves provide the cumulative sediment load as a function of time in years. The slope of the line gives the mean annual sediment load. Mass curves are useful to identify significant changes in flow regime. For example, the effect of the Cochiti Dam on sediment transport is illustrated in Figure 11.10a. Double mass curves plot the cumulative sediment discharge as a function of the cumulative water discharge. The slope of the double mass curve provides average sediment concentration as shown in Figure 11.10b.

The second approach combines a sediment-rating curve between total sediment discharge, or flux-averaged concentration, and water discharge; and a flow-duration curve. This method is referred to as the flow-duration/sediment-rating-curve method. The flow-duration curve states the percentage of time a given river discharge is exceeded. As an example, the flow-duration curve of the Chaudière River

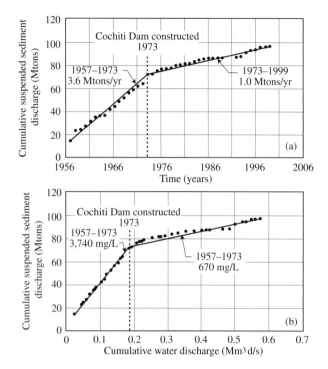

Figure 11.10. a) Mass; and b) double mass curve for the Rio Grande near Albuquerque (after Julien *et al.*, 2005)

is plotted in Figure 11.11a and discrete values are reported in columns (2) and (4) of Table 11.4 respectively. Notice that the selected time percentage intervals are smaller as discharge increases. The sediment-rating curve of the Chaudière River at St-Lambert-de-Lévis is shown in Figure 11.11b, from which, the flux-averaged concentration is approximated by $C_{mg/l} = 3.88 \times 10^{-4} Q_{f^3/s}^{1.3}$. This figure is quite enlightening for several reasons: (1) the highest value of daily sediment load more than 10,000 times the low values, thus high values provide most of the sediment load to this river; (2) sampling over a wide range of discharge will provide a better correlation coefficient in sediment-rating curves; and (3) the average of numerical values is higher than the average of log-transformed values. Sediment concentration calculations for each interval are given in Table 11.4 column (5).

Experience indicates that the flow-duration/sediment-rating curve method is most reliable: (1) when the period of recording is long; (2) when sufficient data at high flows is available; and (3) when the sediment-rating curve shows considerable scatter. Flood flows carry most of the sediments and in order to give full weight to the total sediment inflow, observations must normally concentrate during the flood period. Unfortunately, measurements are rarely available during extreme events,

Table 11.4. *Long-term sediment yield of the Chaudière River from flow-duration–sediment-rating curve method*

Time intervals % (1)	Interval midpoint (%) (2)	Interval Δp (%) (3)	Discharge Q_{cfs} (ft³/s) (4)	Concentration C (mg/l) [a] (5)	$Q_{cfs} \times \Delta p$ (ft³/s) (6)	Sediment Load $Q_s \times \Delta p^*$ (tons/year)[b] (7)
0.00–0.02	0.01	0.02	58,000	607	12	6,280
0.02–0.1	0.06	0.08	52,000	526	42	19,539
0.1–0.5	0.3	0.4	43,000	411	172	63,102
0.5–1.5	1	1	33,000	292	330	85,820
1.5–5.0	3.25	3.5	21,000	162	737	106,213
5–15	10	10	10,640	67	1,064	63,529
15–25	20	10	5,475	29	548	13,782
25–35	30	10	3,484	16	348	4,873
35–45	40	10	2,435	10	244	2,138
45–55	50	10	1,839	7	184	1,121
55–65	60	10	1,375	4.7	138	575
65–75	70	10	1,030	3.2	103	296
75–85	80	10	763	2.1	76	149
85–95	90	10	547	1.4	55	69
95–100	97.5	5	397	0.9	14	12
Total		100			4,067	367,500

Notes: Columns (2) and (4) define the flow-duration curve. Columns (4) and (5) define the sediment-rating curve. The product of columns (3) and (4) is given in Column (6).

[a] The concentration C in mg/l is calculated from $C_{mg/l} = 0.04 Q_{cms}^{1.3} = 3.89 \times 10^{-4} Q_{cfs}^{1.3}$.

[b] The annual sediment yield in metric tons/year is calculated from $Q_s \cong \Delta p = 0.893 \times C \times Q \times \Delta p$ with C in mg/l and Q in ft³/s.

299

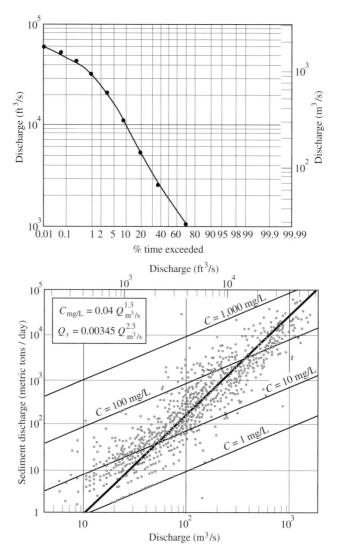

Figure 11.11. a) Flow-duration curve of the Chaudière River b) Sediment-rating curve of the Chaudière River

and must be extrapolated, preferably from measurements during high flows. On an annual basis, the sediment load is given by

$$Q_s(\text{metric tons/year}) = 31.56\, C_{mg/l} Q(\text{in m}^3/\text{s}) \qquad (11.32a)$$

or

$$Q_s(\text{metric tons/year}) = 0.893\, C_{mg/l} Q(\text{in ft}^3/\text{s}) \qquad (11.32b)$$

The total annual sediment load is then given by the sum of all the intervals of the flow-duration curve. In this example, the sum of all numbers in column (7) gives an average annual sediment load of 367,500 metric tons per year for the Chaudière River. The results in Table 11.4 also show that most of the sediment load comes from high flow discharges exceeding 7,000 cfs. Likewise the median of the sediment load corresponds to a discharge of about 27,000 cfs which is exceeded 1.5% of the time or about 5 days per year. The maximum sediment load in column (7) is for discharges around 21,000 cfs, however, it should be considered that the time intervals in column (1) get smaller at high flows.

11.4 Sediment sources and sediment yield

In streams with very coarse bed material, stiff clay, or bedrock control, the sediment transport capacity of fine fractions calculated from sediment transport formulas far exceeds the sediment supply from upstream sources. Sediment transport in such streams can be obtained from an analysis of sediment sources (Section 11.4.1) and sediment yield (Section 11.4.2).

11.4.1 Sediment sources

The analysis of sediment sources aims at estimating the total amount of sediment eroded on the watershed on an annual basis, called annual gross erosion. The annual gross erosion A_T depends on the source of sediments in terms of upland erosion A_U, gully erosion A_G, and local bank erosion A_B, thus $A_T = A_U + A_G + A_B$.

Upland erosion A_U generally constitutes the primary source of sediment, other sources of gross erosion such as mass wasting or bank erosion A_B and gully erosion A_G must be estimated at each specific site. For instance, the annual volume of sediment scoured through lateral migration of the stream and the upstream migration of headcuts can be determined from past and recent aerial photographs and field surveys. In stable fluvial systems, the analysis of sediment sources focuses on upland erosion losses from rainfall and snowmelt.

The impact of raindrops on a soil surface can exert a surface shear stress up to 10 Pa, thus far exceeding the bonding forces between soil particles (Hartley and Julien, 1992). The detached particles are transported through sheet flow into rills and small channels (Alonso *et al.*, 1991). The critical shear stress of cohesive soils can also be very high (Smerdon and Beasley, 1961) for dry soils but decreases on wet soils. With reference to Example 2.2, the unit upland sediment discharge from sheet and rill erosion can be written in the form

$$q_t = e_1 S_o^{e_2} q^{e_3} \qquad (11.33a)$$

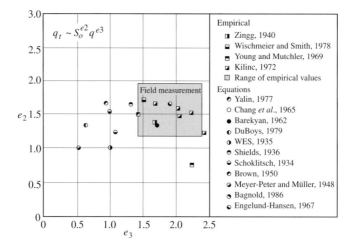

Figure 11.12. Exponents of the equation for sheet flow and ill erosion (after Julien and Simons, 1985a)

The values of the exponents e_2 and e_3 from field observations and from bedload equations from Julien and Simons (1985a) are shown on Figure 11.12. The typical range for field values is $1.2 < e_2 < 1.9$ and $1.4 < e_3 < 2.4$. The equation of Kilinc (1972) for sheet and rill erosion is recommended for bare sandy soils:

$$q_t(\text{in lb/fts}) = 1.24 \times 10^5 \, S_o^{1.66} \, q^{2.035}; (q \text{ in ft}^2/\text{s}) \qquad (11.33b)$$

$$q_t(\text{in t/ms}) = 2.55 \times 10^4 \, S_o^{1.66} \, q^{2.035}; (q \text{ in m}^2/\text{s}) \qquad (11.33c)$$

Note that the coefficient 25,500 in Equation (11.33c) refers to English tons (ton = 2,000 lb). The factor 23,200 can be used for q_t in metric tons per meter per second (1 metric ton = 1,000 kg).

Considering various soil types and vegetation, the annual rainfall erosion losses A_U can be calculated from the Universal Soil-Loss Equation. The USLE computes soil losses at a given site from the product of six major factors:

$$A_U = \hat{R}\hat{K}\hat{L}\hat{S}\hat{C}\hat{P} \qquad (11.34)\blacklozenge$$

when A_U is the soil loss per unit area from sheet and rill erosion normally in tons per acre, $\hat{R}$ is the rainfall erosivity factor, $\hat{K}$ is the soil erodibility factor usually in tons per acre, $\hat{L}$ is the field length factor, $\hat{S}$ is the field slope factor, $\hat{C}$ is the cropping-management factor normalized to a tilled area with continuous fallow, and $\hat{P}$ is the conservation practice factor normalized to straight-row farming up and down the slope. GIS-based computer models have been developed to calculate annual rainfall erosion losses (e.g. Molnar and Julien 1998, Johnson *et al.* 2000, and Kim and Julien 2007).

The rainfall erosivity factor $\hat{R}$ can be calculated from the summation for each storm during the period considered

$$\hat{R} = 0.01 \sum (916 + 331 \ \log I) I \tag{11.35}$$

where I is the rainfall intensity in in/hr. The annual rainfall erosion index in the United States decreases from a value exceeding 500 near the Gulf of Mexico to values less than 100 in the northern States and in the Rockies. The slope length-steepness factor $\hat{L}\hat{S}$ is a topographic factor which can be approximated from the field runoff length X_r in feet and surface slope S_o in ft/ft by:

$$\hat{L}\hat{S} = \sqrt{X_r}\left(0.0076 + 0.53 \ S_o + 7.6 \ S_o^2\right) \tag{11.36}$$

The factor $\hat{L}\hat{S}$ is normalized to a runoff length of 72.6 feet and a 9 percent field slope.

In the more general case of erosion from sheet flow, modifications to Equation (11.33) reflect the influence of soil type, vegetation, and practice factors using the factors $\hat{K}$, $\hat{C}$ and $\hat{P}$ as:

$$q_t(\text{in tons/ms}) = 1.7 \times 10^5 \ S_o^{1.66} \ q^{2.035} \ \hat{K}\hat{C}\hat{P} \tag{11.37}$$

where the surface slope S_o is in m/m, the unit discharge q is in m²/s. The soil erodibility factor $\hat{K}$, the cropping-management factor $\hat{C}$, and the conservation practice factor $\hat{P}$ are obtained from Tables 11.5, 11.6a, b, and c and 11.7, respectively. Equation (11.34) is limited to rainfall erosion losses, Equation (11.37) is applicable to both rainfall and snowmelt erosion losses.

The equivalent upland erosion is then calculated from

$$A_u = \int_{time} \int_{width} q_t \, dw \, dt \tag{11.38}$$

More information on upland erosion methods are detailed in Julien (2002) and US Bureau of Reclamation (2006). Also, GIS-based computer models can simulate the dynamics of surface runoff and sheet erosion during rainstorms. For instance the models CASC2D-SED (Johnson *et al.* 2000, Julien and Rojas 2002, and Rojas *et al.* 2008) and TREX (Velleux *et al.* 2006, 2008) have been successfully applied to several watersheds.

Table 11.5. *Soil erodibility factor $\hat{K}$ in tons/acre*

	Organic matter content (%)	
Textural class	0.5	2
Fine sand	0.16	0.14
Very fine sand	0.42	0.36
Loamy sand	0.12	0.1
Loamy very fine sand	0.44	0.38
Sandy loam	0.27	0.24
Very fine sandy loam	0.47	0.41
Silt loam	0.48	0.42
Clay loam	0.28	0.25
Silty clay loam	0.37	0.32
Silty clay	0.25	0.23

Source: Modified after Schwab *et al.* (1981)

Table 11.6a. *Cropping-management factor $\hat{C}$ for undisturbed forest land*

Percent of area covered by canopy of trees and undergrowth	Percent of area covered by duff at least 2 inches deep	Factor $\hat{C}$
100–75	100–90	0.0001–0.001
70–45	85–75	0.002–0.004
40–20	70–40	0.003–0.009

Source: Modified after Wischmeier and Smith (1978).

Table 11.6b. *Cropping-management factor $\hat{C}$ for construction slopes*

Type of mulch	Mulch rate tons per acre	Factor C
Straw	1.0–2.0	0.06–0.20
Crushed stone	135	0.05
1/4 to 1.5 inch	240	0.02
Wood chips	7	0.08
	12	0.05
	25	0.02

Source: Modified after Wischmeier and Smith (1978)

Table 11.6c. *Cropping-management factor Ĉ for permanent pasture, range, and idle land*

Vegetative canopy–type and height[a]	Type[b]	Cover that contacts the soil surface–percent ground cover					
		0	20	40	60	80	95+
No appreciable canopy	G	0.45	0.20	0.10	0.042	0.013	0.003
	W	0.45	0.24	0.15	0.091	0.043	0.011
Tall weeds or short brush with average drop fall height of 20 inches	G	0.17–0.36	0.10–0.17	0.06–0.09	0.032–0.038	0.011–0.013	0.003
	W	0.17–0.36	0.12–0.20	0.09–0.13	0.068–0.083	0.038–0.042	0.011
Appreciable brush or bushes, with average drop fall height of 6 1/2 feet	G	0.28–0.40	0.14–0.18	0.08–0.09	0.036–0.040	0.012–0.013	0.003
	W	0.28–0.40	0.17–0.22	0.12–0.14	0.078–0.087	0.040–0.042	0.011
Trees, but not appreciable low brush, average drop fall height of 13 feet	G	0.36–0.42	0.17–0.19	0.09–0.10	0.039–0.041	0.012–0.013	0.003
	W	0.36–0.42	0.20–0.23	0.13–0.14	0.084–0.089	0.041–0.042	0.011

Source: Modified after Wischmeier and Smith, (1978)

Note: The listed Ĉ values assume that the vegetation and mulch are randomly distributed over the entire area.

[a] Canopy height is measured as the average fall height of water drops falling from the canopy to the ground. Canopy effect is inversely proportional to drop fall height and is negligible if fall height exceeds 33 feet.

[b] G: cover at surface is grass, grasslike plants, decaying compacted duff, or litter at least 2 inches deep. W; cover at surface is mostly broadleaf herbaceous plants (as weeks with little lateral-root network near the surface) or undecayed residues or both.

Table 11.7. *Conservation practice factor* $\hat{P}$ *for contouring, strip cropping, and terracing*

Land slope (%)	Farming on contour	Contour strip crop	Terracing (a)	Terracing (b)
2 – 7	0.50	0.25	0.50	0.10
8 – 12	0.60	0.30	0.60	0.12
13 – 18	0.80	0.40	0.80	0.16
19 – 24	0.90	0.45	0.90	0.18

(a) For erosion-control planning on farmland.
(b) For prediction of contribution to off-field sediment load
Source: modified after Wischmeier (1972)

11.4.2 Sediment yield

The rate at which sediment is carried by natural streams is much less than the gross erosion on its upstream watershed. Sediment is deposited between the source and the stream cross-section whenever the transport capacity of runoff water is insufficient to sustain transport. The sediment-delivery ratio S_{DR} denotes the ratio of the sediment yield Y at a given stream cross-section to the gross erosion A_T from the watershed upstream of the measuring point. The sediment yield can therefore be written as

$$Y = A_T S_{DR} \qquad\qquad (11.39)\blacklozenge$$

The sediment-delivery ratio depends primarily on the drainage area A_t of the upstream watershed, as shown in Figure 11.13. The sediment-delivery ratio decreases with drainage because sediment from upland areas is trapped in lakes and reservoirs and on flood plains. It can be expected that drainage basins with numerous lakes and reservoirs would have lower values of sediment-delivery ratio than watersheds with streams only.

The sediment yield of watersheds can be obtained from the gross erosion and from the sediment-delivery ratio. An alternative method is to simply measure the accumulation of sediment in reservoirs. Specific degradation refers to the ratio of the sediment yield divided by the drainage area of the watershed.

Specific degradation results as a function of total annual rainfall and drainage area are shown in Figures 11.14a and 11.14b respectively. The results from the field mea-surements of reservoir sedimentation in US reservoirs by Kane and Julien (2007) show a rather wide scatter of the data around mean annual values of several hundred

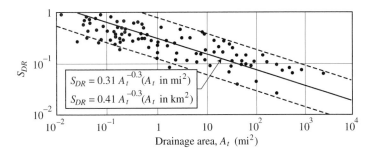

Figure 11.13. Sediment-delivery ratio (after Boyce, 1975)

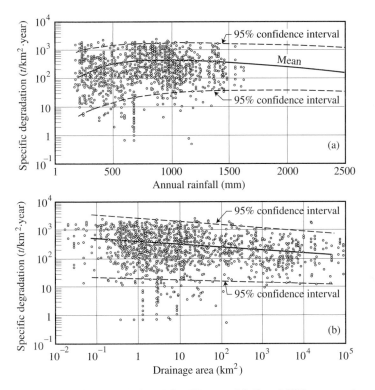

Figure 11.14. Specific degradation (after Kane and Julien, 2007) versus a) annual rainfall; and b) drainage area

tons of sediment per square kilometer. There is a decrease of specific degradation with drainage area.

Case study 11.3 illustrates the application of several concepts for supply-limited transport to the Chaudière watershed in Canada.

Case study 11.3 Chaudière River, Canada

The Chaudière River, Canada, drains a 5,830 km² Appalachian basin to the St Lawrence River near Québec (Figure CS-11.3.1a). Approximately 65% of the

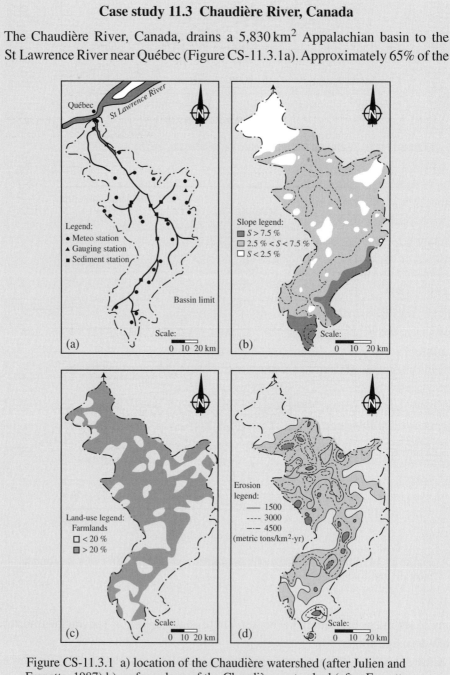

Figure CS-11.3.1 a) location of the Chaudière watershed (after Julien and Frenette, 1987) b) surface slope of the Chaudière watershed (after Frenette and Julien, 1986b) c) land use of the Chaudière watershed (after Frenette and Julien, 1986b) d) specific degradation in metric par km² (after Julien, 1979).

watershed area is still forested and 35% supports agriculture and pasture (Figure CS-11.3.1c). An analysis of soil erosion using the USLE shows that soil losses are primarily a function of the surface slope S_0 and the crop-management factor $\hat{C}$ of the USLE, as shown in Figure CS-11.3.1b and d, the average gross erosion loss is about 770 ton/km^2 year. The mean annual gross rainfall erosion from upland sources calculated from the USLE is of the order of 4.5×10^6 metric tons/year (Figure CS-11.3.1d). Given the mean annual sediment yield of 267 ktons/year from Table 11.4, the sediment-delivery ratio of this watershed is about 0.08 as shown in Figure 11.13.

Short-term simulations of sediment discharge on Figure CS-11.3.2 are possible despite the large scatter on the sediment-rating curve on Figure 11.11b. Monthly simulations of both rainfall and snowmelt erosion losses on Figure CS-11.3.3 obtained from Equation (11.37) demonstrate that about 70% of the mean annual sediment load results from snowmelt.

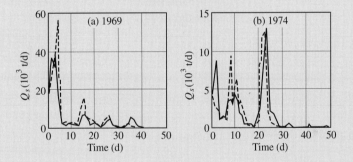

Figure CS-11.3.2 Daily sediment load simulation of the Chaudière River (after Frenette and Julien, 1986b).

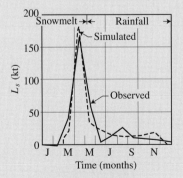

Figure CS-11.3.3 Monthly sediment load distribution of the Chaudière River (after Julien, 1982).

Exercises

♦♦11.1 The particle size distribution of the bed material of the Mississippi River at Tarbert Landing is $d_{10} = 0.072$ mm; $d_{35} = 0.12$ mm; $d_{50} = 0.16$ mm; $d_{65} = 0.18$ mm; and $d_{90} = 0.25$ mm. The suspended samples shown in Figure CS-10.1.1 contain 20–200 mg/l of sand depending on sampling depth, and 150–300 mg/l of silt and clay. Determine: (a) the gradation coefficient of the bed material; (b) what size fraction corresponds to washload; (c) can washload be neglected?; and (d) should the bed material load calculation be done by size fraction?

♦11.2 Given Equation (11.29) of Wu *et al.* (2004) for calculations by size fractions, what would be the ratio of $Q_{s_f}/Q_{s_{50}}$ for the Big Sand Creek example in Case study 11.1? Also, estimate the median grain diameter of sediment in suspension.

Problems
♦*Problem 11.1*

Compute the average sediment concentration C_{ppm} in an alluvial canal using the following methods: Engelund and Hansen, Ackers and White, and Yang. The canal carries a discharge of 105 m³/sec with the water temperature of 15°C. The channel has a slope of 0.00027, an alluvial bed width of 46 m, a flow depth of 2.32 m, and a sideslope of 2 to 1. The bed material (specific gravity $G = 2.65$) has the following particle size distribution:

Fraction diameter (mm)	Geometric mean (mm)	Fraction by weight (Δp_i)
0.062–0.125	0.088	0.04
0.125–0.25	0.177	0.23
0.25–0.50	0.354	0.37
0.50–1.0	0.707	0.27
1.0–2.0	1.414	0.09

(*Answer:* Engelund-Hansen $C = 356$ ppm; Ackers-White $C = 866$ ppm; Yang $C = 140$ ppm)

♦♦*Problem 11.2*

The Conca de Tremp watershed covers 43.1 km² in Spain. The elevation ranges from 530–1,460 m above sea level, the climate is typically Mediterranean with

690 mm of mean annual precipitation and a 12.5°C mean annual temperature. The Mediterranean forest has been depleted and the region has been intensively farmed for centuries. Figure P-11.2 shows the upland erosion map in metric tons per hectare per year (from Julien and Gonzalez del Tanago, 1991): (a) estimate the gross upland erosion and the sediment yield of the watershed; (b) how does the erosion rate compare with the geologic erosion rate 0.1 tons/acre/year (1 acre = 0.4045 hectare)? (c) compare with the accelerated erosion rates for pasture–5 tons/acre/year? (d) compare with the erosion rate of urban development–50 tons/acre/year? and (e) plot the estimated sediment yield from (a) and the estimated sediment-delivery ratio with specific degradation rates on Figure 11.14.

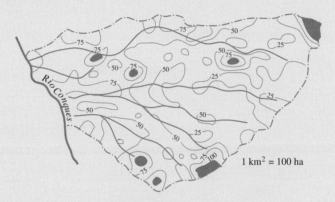

Figure P-11.2 Upland erosion map in metric tons/ha·yr

♦Problem 11.3

Consider sediment transport in the Elkhorn River at Waterloo, Nebraska, given the total drainage area of 6,900 mi^2. The flow-duration curve and the sediment-rating curve are detailed in the following tabulations:

Flow-duration curve

% time exceeded	discharge (ft^3/s)
0.05	37,000
0.3	15,000
1	9,000
3.25	4,500
10	2,100
20	1,200

% time exceeded	discharge (ft³/s)
30	880
40	710
50	600
60	510
70	425
80	345
90	260
96.75	180

Sediment-rating curve

discharge (ft³/s)	suspended load (thousand tons/day)
280	0.25
500	0.6
800	1.0
1,150	3.0
1,800	8.0
2,300	18.0
4,200	40.0
6,400	90.0
8,000	300.0
10,000	500.0

Calculate: (a) the mean annual suspended sediment load using the flow duration–sediment rating curve method; (b) the sediment yield per square mile; and (c) plot the results on Figure 11.14.

(*Answer:* (a) 4.9×10^6 tons/year; (b) 710 tons/mi$^2 \cdot$ year).

Problem 11.4

Compare the bed sediment discharge equation proposed by Julien (2002) as $q_{bv} \approx 18\sqrt{gd_s^3}\tau_*^2$, applicable where $0.1 < \tau_* < 1$, with: (a) the Big Sand Creek in Case study 11.1; (b) the Colorado River in Case study 11.2; and (c) the example in Figure 11.6.

◆◆*Problem 11.5*

With reference to Case study 11.3 on the Chaudière River: (a) use the data of Table 11.4 to plot the entire flow-duration curve; (b) calculate the cumulative sediment load from Table 11.4, column (7), from the bottom up; (c) divide the cumulative sediment load by 367,500 ton/year and plot as a function of the exceedence probability (column (2) of Table 11.4); and (d) determine the sediment load exceeded 1% of the time.

◆◆*Problem 11.6*

An example of suspended sediment data on the Mississippi River at Tarbert Landing is given in the table below (Akalin, 2002). The discharge on April 10, 1998 was 847,658 ft^3/s, the water temperature was 16°C, and the average slope is 3.78 cm/km. The average gradation of the bed material at Tarbert Landing is 1% very fine sand, 37% fine sand, 57% medium sand, and 5% coarse sand. Figure P-11.6 also shows how the sediment concentration of fine sand decreases with temperature.

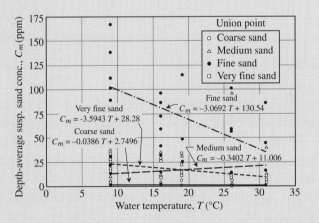

Figure P-11.6 Sand concentration versus temperature (after Akalin, 2002)

Vertical (ft)	Depth (ft)	Sampling depth	Velocity (ft/s)	Suspended sediment percentage finer				Point concentration (ppm)		
				0.425 (mm)	0.25 (mm)	0.125 (mm)	0.0625 (mm)	Sand	Fine	Total
1,371	62	6.2	4.85	100	91.4	31.7	4.3	29	162	191
		18.6	4.11	100	90.0	23.7	2.4	49	151	200
		31.0	4.04	100	92.7	31.5	4.0	40	135	175

Vertical (ft)	Depth (ft)	Sampling depth	Velocity (ft/s)	Suspended sediment percentage finer 0.425 (mm)	0.25 (mm)	0.125 (mm)	0.0625 (mm)	Point concentration (ppm) Sand	Fine	Total
		43.4	4.25	100	85.0	16.3	1.7	88	148	236
		55.8	3.57	97.7	80.4	14.8	1.2	107	158	265
		60.8	2.36	95.7	74.9	13.0	1.2	142	157	299
2,129	63	6.3	5.99	100	90.0	25.4	2.3	33	154	187
		18.9	4.94	100	87.3	19.2	1.5	56	138	194
		31.5	3.9	98.4	80.6	15.3	1.2	95	153	248
		44.1	3.5	98.2	72.3	9.1	0.6	194	159	353
		56.7	2.9	98.4	55.4	4.6	0.1	390	141	531
		61.7	2.14	96.2	70.1	6.2	0.5	296	150	446
2,730	57	5.7	6.42	99.2	96.9	39.3	5.4	38	161	119
		17.1	6.47	100	98.5	34.6	1.9	62	154	216
		28.5	5.88	99.0	96.9	18.7	0.6	116	162	278
		39.9	4.75	98.8	95.5	14.2	0.8	185	171	356
		51.3	4.71	99.0	96.0	13.0	0.5	264	167	431
		55.9	3.28	98.5	95.4	11.5	0.2	407	171	578

Determine the following: (a) what is the settling velocity of these size fractions?; (b) what is the shear velocity?; (c) which of washload or bed material load is predominant?; (d) the washload concentration variability over depth and width; (e) the sediment concentration of medium, fine, and very fine sand; (f) plot the Rouse diagrams for each size fraction and for the total sand fraction; (g) extrapolate the concentration to $a = 2d_{50}$ from the Rouse diagrams and compare the results; (h) extrapolate the respective values of the Rouse number and compare with the theoretical values assuming $\kappa = 0.4$; (i) plot the logarithmic velocity profiles and determine the values of κ; (j) plot the particle size distribution of the bed material and the suspended sediment; and (k) plot the d_{50t}/d_{50} on Figure 11.7b.

♦Problem 11.7

With reference to Case study 11.2, calculate the unit sediment discharge using the bedload methods of DuBoys, Meyer-Peter and Müller; and Einstein–Brown in Chapter 9. Plot the results of calculations based on d_{50} for $h = 4$ ft and $h = 10$ ft in Figure CS-11.2.2.

◆Problem 11.8

Consider the canal data compiled by Kodoatie (1999). Plot the resistance to flow measurements in Figure 6.6a. Examine the range of u_*/ω values. Compare the measured concentration with those predicted from the methods of this chapter.

Canal data	Water discharge (m³/s)	Channel width W (m)	Flow depth h (m)	Flow velocity V (m/s)	Median bed diameter d_{50} (mm)	Surface slope S_w (m/m)	Water temp. (°C)	Measured concentration C (ppm)
American canal								
	12.59	11.73	1.83	0.59	0.096	0.000063	23	370
	1.56	3.49	0.80	0.56	0.173	0.000253	21	249
	1.22	3.19	0.80	0.47	0.229	0.000294	21	406
	29.18	22.19	2.53	0.52	0.253	0.000058	22	115
	29.40	14.81	2.59	0.77	0.311	0.000120	22	185
Indian canal								
	156.05	56.27	3.39	0.82	0.020	0.000060	20	2,601
	59.16	25.49	2.44	0.95	0.021	0.000084	20	5,759
	153.25	56.02	3.37	0.81	0.024	0.000060	20	2,887
	60.72	25.56	2.49	0.95	0.025	0.000084	20	5,182
	157.41	56.47	3.35	0.83	0.030	0.000070	20	2,316
Pakistani canal								
	158.09	118.87	2.23	0.60	0.083	0.000070	28	369
	94.27	88.39	1.46	0.73	0.084	0.000137	28	190
	29.59	35.66	1.68	0.49	0.085	0.000085	31	103
	76.94	69.49	1.83	0.61	0.108	0.000132	27	125
	52.13	35.66	2.29	0.64	0.110	0.000075	32	156

◆◆Problem 11.9

Consider the Amazon River data below from Posada-Garcia (1995). Compare the concentration measurements with those predicted from the methods of this chapter.

Water discharge (m³/s)	Channel width (m)	Flow depth (m)	Flow velocity (m/s)	Median bed diameter (mm)	Surface slope (m/m)	Water temp. (°C)	Suspended sed. conc. (mg/l)
38,100	970	23.00	1.71	0.255	0.0000686	26	351
43,600	1,080	23.80	1.70	0.220	0.0000691	27	398
63,600	1,400	23.70	1.92	0.233	0.0000580	27	276
57,100	1,360	22.70	1.85	0.196	0.0000641	27	606
86,100	2,100	21.90	1.87	0.238	0.0000560	26	238
83,700	2,129	22.00	1.79	0.198	0.0000560	27	243
65,400	2,160	19.00	1.59	0.244	0.0000641	27	499

Water discharge (m³/s)	Channel width (m)	Flow depth (m)	Flow velocity (m/s)	Median bed diameter (mm)	Surface slope (m/m)	Water temp. (°C)	Suspended sed. conc. (mg/l)
66,600	1,720	21.80	1.78	0.212	0.0000641	27	548
62,700	1,530	23.90	1.71	0.343	0.0000452	27	555
71,300	3,020	16.90	1.40	0.192	0.0000452	27	501
151,000	1,000	62.30	2.42	0.409	0.0000370	27	156
80,800	1,400	36.10	1.60	0.154	0.0000452	27	461
85,200	1,418	37.30	1.61	0.120	0.0000343	27	376
75,700	1,890	21.80	1.84	0.331	0.0000257	27	481
140,000	3,100	28.10	1.61	0.244	0.0000330	26	186
133,000	3,130	27.60	1.54	0.216	0.0000330	27	181
90,600	1,510	44.60	1.35	0.259	0.0000138	27	265
155,000	2,400	45.00	1.44	0.171	0.0000189	27	151
120,000	2,300	38.90	1.34	0.141	0.0000138	27	290
235,000	2,600	48.90	1.85	0.243	0.0000200	26	207
230,000	2,340	55.80	1.76	0.237	0.0000200	27	216

♦Problem 11.10

Consider the Niobrara River data at comparable flow depth from Colby and Hembree (1955). Plot the concentration as a function of channel slope, temperature, and u_*/ω. Carry out calculations using methods from this chapter and compare the results.

Q (m³/s)	W (m)	h (m)	S (m/m)	$T°$ (°C)	d_{50} (mm)	d_{65} (mm)	σ_g	C_T (ppm)
11.35	21.34	0.49	0.001705	5.0	0.30	0.41	2.345	1,890
11.72	21.64	0.49	0.001705	6.7	0.28	0.34	1.643	2,000
16.05	21.95	0.58	0.001799	11.7	0.21	0.25	1.573	2,220
9.74	21.34	0.47	0.001420	15.6	0.26	0.32	1.653	1,780
9.42	21.03	0.48	0.001402	16.1	0.32	0.38	1.514	1,490
12.88	21.64	0.53	0.001686	14.4	0.31	0.38	1.669	1,900
5.91	21.03	0.42	0.001250	28.3	0.32	0.38	1.631	392
7.53	21.34	0.49	0.001155	22.2	0.29	0.37	1.824	820
5.86	21.34	0.44	0.001212	22.8	0.34	0.42	1.956	429
6.65	21.18	0.47	0.001136	16.1	0.30	0.36	1.638	736

♦Problem 11.11

Consider the Middle Loup River data from Hubbell and Matejka (1959). At comparable discharge, plot the sediment concentration as a function of water temperature. Carry out calculations using methods from this chapter and compare the results.

Q (m³/s)	W (m)	h (m)	S (m/m)	$T°$ (°C)	d_{50} (mm)	d_{65} (mm)	σ_g	C_T (ppm)
10.31	43.89	0.32	0.001458	24.4	0.317	0.399	1.980	632
10.45	43.28	0.36	0.001250	21.7	0.424	0.586	2.403	687
10.22	44.20	0.33	0.001345	10.0	0.339	0.416	1.849	1,410
10.39	44.81	0.37	0.001288	31.1	0.383	0.476	2.301	548
10.93	45.11	0.33	0.001326	26.1	0.334	0.424	2.095	686
10.36	45.11	0.33	0.001307	18.3	0.274	0.344	1.687	1,020

♦♦Computer problem 11.1 Sediment-rating curve of the Niobrara River

The Niobrara River, Nebraska, carries significant volumes of sand. The bed material size has a geometric mean of 0.283 mm and standard deviation $\sigma_g = 1.6$, $d_{35} = 0.233$ mm, $d_{50} = 0.277$ mm, $d_{65} = 0.335$ mm, and $d_{90} = 0.53$ mm. The detailed sieve analysis is given below.

Sieve opening (mm)	% Finer
0.062	0.05
0.125	4.2
0.25	40.0
0.5	89.0
1.0	96.5
2.0	98.0
4.0	99.0

The stage–discharge relationship for the Niobrara River is shown in Figure CP-11.1a at flow depths ranging from 0.7 to 1.3 feet and unit discharges from 1.7 to 5 ft²/s. The washload consisted of particles finer than 0.125 mm. Given the channel width 110 ft, the channel slope 0.00129, and the water temperature at 60°F: (a) calculate the unit sediment discharge by size fractions using three appropriate methods from this chapter; (b) plot the results on the sediment-rating curve in Figure CP-11.1b; and (c) compare the calculated transport rates with the field measurements.

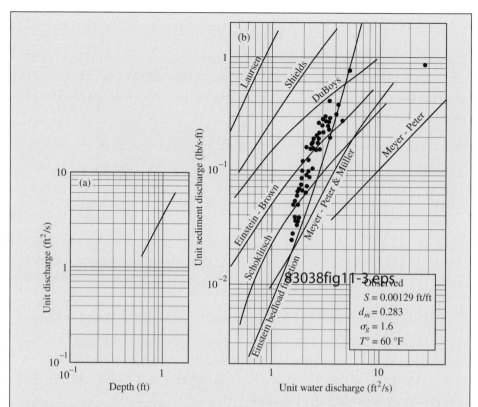

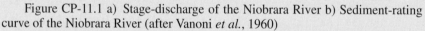

Figure CP-11.1 a) Stage-discharge of the Niobrara River b) Sediment-rating curve of the Niobrara River (after Vanoni *et al.*, 1960)

♦♦*Computer problem 11.2 Total bed sediment discharge*

Consider the channel reach analyzed in Computer problems 3.1 and 8.2. Select one appropriate bed-sediment discharge relationship to calculate the bed-sediment discharge in metric tons/m·day by size fractions. Plot the total sediment transport capacity along the 25 km reach, discuss the methods, assumptions and results.

12

Reservoir sedimentation

As natural streams enter reservoirs, the stream flow depth increases and the flow velocity decreases. This reduces the sediment transport capacity of the stream and causes settling. The pattern of deposition generally begins with a delta formation in the reservoir headwater area. Density currents may transport finer sediment particles closer to the dam. Figure 12.1 depicts a typical reservoir sedimentation pattern. Aggradation in the upstream backwater areas may increase the risk of flooding over long distances above the reservoir.

The rate of sedimentation in reservoirs varies with sediment production on the watershed, the rate of transportation in streams, and the mode of deposition. Reservoir sedimentation depends on the river regime, flood frequencies, reservoir geometry and operation, flocculation potential, sediment consolidation, density currents, and possible land use changes over the life expectancy of the reservoir. In the analysis of reservoir sedimentation, storage losses in terms of live and dead storage, trap efficiency, control measures and the reservoir operations must be considered given: the inflow hydrograph, the sediment inflow, the sediment characteristics, the reservoir configuration, the regional geography, and land use.

The concept of life expectancy of reservoirs describes the time at which a reservoir is expected to become entirely filled with sediment. Its evaluation represents a challenge since the sediment sources arise from various geological formations, cutting and burning of brushland and forest, over-grazed grasslands, natural hazards including landslides, typhoons, and volcanoes, and changes in land use are all likely to occur during the expected life of the reservoir. Once the incoming sediment load has been determined from Section 12.1, the analysis of backwater profiles in Section 12.2 is combined with an analysis of sediment transport capacity to calculate the aggradation rate and the trap efficiency from Section 12.3. Sediment deposits consolidate over time (Section 12.4) and the life expectancy of a reservoir can be estimated from Section 12.5. The analysis of density currents (Section 12.6) is sometimes conducive to sediment management techniques

319

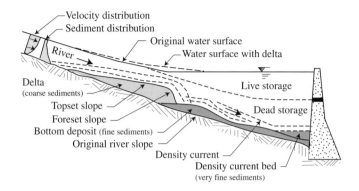

Figure 12.1. Typical reservoir sedimentation pattern (after Frenette and Julien, 1986a)

such as flushing. Reservoir sedimentation surveys and sediment control measures are briefly outlined in Sections 12.7 and 12.8 respectively. Two examples and two case studies provide more details on the application of methods covered in this chapter.

12.1 Incoming sediment load

The incoming sediment load must be measured at appropriate gauging stations over several years prior to construction. Flow and sediment measurements define the long-term sediment load as described in Section 11.3. The annual sediment yield can also be obtained from estimates of sediment sources and sediment yield as shown in Section 11.4.

When washload is dominant, mathematical models can be used to predict soil losses by overland flow (e.g. Case study 11.3). Watershed models using the universal soil-loss equation among others can be used when sufficient data are available from topographic, land use, and agricultural maps, geographic information systems, aerial photographs, and field surveys. Sediment yield estimates based on physical characteristics of the watershed are extremely valuable because the rate of sediment transport can then be predicted for alternative watershed conditions. For instance, Kim and Julien (2006) provided a method to estimate the sediment load into the Imha reservoir in South Korea on a mean annual basis and for extreme single events like typhoon Maemi in 2003.

12.2 Reservoir hydraulics

As a stream enters a reservoir, the flow depth increases and the velocity and friction slope decrease, as generally described by M-1 backwater curves (Chapter 3).

Most streams can be analyzed by one-dimensional approximations of the equations of conservation of mass and momentum. The resulting backwater equation has been derived in Example 3.10, and the water surface elevation h varies with downstream distance x as a function of the bed slope S_o, the friction slope S_f, and the Froude number $\text{Fr} = V/\sqrt{gh}$, given the mean flow velocity V and the gravitational acceleration g.

$$\frac{dh}{dx} = \frac{S_o - S_f}{1 - \text{Fr}^2} \qquad \text{(E-3.10.1)}$$

Equation (E-3.10.1) can be solved numerically from the point of maximum depth at the downstream end. The flow depth change Δh corresponding to reach length Δx is calculated by solving Equation (E-3.10.1) as:

$$\Delta h = \frac{(S_o - S_f)\Delta x}{1 - \text{Fr}^2} \qquad \text{(12.1)}$$

At the upstream end, the normal flow depth corresponds to $\Delta h \to 0, \Delta x \to \infty$ and $S_f = S_o$; while in the reservoir, $\text{Fr}^2 \to 0$, $S_f \to 0$ and $\Delta x = \Delta h/S_o$. Solution to Equation 12.1 is sought while satisfying the continuity relationship $Q = VhW$, given the channel width W. The resistance equation is also required in terms of hydraulic radius and friction slope as compiled in Table 6.1. As a delta forms from the aggradation of bed material load (sand and gravels), the topset bed slope will decrease to about half the original stream slope, as shown in Figure 12.2.

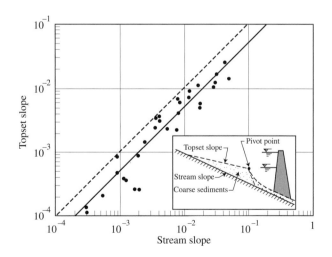

Figure 12.2. Topset slope versus stream slope (after US Bureau of Reclamation, 2006)

12.3 Trap efficiency and aggradation

Owing to conservation of sediment mass, part of the total load deposits on the channel bed as the sediment transport capacity decreases in the downstream direction. With reference to the sediment continuity relationship (Equation (10.3)) without sediment source ($\dot{C} = 0$):

$$\frac{\partial C}{\partial t} + \frac{\partial \hat{q}_{tx}}{\partial x} + \frac{\partial \hat{q}_{ty}}{\partial y} + \frac{\partial \hat{q}_{tz}}{\partial z} = 0 \tag{12.2}$$

where the mass fluxes per unit area $\hat{q}_{tx}, \hat{q}_{ty}$, and $\hat{q}_{tz}$ were defined in Equation (10.4).

Assuming a steady supply of sediment ($\partial C/\partial t = 0$), Equation (12.2) for one-dimensional flow ($\partial \hat{q}_{ty}/\partial y = 0$) reduces to

$$\frac{\partial \hat{q}_{tx}}{\partial x} + \frac{\partial \hat{q}_{tz}}{\partial z} = 0 \tag{12.3}$$

It is further assumed that the diffusive and mixing fluxes from Equation 10.4 are small compared to the advective fluxes in a reservoir. Considering settling as the dominant advective flux in the vertical direction $v_z = -\omega$, one obtains from $\hat{q}_{tx} = v_x C$ and $\hat{q}_{tz} = -\omega C$:

$$\frac{\partial v_x C}{\partial x} - \frac{\partial \omega C}{\partial z} = 0 \tag{12.4}$$

A practical approximation is obtained for gradually varied flow ($\partial v_x/\partial x \to 0$), constant fall velocity ω and $\partial C/\partial z = -C/h$, thus

$$v_x \frac{\partial C}{\partial x} + \frac{\omega C}{h} = 0 \tag{12.5}$$

The solution for grain sizes of a given sediment fraction i (constant fall velocity) at a constant unit discharge $q = Vh$, given $v_x = V$, is a function of the upstream sediment concentration C_{oi} of fraction i at $x = 0$:

$$C_i = C_{oi} e^{-\frac{X\omega_i}{hV}} \tag{12.6}$$

As shown in Figure 12.3, this relationship is in good agreement with the relationship of Borland (1971) and the measurements of Cecen *et al.* (1969).

This shows that the concentration left in suspension is negligible ($C_i/C_{oi} = 0.01$) at a distance X_{C_i}.

$$X_{C_i} = 4.6\frac{hV}{\omega_i} = \frac{4.6q}{\omega_i} \tag{12.7}$$

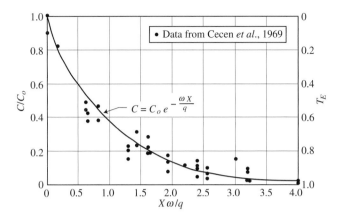

Figure 12.3. Relative concentration C/C_o and trap efficiency T_E versus $X\omega/q$

The percentage of sediment fraction i that settles within a given distance X defines the trap efficiency, T_{Ei} as:

$$T_{Ei} = \frac{C_{oi} - C_i}{C_{oi}} = 1 - e^{-\frac{X\omega_i}{hV}} \qquad (12.8)\blacklozenge\blacklozenge$$

Without resuspension, 99% of the sediment in suspension settles within a distance $X_{C_i} = 4.6\frac{hV}{\omega_i}$. When calculating the trap efficiency of silt and clay particles, careful consideration must also be given to density currents (Section 12.6) and possible flocculation, in which case the flocculated settling velocity ω_{fi} from Section 5.4.3 must be used instead of ω_i.

Another way to estimate the percentage of sediment trapped in a reservoir is from the ratio of reservoir volumetric capacity to the annual volumetric inflow of water. The relationships from Brune and Churchill are shown in Figure 12.4 for comparison with field measurements.

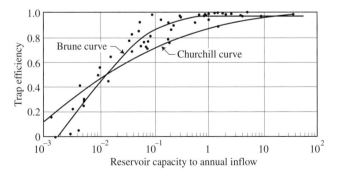

Figure 12.4. Trap efficiency versus reservoir capacity to annual inflow ratio (modified after US Bureau of Reclamation, 2006)

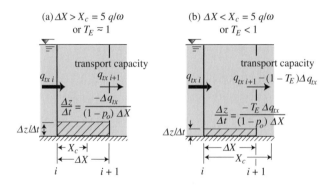

Figure 12.5. Aggradation–degradation scheme

The settling sediment flux in the z direction for a given size fraction i causes a change in bed surface elevation z. Given the porosity $p_o = \forall_v/\forall_t = 1 - C_v$ of the bed material, the integrated form of Equation (12.3) over the depth h is a function of the unit sediment discharge by volume q_{tx} $\left(\text{in } L^2/T\right)$:

$$T_{Ei}\frac{\partial q_{txi}}{\partial x} + (1 - p_o)\frac{\partial z_i}{\partial t} = 0 \qquad (12.9\text{a})$$

or

$$\frac{\partial z_i}{\partial t} = -\frac{T_{Ei}}{(1 - p_o)}\frac{\partial q_{txi}}{\partial x} \qquad (12.9\text{b})\blacklozenge$$

Values of porosity p_o depend on the specific weight of sediment deposits covered in Section 12.4. For model grid size $\Delta X > X_C$, the trap efficiency is unity and aggradation responds directly to changes in the sediment transport capacity of the stream. For $\Delta X < X_c$, only part of the sediment load in suspension will settle within the given reach. The sediment load at the downstream end can then exceed the sediment transport capacity of the stream. For instance, in numerical models of sedimentation of fine sediment in reservoirs with $\Delta X < X_c$, or $T_E < 1$, the downstream sediment transport will be calculated, as shown in Figure 12.5b, by adding the excess sedimentation flux to the calculated transport capacity

$$\underbrace{q_{txi+1}}_{actual} = \underbrace{q_{txi+1}}_{capacity} - \underbrace{(1 - T_{Ei})\Delta q_{txi}}_{non\ settled} \qquad (12.9\text{c})$$

where, $\Delta q_{txi} = q_{txi+1} - q_{txi}$ is negative when sediment transport capacity decreases in the x direction. This correction is important for a better distribution of silts and clays in reservoirs where the settling velocities are very small.

12.4 Dry specific weight of sediment deposits

The conversion of the incoming weight of sediment to volume necessitates knowledge of the average dry specific weight of a mixture γ_{md}, defined in Chapter 2 as the dry weight of sediment per unit total volume including voids. For material coarser than 0.1 mm, the specific dry weight of the mixture remains practically constant around $\gamma_{md} = 14.75$ kN/m^3 or 93 lb/ft^3. The corresponding dry mass density of the mixture $\rho_{md} = \gamma_{md}/g = 1,500$ kg/m^3 or 2.9 slug/ft^3. The porosity p_o of sand material is then obtained from $p_o = 1 - \gamma_{md}/\gamma_s = 0.43$. The volumetric sediment concentration $C_V = 1 - p_o$ and the void ratio is $e = p_o/(1 - p_o)$. The initial dry specific mass of sediment deposits varies with the median grain size of the deposit. The measurements in Figure 12.6 show that

$$\rho_{md} \text{ in kg/m}^3 \cong 1,600 + 300\log d_{50}(\text{mm}) \qquad (12.10a)$$

Under pressure, the dry specific weight of finer sediment fractions varies in time due to the consolidation of the material and exposure to the air. After T_c years the dry specific weight of a mixture γ_{mdT} increases as a function of time from the initial dry specific weight γ_{md1} after $T_c = 1$ year according to

$$\gamma_{mdT} = \gamma_{md1} + K\log T_c \qquad (12.10b)\blacklozenge$$

Values of the initial dry specific weight γ_{md1} and consolidation factor K in lb/ft^3 are compiled in Table 12.1. Assuming continuous uniform settling during a period of T_c years, the depth-averaged dry specific weight of the mixture after T_c years $\overline{\gamma}_{mdT}$ is given by the formula of Miller (1953):

$$\overline{\gamma}_{mdT} = \gamma_{md1} + 0.43K\left(\frac{(T_c\ln T_c)}{T_c - 1} - 1\right) \qquad (12.11)$$

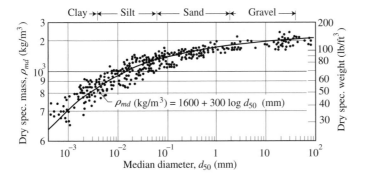

Figure 12.6. Dry specific mass of sediment deposits versus median diameter (modified after Wu and Wang, 2006)

Table 12.1. *Dry specific weight of sediment deposits (γ_{md1} and K in lb/ft³)*

	Lane and Koelzer						Trask					
	Sand		Silt		Clay		Sand		Silt		Clay	
	γ_{md1}	K	γ_{md1}	K	γ_{md1}	K	γ_{md1}	K	γ_{md1}	K	γ_{md1}	K
Sediment always submerged or nearly submerged	93	0	65	5.7	30	16.0	88	0	67	5.7	13	16.0
Normally moderate reservoir drawdown	93	0	74	2.7	46	10.7	88	0	76	2.7	–	10.7
Normally considerable reservoir drawdown	93	0	79	1.0	60	6.0	88	0	81	1.0	–	6.0
Reservoir normally empty	93	0	82	0.0	78	0.0	88	0	84	0.0	–	0.0

Note: 62.4 lb/ft³ $= 9,810$ N/m³

where:

$$\bar{\gamma}_{mdT} = \text{dry specific weight after } T_c \text{ years (lb/ft}^3)$$
$$\gamma_{md1} = \text{initial dry specific weight (lb/ft}^3)$$
$$K = \text{consolidation factor (lb/ft}^3)$$
$$T_c = \text{consolidation time (years)}$$

In the case of heterogeneous sediment mixtures, the specific weight of a mixture is calculated using weight-averaged values for each size fraction, as shown in Example 12.1.

Example 12.1 Density of sediment deposits

The source of sediment entering a large reservoir contains 20% sand, 65% silt, and 15% clay:

(1) Determine the dry specific weight of the mixture after 1 year, and after 100 years, considering nearly submerged conditions; refer to the following tabulation

Size class	Fraction Δp_i	$\overline{\gamma}_{md1}$ lb/ft^3	$\Delta p_i \cdot \overline{\gamma}_{md1}$ lb/ft^3	K lb/ft^3	$\overline{\gamma}_{md100}$ lb/ft^3	$\Delta p_i \cdot \overline{\gamma}_{md100}$ lb/ft^3
Sand	0.20	93	18.6	0	93	18.6
Silt	0.65	65	42.2	5.7	74	48.1
Clay	0.15	30	4.5	16.0	55	8.2
Total	1.00		$\overline{\gamma}_{md1} = 65.3$ lb/ft^3 after 1 year		$\overline{\gamma}_{md100} = 75$ lb/ft^3 after 100 years	

(2) Calculate the porosity after 1 year and after 100 years.

$$P_{o(1\ year)} = 1 - \frac{\overline{\gamma}_{md1}}{\gamma_s} = 1 - \frac{65.3}{2.65 \times 62.4} = 0.60$$

$$P_{o(100\ years)} = 1 - \frac{\overline{\gamma}_{md100}}{\gamma_s} = 1 - \frac{75}{2.65 \times 62.4} = 0.55$$

12.5 Life expectancy of reservoirs

The life expectancy of a reservoir is the expected time at which the reservoir will be completely filled with sediments. Its determination requires knowledge of the storage capacity or volume of the reservoir $\forall_R$, the mean annual incoming total sediment discharge Q_t in weight per year, the sediment size distribution, the trap efficiency of the reservoir T_E, and the dry specific weight of sediment deposits γ_{md}. After transforming the incoming mean annual sediment discharge into volume of sediment trapped in the reservoir, the life expectancy T_R is

$$T_R = \frac{\forall_R\, \gamma_{mdT}}{\sum_i T_{Ei}\, \Delta p_i\, Q_{ti}} \qquad (12.12)\blacklozenge$$

The life expectancy represents an average duration upon which the economic feasibility of the reservoir can be based. The accuracy of life expectancy calculations depends on the annual sediment discharge. Mean annual values are useful for long-term estimates. However, the analysis of extreme events is also important when the life expectancy is short and the variability in mean annual sediment yield is large (e.g. arid areas). The probability of occurrence of one or several severe events that may fill the reservoir before the expected life duration must then be considered. For instance, the risk incurred by the occurrence of extreme floods in the next five years, and its impact on the life expectancy must be examined. The risk of an extreme event causing severe depletion of the live storage would detrimentally impact further use of the reservoir for hydropower generation and flood control.

In reservoirs having a reduced capacity–inflow ratio, it is important to consider aspects of the incoming sediment load and the settling capacity of fine particles: (1) What are the likely changes in land use of the watershed in terms of potential increase in sediment production from upland areas during the life expectancy of the reservoir? (2) What is the possible impact caused by an extreme event such as a 1,000 year flood on the storage capacity and operation of the reservoir? (3) What is the potential effect of flocculation on changes in trap efficiency of the reservoir?

12.6 Density currents

In general, engineering problems associated with density currents are concerned with the passage of sediment through the reservoir and with the water quality issues of the sediment-laden waters. A density current may be defined as the movement under gravity of fluids of slightly different density. For density currents in reservoirs, the density difference is caused by sediment-laden river water with a specific weight different from clear water in the reservoir.

Table 12.2 shows the mass density of sediment mixtures at different temperatures. It takes a sediment concentration of approximately 330 mg/l to compensate a temperature decrease of 1°C. For instance, at a concentration of 2,500 mg/l, the density of a mixture at 25°C is less than that of clear water at 10°C. In thermally stratified reservoirs during summer months, warm sediment-laden water would flow along the thermocline, to result in interflow, as sketched in Figure 12.7.

Density currents consist chiefly of particles in suspension of less than 20 μm in diameter. Stokes' law gives a settling rate for particles of 10 μm diameter of approximately 0.0001 ft/s, thus, they need about three hours to settle over a distance of one foot. Also, transverse turbulent fluctuations of the order of 1 percent in a current having a mean velocity of only 0.1 ft/s would be sufficient to keep such particles in suspension. Sediment particles found in most density currents are commonly referred to as "washload," originating from erosion on the land slopes of the drainage area rather than from the streambed. In the river, the concentration of such material is practically constant from bed to surface, and is relatively independent of major changes in flow conditions that occur at the plunge point of the reservoir.

The relation between the mass density difference $\Delta\rho$, or the specific weight difference $\Delta\gamma$, and the volumetric concentration is described by

$$\frac{\Delta\rho}{\rho} = \frac{\Delta\gamma}{\gamma} = C_v(G-1) = 10^{-6}\frac{(G-1)}{G}C_{mg/l} \qquad (12.13)$$

Table 12.2. *Mass density of sediment mixtures ρ_m in kg/m^3*

Temp. °C	Clear water	Sediment concentration mg/l					
		500	1,000	2,500	5,000	10,000	25,000
4	999.97	1,000.28	1,000.59	1,001.53	1,003.09	1,006.20	1,015.54
5	999.96	1,000.28	1,000.59	1,001.52	1,003.08	1,006.19	1,015.53
7	999.90	1,000.21	1,000.52	1,001.46	1,003.01	1,006.13	1,015.47
10	999.69	1,000.01	1,000.32	1,001.26	1,002.81	1,005.92	1,015.26
15	999.10	999.41	999.72	1,000.66	1,002.21	1,005.32	1,014.65
20	998.20	998.52	998.83	999.76	1,001.31	1,004.42	1,013.75
25	997.05	997.37	997.68	998.61	1,000.16	1,003.27	1,012.58
30	995.68	995.99	996.30	997.23	998.78	1,001.88	1,011.18
35	994.12	994.43	994.74	995.67	997.22	1,000.31	1,009.60
40	992.42	992.73	993.04	993.97	995.51	998.60	1,007.87

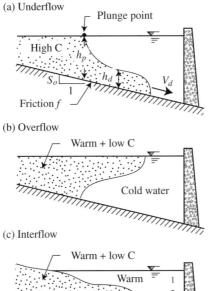

1: Epilimnion
2: Thermocline
3: Hypolimnion

Figure 12.7. Types of density currents a) underflow b) overflow c) interflow

Per Figure 12.7, underflows are the most common type of density currents. High concentrations of washload ($C > 5,000$ mg/l) can overcome density differences due to water temperature (e.g. Table 12.2). The plunge point usually appears when the densimetric Froude number at the plunge point is $\mathrm{Fr}_p = V_p/\sqrt{gh_p\Delta\rho/\rho} \cong 0.5 - 0.78$. The flow depth at the plunge point h_p is calculated from the unit discharge q as $h_p = (\rho q^2/\mathrm{Fr}_{ps}^2 g \Delta\rho)^{1/3}$.

By analogy with open-channel flow, the average velocity V_d of the density current of thickness h_d given the bed slope S_o of the reservoir is a function of the densimetric friction factor f_d

$$V_d = \sqrt{\frac{8\Delta\rho g h_d S_o}{\rho f_d(1+\alpha_d)}} \tag{12.14}$$

the density current unit discharge corresponds to $q_d = V_d h_d$ and the sediment flux by volume q_{tv} is calculated by $q_{tv} = C_v q_d$ as the sediment concentration is relatively uniform. In the case of turbulent flow, $\mathrm{Re}_d = V_d h_d/v > 1000$, the densimetric friction factor can be approximated by $f_d \simeq 0.01$ and $\alpha_d \simeq 0.5$.

Any attempt to refine the analysis of turbulent flows is hindered by the fact that as the degree of turbulence increases, the interface becomes increasingly difficult to define, due to mixing and resulting vertical density variations. The interface of a density current at very low velocities is smooth and distinct, and consists of a sharp discontinuity of density across which the velocity variation is continuous. As the relative velocity between the two layers is increased, waves are formed at the interface, and at a certain critical velocity the mixing process begins by the periodic breaking of the interfacial waves.

Various criteria for determining the flow conditions at which mixing begins have been proposed. Keulegan (1949) defined a mixing stability parameter ϑ from viscous and gravity forces as

$$\vartheta = \left(\frac{vg\Delta\rho}{V_d^3\rho}\right)^{1/3} = \left(\frac{1}{\mathrm{Fr}_d^2\,\mathrm{Re}_d}\right)^{1/3} \tag{12.15}$$

in which the densimetric Froude number $\mathrm{Fr}_d = V_d/\sqrt{\frac{\Delta\rho}{\rho}gh_d}$, and the densimetric Reynolds number, $\mathrm{Re}_d = V_d h_d/v$. The average experimental value of ϑ_c for turbulent flow is $\vartheta_c = 0.18$. The flow is stable when $\vartheta > \vartheta_c$ and mixing occurs when $\vartheta < \vartheta_c$.

The corresponding mean flow velocity of a stable density current V_d is calculated from Equation (12.15) and is usually of the order of $V_c \simeq 0.03$ m/s, or 3 km/day. Given the low friction factor $f \simeq 0.01$, the corresponding shear velocity is of the order of $u_* \simeq 1$ mm/s. With suspensions maintained when $u_* > 2\omega$ (Figure 10.10), the corresponding diameter of sediment particles in suspension is therefore less than about 30 μm. It is concluded that silts and clays contribute to density currents. Another consideration for density currents stems from $\text{Fr}_d^2 = 8\,S_o/f\,(1+\alpha)$, which requires a steep slope, typically $S_o > 50$ cm/km.

In summary, underflow occurs when the silt-laden flows have a higher density than the deep reservoir water. Underflow density currents require: (1) large sediment concentrations; (2) sediment particles are typically finer than about 30 μm; and (3) steep bed slope of the reservoir, typically $S_o > 50$ cm/km. Example 12.2 provides an application of this method to the density currents observed in Lake Mead.

Overflow and interflow density currents occur when the reservoir is thermally stratified (shown in Figure 12.7). During the summer months, the surface temperature in the epilimnion may reach 20°–25°C and the density difference with the colder water of the hypolimnion can be of the order of 2 kg/m^3 (see Table 12.2). River flows may carry up to a couple of thousand mg/l of washload (silts and clays) and still be lighter then the colder reservoir waters. In such cases, river waters will spread out as overflow in the shallow epilimnion and interflow when the thermocline is deep. A recent example is described in DeCesare *et al.* (2006).

Example 12.2 Density currents at Lake Mead, United States

Surveys of Lake Mead indicated the existence of underflows when the density difference between the current and the surrounding water is of the order of $\Delta\rho/\rho = 0.0005$. The average bed slope of the reservoir is about 5 ft/mile, or $S_o = 0.00095$. If measurements indicate a density current depth of approximately 15 ft, the magnitude of the velocity can be obtained from Equation (12.14) after assuming $f_d = 0.010$ and $\alpha_d = 0.5$ for turbulent flow.

$$V_d = \sqrt{\frac{8(0.0005)\,32.2\text{ft}}{0.010(1.5)\text{s}^2}} \times \sqrt{15\text{ ft }(0.00095)} = 0.35\text{ft/s}$$

The density current Reynolds number is therefore

$$\text{Re}_d = \frac{V_d h_d}{\nu} = 5 \times 10^5$$

The computed velocity of 0.35 ft/s (approximately 6 miles/day) is consistent with field measurements. The stability of the density current is calculated from Equation (12.15)

$$\vartheta = \left(\frac{1.2 \times 10^{-5}\,\text{ft}^2}{\text{s}} \times \frac{32.2\,\text{ft}}{\text{s}^2} \times \frac{0.0005\,\text{s}^3}{(0.35)^3\,\text{ft}^3} \right)^{1/3} = 0.0165$$

Mixing will occur because $\vartheta < 0.18$ for turbulent flow.

The rate of sediment transport by the density current can also be estimated. From Equation (12.13), $C_v = 0.0003$ or $C = 803$ mg/liter, and if the average width W_d of the density current is taken as 500 ft, then the discharge represented by the density current is $Q_d = Vh_d W_d = 2,625$ ft^3/s, and that of the sediment $Q_{sd} = Q_d C_v = 0.78$ ft^3/s, or 130 lb/s, which is approximately equivalent to a sediment transport rate of 5,600 tons/day.

12.7 Reservoir sedimentation surveys

Reservoir surveys obtained after the closure of a dam provide useful information on the deposition pattern, trap efficiency, and density of deposited sediment. This information may be necessary for efficient reservoir operation. Before undertaking a new survey, the original and/or preceding ones should be studied. Subsequently a decision should be made on which technique to use for the forthcoming survey. There are three basic survey techniques:

(1) For filled reservoirs; sonic equipment measuring primarily the reservoir bottom elevation, and terrestrial or aerial photogrammetric surveying measuring the water surface area at a given stage;
(2) For empty reservoirs; aerial photogrammetric survey of the reservoir topography; and
(3) For partly empty reservoirs; a combination of the above.

In addition it is essential to sample the deposited sediments, to determine specific mass density and particle size distribution.

Electronic surveying equipment can be used quite effectively. Airborne or satellite-based methods prove to be economically attractive; they require a minimum of ground control and with stereoscopic equipment a contour map can easily be drawn. Global Positioning Systems (GPS) and differential GPS are recommended, and a 1–2 meter position is currently considered acceptable for reservoir mapping and sediment volume estimates (US Bureau of Reclamation, 2006).

Sonic sounders or fathometers, operating from a boat crossing the reservoir, detect the reservoir bottom continuously. The sonic sounder consists of a sensor unit containing the transmitting and receiving transducer and a recording unit, which continuously records the water depth on a chart. Sonic sounders are sometimes capable of delineating the differences in densities of submerged deposits.

Sediment samplers, such as gravity core samplers, piston core samplers, and spud-rod samplers, are used to take undisturbed volumetric samples of deposited sediments. As a last resort, radioactive gamma probes determine the specific weight of the deposit in situ without removing a sample.

12.8 Control measures

Because of the large number of variables involved in reservoir sedimentation problems, no single control measure can be suggested. To simplify the discussion, the control measures can be grouped into categories: watershed control, inflow control, and deposition control.

Control of the watershed may be the most effective sediment control measure, if at all possible, since it reduces the sediment production from the watershed. Such a control measure is certainly feasible on small watersheds; however, it may be an expensive long-term undertaking in larger ones. Proper soil conservation practices provide an effective means of reducing erosion. Measures to increase the vegetative cover of a watershed are very important to control sheet, gully, and channel erosion.

The control of sediment inflow into a reservoir can be achieved by proper watershed management supplemented with sediment-retarding structures throughout the watershed. Stream-channel improvement and stabilization should be considered, including the building of settling (or debris) basins, sabo dams, off-channel reservoirs, utilization of existing or new by-pass canals, and vegetation screens (as dense plant growth reduces the flow velocity and causes deposition).

Control of deposition begins with the proper design of a reservoir, particularly the position and operation of outlets, spillways, and possibly sluice gates drawing directly sediment-laden density currents. After deposition has occurred, various methods exist for the removal of sediments. Dredging should be minimized because it is expensive. Flushing and sluicing can be examined as viable alternatives for low dams (Ji, 2006). In reservoirs with a low capacity–inflow ratio, flushing for several days during the flood season is sometimes sufficient to wash out several years' of accumulated sediments. Flushing reservoirs with a large capacity–inflow ratio is generally counter-productive. Two case studies illustrate various aspects of reservoir sedimentation.

Case study 12.1 Tarbela Dam, Pakistan

Tarbela Dam is located on the Indus River near Islamabad in Pakistan. It is the world's largest rock and earth-fill dam, 2.74 km long and 143 m high. Built at a cost of about 1.5 billion US dollars, it has the capacity to generate 3,750 MW of hydropower. The reservoir spans over 80 km upstream of the dam (Figure CS-12.1.1) with a total storage capacity of 11.3 million acre-feet, or 13.9 km^3. The upstream watershed covers 171,000 km^2, of which only 6% receives monsoon rain, the annual precipitation on the remainder of the watershed does not exceed 10 cm (Lowe and Fox, 1982).

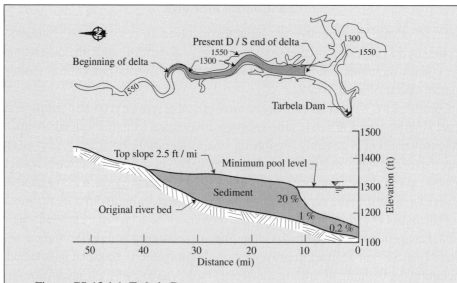

Figure CS-12.1.1 Tarbela Dam

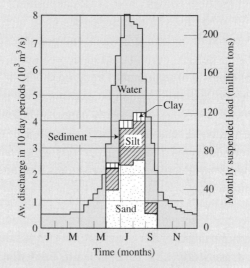

Figure CS-12.1.2 Discharge and sediment load, Indus River

Surface runoff and incoming sediment load depend primarily on melting snow (Figures CS-12.1.2 and 3). The mean annual flow based on a 115-year record is 78.9 km³/year, which is about 5 to 6 times the reservoir storage capacity. The average annual sediment load is of the order of 200 million tons per year. With a sediment size distribution of 60% sand, 33% silt and 7% clay, a dry specific mass of about 1,350 kg/m³ is estimated. With a trap efficiency estimated at 90%,

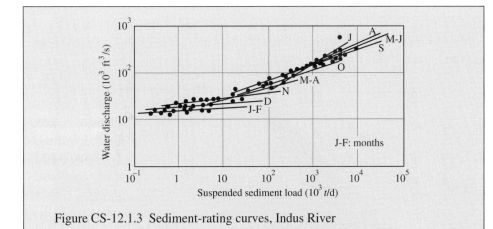

Figure CS-12.1.3 Sediment-rating curves, Indus River

the depletion in storage capacity due to sedimentation is about 0.135 km³ per year, from which the life expectancy is about 100 years.

Case study 12.2 Molineros Reservoir, Bolivia

The projected Molineros Reservoir is located on the Caine River in the Bolivian Andes. In 1983, the proposed project included a 200 m-high dam with hydropower production reaching 132 MW. The 45 km-long reservoir in a narrow gorge has a live storage capacity of 2.98 km³.

The steep watershed is sparsely covered with ground vegetation on very erodible sandstone. About 80% of the 630 mm mean annual precipitation on the 9,530 km² watershed occurs during the wet season extending from December to March. The mean annual flow is 47.2 m³/s with possible flood flows reaching 3,750 m³/s and 4,150 m³/s at a period of return of 1,000 and 10,000 years respectively.

The sediment-rating curve and the flow duration are shown on Figure CS-12.2.1; sediment concentrations as high as 150,000 mg/l have been measured. The analysis of annual suspended sediment load gives 92×10^6 tons/year, shown in Figure CS-12.2.2. Assuming that about 15% of the total load is bedload, a mean annual sediment load of 108×10^6 tons/year is estimated. Given that the sediment size distribution is about 25% sand, 65% silt, and 10% clay, the dry specific mass of reservoir deposits range from 1,100 kg/m³ to 1,300 kg/m³. The trap efficiency of this large reservoir approaches 100% for all size fractions. Considering a reservoir capacity close to 3 km³, the life expectancy varies between 19 and 35 years depending on various alternatives selected. In Figure CS-12.2.2 it is shown that in this case, the inflow of sediment during a 100-year flood is equivalent to about four times the mean annual sediment inflow. Large sediment loads in the river have postponed the otherwise viable project since 1972.

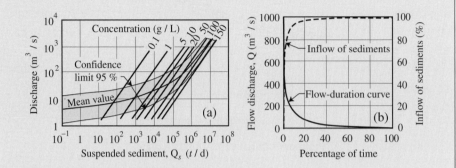

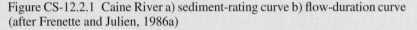

Figure CS-12.2.1 Caine River a) sediment-rating curve b) flow-duration curve (after Frenette and Julien, 1986a)

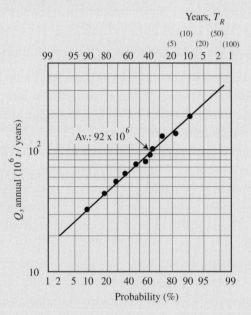

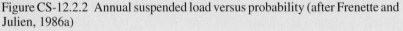

Figure CS-12.2.2 Annual suspended load versus probability (after Frenette and Julien, 1986a)

Exercises

♦♦12.1 Estimate the mass density of a sediment mixture when $C = 17,000$ mg/l at $0°$ C $< T° < 40°$C.

♦12.2. Determine the variability in porosity and C_v as a function of grain size from Equation (12.10a).

Problems
♦♦*Problem 12.1*

From the data given in Case study 12.2 on the Molineros Reservoir Project; (a) determine the trap efficiency and the specific weight of sediment deposits after 10 years; (b) use the flow-duration–sediment-rating curve method to estimate the annual sediment load; (c) calculate the life expectancy of the reservoir; and (d) examine the impact of possible occurrence of a 100-year flood in the next five years on the life expectancy of the reservoir.

♦♦*Problem 12.2*

From the data shown for the interflow of the Imha Reservoir after typhoon Maemi on 9/12/2003 (Kim, 2006) shown in Figure P-12.2, estimate the following: (a) position of the thermocline and epilimnion depth; (b) maximum concentration of suspended solids; (c) mass density of the epilimnion with the sediment and clear water of the hypolimnion; (d) explain why interflow exists in this case; (e) estimate the particle size from the rate of settling; and (f) if the reservoir bottom is at elevation 110 m, how long will it take for the remaining sediment in suspension to settle?

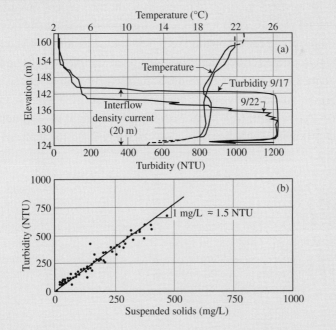

Figure P-12.2 Turbidity and temperature profiles for the Imha Reservoir (from Kim, 2006)

♦♦*Review Problem 12.3: this is a typical review problem for this entire book*

Consider the Missouri River data at a discharge of 2,980 m^3/s, slope 15.5 cm/km, depth 3.08 m, velocity 1.37 m/s, $d_{50} = 0.2$ mm, $d_{90} = 0.26$ mm at a temperature of 22°C. Determine the following in SI: (1) kinematic viscosity; (2) shear velocity; (3) dimensionless grain diameter; (4) Reynolds number of the particle; (5) drag coefficient of settling particles; (6) settling velocity; (7) angle of repose; (8) critical Shields parameter; (9) critical shear stress; (10) critical shear velocity; (11) grain shear Reynolds number; (12) laminar sublayer thickness; (13) calculate z_o; (14) hydraulic radius; (15) Manning n; (16) Darcy–Weisbach f; (17) Chézy C; (18) Froude number; (19) grain roughness height; (20) transport-stage parameter; (21) grain Shields parameter; (22) grain resistance f' and C'; (23) form resistance f'' and C''; (24) bedform type from Simons and Richardson; (25) dune height; (26) dune length; (27) dominant mode of sediment transport (bedload or suspended load); (28) bedload thickness; (29) bedload using MPM; (30) bedload using Einstein–Brown; (31) near-bed concentration Ca; (32) concentration Ca by volume and in ppm; (33) Rouse number assuming $\kappa = 0.4$; (34) sediment concentration at mid-depth; (35) flux-averaged concentration from Yang's equation; (36) bed material discharge from Yang's equation; (37) vertical mixing coefficient; (38) lateral mixing coefficient; (39) dispersion coefficient; (40) time scale for vertical mixing; (41) length scale for vertical mixing; (42) time scale for lateral mixing; (43) length scale for lateral mixing; (44) what sampler would be appropriate for suspended sediment measurements; (45) what sampler would be appropriate for bed material; (46) what sampler would be appropriate for bedload measurements; (47) what laboratory technique would be appropriate for the particle size distribution of the bed material; (48) what is the trap efficiency over $\Delta x = 1$ km; (49) what is the specific weight of sediment after 100 years; and (50) what would be the porosity of the bed sediment after 1,000 years.

Appendix A

Einstein's Sediment Transport Method

Einstein's method (1950) is herein presented using the original notations. Einstein's bed sediment discharge function gives the rate at which flow of any magnitude in a given channel transports the individual sediment sizes found in the bed material. His equations are extremely valuable in many studies for determining the time change in bed material when each size moves at its own rate. For each size D_s of the bed material, the bedload discharge is given as

$$i_B \, q_B \qquad\qquad (A.1)$$

and the suspended sediment discharge is given by

$$i_s \, q_s \qquad\qquad (A.2)$$

and the total bed sediment discharge is

$$i_T \, q_T = i_s \, q_s + i_B \, q_B \qquad\qquad (A.3)$$

and finally

$$Q_T = \sum i_T \, q_T \qquad\qquad (A.4)$$

where i_T, i_s, and i_B are the fractions of the total, suspended and contact bed sediment discharges, q_T, q_s, and q_B for a given grain size D_s. The term Q_T is the total bed sediment discharge. The suspended sediment discharge is related to the bed sediment discharge because there is a continuous exchange of particles between the two modes of transport.

With suspended sediment discharge related to the bed sediment discharge, Equation A.3 becomes

$$i_T q_T = i_B q_B (1 + P_E I_1 + I_2) \qquad\qquad (A.5)$$

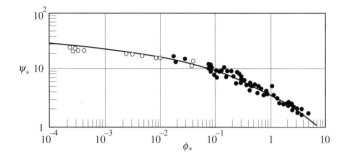

Figure A.1. Einstein's $\varphi_* - \psi_*$ bedload function (Einstein, 1950)

where

$$i_B \, q_B = \frac{\varphi_* i_B \gamma_s}{\left(\dfrac{\rho}{\rho_s - \rho} \dfrac{1}{g \, D_s^3} \right)^{1/2}} \tag{A.6}$$

and

γ_s = the unit weight of sediment;

ρ = the density of water;

ρ_s = the density of sediment;

g = gravitational acceleration;

φ^* = dimensionless sediment transport function $f(\psi_*)$ given in Figure A.1

$$\psi_* = \xi Y (\log 10.6 / B_x)^2 \psi \tag{A.7}$$

$$\psi = \left(\frac{\rho_s - \rho}{\rho} \right) \frac{D_s}{R_b' S_f} \tag{A.8}$$

ξ = a correction factor given as a function of $D_s / \overline{X}$ in Figure A.2;

$$\overline{X} = 0.77\Delta, \text{ if } \Delta/\delta' > 1.8; \tag{A.9a}$$

$$\overline{X} = 1.39\delta', \text{ if } \Delta/\delta' < 1.8 \tag{A.9b}$$

Δ = the apparent roughness of the bed, k_s / X;
X = a correction factor in the logarithmic velocity distribution equation given as a function of k_s / δ' in Figure A.3a

$$\delta' = 11.6 \upsilon / V_*' \tag{A.10}$$

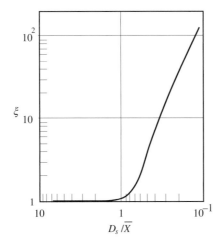

Figure A.2. Hiding factor (Einstein, 1950)

V/V_*' = Einstein's velocity distribution equation

$$V/V_*' = 5.75 \log(30.2 y_o/\Delta) \tag{A.11}$$

V_*' = the shear velocity due to grain roughness

$$V_*' = \sqrt{g\, R_b' S_f} \tag{A.12}$$

R_b' = the hydraulic radius of the bed due to grain roughness,

$R_b' = R_b - R_b''$;

R_b'' = the hydraulic radius of the bed due to channel irregularities

S_f = the slope of the energy grade line normally taken as the slope of the water surface;

Y= another correction term given as a function of D_{65}/δ' in Figure A.3b; and

$B_x = \log (10.6 \overline{X}/\Delta)$.

The preceding equations are used to compute the fraction i_B of the load. The other terms in Equation A.5 are

$$P_E = 2.3 \log 30.2 y_o/\Delta \tag{A.13}$$

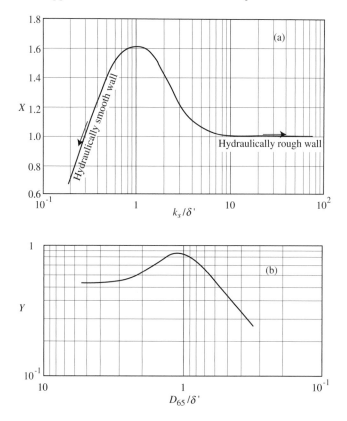

Fig. A.3 a) Einstein's multiplication factor X in the logarithmic velocity equations (Einstein, 1950) b) Pressure correction (Einstein, 1950)

I_1 and I_2 are integrals of Einstein's form of the suspended sediment Equation A.2

$$I_1 = 0.216 \frac{E^{Z-1}}{(1-E)^Z} \int_E^1 \left(\frac{1-y}{y}\right)^Z dy \qquad (A.14)$$

$$I_2 = 0.216 \frac{E^{Z-1}}{(1-E)^Z} \int_E^1 \left(\frac{1-y}{y}\right)^Z \ln y\, dy \qquad (A.15)$$

where

$Z = \omega/0.4V_*'$;

$\omega =$ the fall velocity of the particle of size D_s;

$E =$ the ratio of bed layer thickness to flow depth, a/y_o;

$y_o =$ depth of flow; and

$a =$ the thickness of the bed layer, $2D_{65}$.

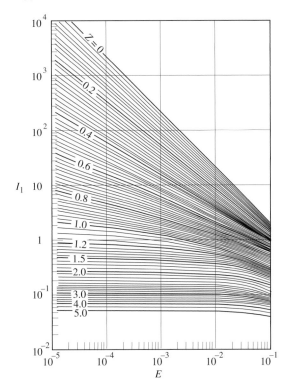

Figure A.4. Integral I_1 in terms of E and Z (Einstein, 1950)

The two integrals I_1 and I_2 are given in Figures A.4 and A.5, respectively, as a function of Z and E.

In the preceding calculations for the total load, the shear velocity is based on the hydraulic radius of the bed due to grain roughness R_b'. Its computation is explained in the following paragraph.

Total resistance to flow is composed of two parts, surface drag and form drag. The transmission of shear to the boundary is accompanied by a transformation of flow energy into turbulence. The part of energy corresponding to grain roughness is transformed into turbulence which stays at least for a short time in the immediate vicinity of the grains and has a great effect on the bedload motion; whereas, the other part of the energy which corresponds to the form resistance is transformed into turbulence at the interface between wake and free stream flow, or at a considerable distance away from the grains. This energy does not contribute to the bedload motion of the particles and may be largely neglected in the sediment transportation.

Einstein's equation for mean flow velocity V in terms of V'_* is

$$V/V'_* = 5.75 \log(12.26 R_b'/\Delta) \tag{A.16a}$$

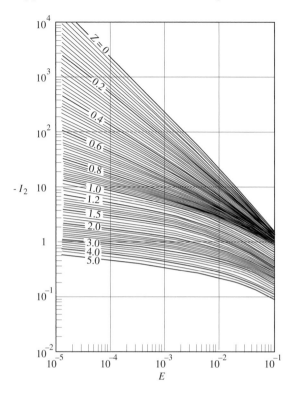

Figure A.5. Integral I_2 in terms of E and Z (Einstein, 1950)

or

$$V/V_*' = 5.75 \log(12.26 R_b'/k_s X) \qquad \text{(A.16b)}$$

Furthermore, Einstein suggested that

$$V/V_*'' = f[\psi'] \qquad \text{(A.17)}$$

where

$$\psi' = \left(\frac{\rho_s - \rho}{\rho}\right) \frac{D_{35}}{R_b' S_f} \qquad \text{(A.18)}$$

The relation for Equation A.18 is given in Figure A.6. The procedure to follow in computing R_b' depends on the information available. If mean velocity V, friction slope S_f, hydraulic radius R_b and bed material size are known, then R_b is computed by trial and error using Equation A.16 and Figure A.6.

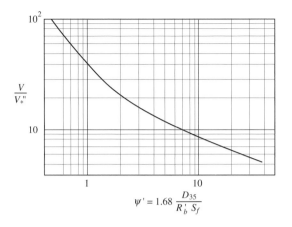

Figure A.6. V/V_*'' vs ψ' (Einstein, 1950)

The procedure for computing total bed sediment discharge in terms of different size fractions of the bed material is:

(1) Calculate ψ^* using Equation A.7 for each size fraction;
(2) Find φ_* from Figure A.1 for each size fraction;
(3) Calculate i_B q_B for each size fraction using Equation A.6;
(4) Sum up the q_B across the flow to obtain i_B Q_B; and
(5) Sum the size fractions to obtain Q_B;

For the suspended sediment discharge:

(6) Calculate Z for each size fraction using the equation below Equation (A-15);
(7) Calculate $E = 2 D_s/y_o$ for each fraction;
(8) Determine I_1 and I_2 for each fraction from Figures A.4 and A.5;
(9) Calculate P_E using Equation A.13;
(10) Compute the suspended discharge from i_B q_B $(P_E I_1 + I_2)$; and
(11) Sum all the q_B and all the i_B to obtain the total suspended discharge Q_{ss}.

Thus, the total bed sediment discharge:

(12) Add the results of Step 5 and 11.

A sample problem showing the calculation of the total bed sediment discharge using Einstein's procedure is presented in Example A.1.

Example A.1 Total bed sediment discharge calculation from Einstein's method

A test reach, representative of the Big Sand Creek near Greenwood, Mississippi, was used by Einstein (1950) as an illustrative example for applying his bed-load function. His example is considered here. For simplicity, the effects due to bank friction are neglected. The reader can refer to the original example for the construction of the representative cross-section. The characteristics of this cross-section are as follows.

The channel slope was determined as $S_f = 0.00105$. The relation between cross-sectional area, hydraulic radius, and wetted perimeter of the representative cross-section and stage are given in Figure A.7. In the case of this wide and shallow channel, the wetted perimeter is assumed to equal the surface width. The averaged values of the four bed-material samples are given in Table A.1. Ninety-six percent of the bed material is between 0.147 and 0.589 mm, which is divided into four size fractions.

The sediment transport calculations are made for the individual size fraction which has the representative grain size equal to the geometric mean grain diameter of each fraction. The water viscosity is $v = 1.0 \times 10^{-5}$ ft^2/sec and the specific gravity of the sediment is 2.65.

The calculation of important hydraulic parameters is performed in Table A.2. The table heading, its meaning, and calculation are explained with footnotes. The bed sediment transport is then calculated for each grain fraction of the bed material at each given flow depth. It is convenient to summarize the calculations in the form of tables. The procedure is given in Table A.3.

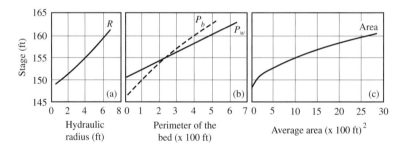

Figure A.7. Description of the average cross-section (Einstein, 1950)

Table A.1. *Bed sediment information for sample problem*

Grain size distribution mm	Average grain size			Settling velocity	
	mm	ft	i_B in %	cm/sec	ft/s
$D_s > 0.60$	–	–	2.4	–	–
$0.60 > D_s > 0.42$	0.50	0.00162	17.8	6.2	0.205
$0.42 > D_s > 0.30$	0.36	0.00115	40.2	4.5	0.148
$0.30 > D_s > 0.21$	0.25	0.00081	32.0	3.2	0.106
$0.21 > D_s > 0.15$	0.18	0.00058	5.8	2.0	0.067
$0.15 > D_s$	–	–	1.8	–	–

$D_{16} = 0.24$ mm $\sigma_g = 1.35$
$D_{35} = 0.29$ mm $Gr = 1.35$
$D_{50} = 0.34$ mm
$D_{65} = 0.37$ mm
$D_{84} = 0.44$ mm

Table A.2. *Hydraulic calculations for sample problem in applying the Einstein procedure (after Einstein, 1950)*

R_b' (1)	V_*' (2)	δ' (3)	κ_s/δ' (4)	X (5)	Δ (6)	V (7)	ψ' (8)	V/V_*'' (9)	V_*'' (10)	R_b'' (11)
0.5	0.129	0.00095	1.21	1.59	0.00072	2.92	2.98	16.8	0.17	0.86
1.0	0.184	0.00067	1.72	1.46	0.00079	4.44	1.49	27.0	0.16	0.76
2.0	0.259	0.00047	2.44	1.27	0.00090	6.63	0.75	51.0	0.13	0.50
3.0	0.318	0.00039	2.95	1.18	0.00097	8.40	0.50	87.0	0.10	0.30
4.0	0.368	0.00033	3.50	1.14	0.00102	9.92	0.37	150.0	0.07	0.14
5.0	0.412	0.00030	3.84	1.10	0.00104	11.30	0.30	240.0	0.05	0.07
6.0	0.450	0.00027	4.26	1.08	0.00107	12.58	0.25	370.0	0.03	0.03

R_b (12)	y_0 (13)	Z_0 (14)	A (15)	P_b (16)	Q (17)	$\overline{X}$ (18)	Y (19)	B_x (20)	$(B/B_x)^2$ (21)	P_E (22)
1.36	1.36	150.2	140	103	409	0.00132	0.84	1.29	0.63	10.97
1.76	1.76	150.9	240	136	1,065	0.00093	0.68	1.19	0.85	11.10
2.50	2.50	152.1	425	170	2,820	0.00069	0.56	0.91	1.27	11.30
3.30	3.30	153.3	640	194	5,380	0.00076	0.55	0.91	1.27	11.50
4.14	4.14	154.9	970	234	9,620	0.00079	0.54	0.91	1.27	11.70
5.07	5.07	156.9	1,465	289	16,550	0.00080	0.54	0.91	1.27	11.90
6.03	6.03	159.5	2,400	398	30,220	0.00082	0.54	0.91	1.27	12.04

(1) R_b' = bed hydraulic radius due to grain roughness, ft.
(2) V_*' = shear velocity due to grain roughness, fps = $\sqrt{gR_b'S_f}$
(3) δ' = thickness of the laminar sublayer, ft = $11.62\, v/V_*'$
(4) κ_s = roughness diameter, ft = D_{65}
(5) X = correction factor in the logarithmic velocity distribution, given in Figure A.3a
(6) Δ = apparent roughness diameter, ft = κ_s/X
(7) V = average flow velocity, fps = $5.75\, V_*'\,\log(12.27\, R_b'/\Delta)$
(8) ψ' = intensity of shear on representative particles
$$= \frac{\rho_s - \rho}{\rho}\,\frac{D_{35}}{R_b'S_f}$$
(9) V/V_*'' = velocity ratio, given in Figure A.6
(10) V_*'' = shear velocity due to form roughness, fps
(11) R_b'' = bed hydraulic radius due to form roughness, ft = $V_*''^2/gS$
(12) R_b = bed hydraulic radius, ft = R, the total hydraulic radius if there is no additional friction = $R_b' + R_b''$
(13) y_0 = average flow depth, ft = R for wide, shallow streams
(14) Z_0 = stage, ft from Figure A.7
(15) A = cross-sectional area, ft^2
(16) P_b = bed wetter perimeter, ft
(17) Q = flow discharge = AV
(18) $\overline{X}$ = characteristic distance, from Equation A.9

Table A.3. *Bed sediment load calculations for sample problem in applying the Einstein procedure (Einstein, 1950)*

D (1)	i_B (2)	R_b' (3)	Ψ (4)	$D/\overline{X}$ (5)	ξ (6)	Ψ_* (7)	φ_* (8)	$i_B q_B$ (9)	$i_B Q_B$ (10)	$\Sigma i_B Q_B$ (11)	$10^3 E$ (12)	Z (13)	I_1 (14)	$-I_2$ (15)	$P_E I_1 + I_2 + 1$ (16)	$i_T q_T$ (17)	$i_T Q_T$ (18)	$\Sigma i_T Q_T$ (19)
0.00162	0.178	0.5	5.08	1.23	1.08	2.90	1.90	0.0267	119.00	400	2.38	3.78	0.078	0.44	1.42	0.03800	168	670
		1.0	2.54	1.74	1.00	1.73	4.00	0.0561	330.00	1335	1.84	2.65	0.131	0.74	1.71	0.09580	561	3928
		2.0	1.27	2.35	1.00	0.90	8.20	0.1150	845.00	3771	1.30	1.88	0.240	1.27	2.44	0.28100	2050	30500
		3.0	0.85	2.16	1.00	0.60	12.80	0.1800	1510.00	6496	0.98	1.53	0.385	2.01	3.44	0.61700	5170	113000
		4.0	0.63	2.05	1.00	0.43	18.00	0.2530	2560.00	10745	0.78	1.33	0.560	2.80	4.75	1.20000	12100	324000
		5.0	0.51	2.03	1.00	0.35	22.50	0.3160	3950.00	16333	0.63	1.18	0.810	3.85	6.78	2.13000	26500	800000
		6.0	0.42	1.98	1.00	0.29	27.00	0.3800	6350.00	27142	0.54	1.08	1.090	4.90	9.20	3.48000	59800	1940000
0.00115	0.402	0.5	3.38	0.82	1.36	2.44	2.45	0.0471	210.00		1.69	2.88	0.117	0.68	1.60	0.07540	335	
		1.0	1.69	1.16	1.10	1.27	5.50	0.1060	623.00		1.31	2.02	0.210	1.19	2.14	0.22700	1330	
		2.0	0.85	1.57	1.01	0.61	12.60	0.2420	1780.00		0.92	1.44	0.450	2.33	3.76	0.91000	6660	
		3.0	0.56	1.44	1.04	0.41	19.00	0.3640	3050.00		0.70	1.17	0.830	3.85	6.73	2.44000	20400	
		4.0	0.42	1.37	1.05	0.30	26.00	0.5000	5050.00		0.56	1.01	1.370	5.70	11.30	5.65000	57100	
		5.0	0.34	1.35	1.05	0.25	31.50	0.6040	7540.00		0.45	0.90	2.120	8.10	17.20	10.40000	129000	
		6.0	0.28	1.32	1.05	0.20	39.00	0.7490	12900.00		0.38	0.83	2.950	10.50	26.00	19.60000	335000	
0.00081	0.320	0.5	2.54	0.61	2.25	3.03	1.75	0.0155	69.00		1.19	1.94	0.230	1.29	2.23	0.03450	153	
		1.0	1.27	0.87	1.26	1.09	6.80	0.0600	353.00		0.92	1.36	0.520	2.60	4.16	0.25000	1460	
		2.0	0.63	1.17	1.10	0.49	15.80	0.1390	1020.00		0.65	0.97	1.530	6.10	12.20	1.70000	12500	
		3.0	0.42	1.08	1.12	0.33	23.50	0.2070	1730.00		0.49	0.79	3.350	11.00	28.70	5.95000	49700	
		4.0	0.32	1.04	1.15	0.25	31.50	0.2790	2820.00		0.39	0.68	6.200	17.50	56.00	15.00000	157000	
		5.0	0.25	1.01	1.17	0.20	39.50	0.3490			0.32	0.61	9.800	25.50	92.00	32.00000	397000	
		6.0	0.21	0.99	1.19	0.17	46.00	0.4060	6980.00		0.27	0.55	15.000	36.00	146.00	59.50000	1020000	
0.00057	0.058	0.5	1.80	0.43	5.40	5.15	0.58	0.00056	2.49		0.85	1.21	0.720	3.35	5.55	0.00312	14	
		1.0	0.90	0.61	2.28	1.39	5.10	0.00500	29.40		0.65	0.86	2.440	8.10	20.00	0.10000	587	
		2.0	0.45	0.83	1.37	0.44	17.50	0.01710	126.00		0.46	0.61	8.400	21.50	74.40	1.26000	9350	
		3.0	0.30	0.76	1.52	0.32	25.00	0.02460	206.00		0.35	0.49	19.300	41.00	183.00	4.50000	37600	
		4.0	0.22	0.72	1.60	0.25	31.50	0.03100	313.00		0.28	0.43	32.000	63.00	312.00	9.68000	97800	
		5.0	0.18	0.71	1.65	0.20	39.50	0.03870	483.00		0.23	0.38	51.000	91.00	516.00	20.00000	248000	
		6.0	0.15	0.70	1.70	0.18	43.50	0.04260	752.00		0.19	0.35	70.000	122.00	722.00	30.80000	526000	

(19) Y = pressure correction term, given in Figure A.3b
(20) B_x = coefficient = log $(10.6 \, \overline{X}/\Delta)]$
(21) B = coefficient = log 10.6
(22) P_E = Einstein's transport parameter
$$= 2.303 \log \frac{30.2 y_o}{\Delta}$$

See the following for explanation of symbols in Table A.3, column by column:

(1) D = representative grain size, ft, given in Table A.1
(2) i_B = fraction of bed material given in Table A.1
(3) R_b' = bed hydraulic radius due to grain roughness, ft, given in Table A.2
(4) ψ = intensity of shear on a particle = $\dfrac{\rho_s - \rho}{\rho} \dfrac{D}{R_b' S_f}$
(5) $D/\overline{X}$ = dimensionless ratio, $\overline{X}$ given in Table A.2
(6) ξ = hiding factor, given in Figure A.2
(7) ψ_* = intensity of shear on individual grain size = $\xi Y (B/B_x)^2 \psi$, (values of Y and $(B/B_x)^2$ are given in Table A.2
(8) φ_* = intensity of sediment transport for individual grain from Figure A.1
(9) $i_B q_B$ = bedload discharge per unit width for a size fraction, lb/sec/ft = $i_B \varphi_* \rho_s (gD)^{3/2} \sqrt{(\rho_s/\rho) - 1}$
(10) $i_B Q_B$ = bedload discharge for a size fraction for entire cross-section, tons/day = 43.2 $W i_B q_B$, $W = P_b$ given in Table A.2 (1 ton = 2000 lbs)
(11) $\Sigma i_B Q_B$ = total bedload discharge for all size fractions for entire cross-section, tons/day
(12) E = ratio of bed layer thickness to water depth = $2 D/y_o$, for values of y_o, see Tables A.2
(13) Z = exponent for concentration distribution = $\omega/(0.4 U_*')$, for values of ω and U_*' see Tables A.1 and A.2
(14) I_1 = integral, given in Figure A.4
(15) $-I_2$ = integral, given in Figure A.5
(16) $P_E I_1 + I_2 + 1$ = factor between bedload and total load, using P_E in Table A.2
(17) $i_T q_T$ = bed material load per unit width of stream for a size fraction, lb/sec-ft = $i_B q_B (P_E I_1 + I_2 + 1)$
(18) $i_T Q_T$ = bed material load for a size fraction of entire cross-section, tons/day = 43.2 $P_b i_B q_B$, P_b is given in Table A.2 (1 ton = 2000 lbs)
(19) $\Sigma i_T Q_T$ = total bed material load for all size fractions, tons/day

Appendix B

Useful mathematical relationships

$$b^x = y, \quad x = \log_b y$$
$$\log(ab) = \log a + \log b$$
$$\log(a/b) = \log a - \log b$$

Quadratic equation

$$ax^2 + bx + c = 0$$

$$x = \frac{-b \pm \sqrt{b^2 - 4ac}}{2a}; \ b^2 > 4ac \text{ for real roots}$$

Determinants

$$\begin{vmatrix} a_1 & b_1 \\ a_2 & b_2 \end{vmatrix} = a_1 b_2 - a_2 b_1$$

$$\begin{vmatrix} a_1 & b_1 & c_1 \\ a_2 & b_2 & c_2 \\ a_3 & b_3 & c_3 \end{vmatrix} = a_1 (b_2 c_3 - b_3 c_2) - b_1 (a_2 c_3 - a_3 c_2) + c_1 (a_2 b_3 - a_3 b_2)$$

Trigonometry

$$\sin^2 \theta + \cos^2 \theta = 1$$
$$1 + \tan^2 \theta = \sec^2 \theta$$
$$1 + \cot^2 \theta = \mathrm{cosec}^2 \theta$$

$$\sin(\theta/2) = \sqrt{\frac{1}{2}(1-\cos\theta)}$$

$$\cos(\theta/2) = \sqrt{\frac{1}{2}(1+\cos\theta)}$$

$$\sin(2\theta) = 2\sin\theta\cos\theta$$

$$\cos(2\theta) = \cos^2\theta - \sin^2\theta$$

$$\sin(a\pm b) = \sin a\cos b \pm \cos a\sin b$$

$$\cos(a\pm b) = \cos a\cos b \mp \sin a\sin b$$

Series

$$\sin x = x - \frac{x^3}{3!} + \frac{x^5}{5!} - \frac{x^7}{7!} \cdots$$

$$\cos x = 1 - \frac{x^2}{2!} + \frac{x^4}{4!} - \frac{x^6}{6!} \cdots$$

$$\sinh x = x + \frac{x^3}{3!} + \frac{x^5}{5!} + \frac{x^7}{7!} \cdots$$

$$\cosh x = 1 + \frac{x^2}{2!} + \frac{x^4}{4!} + \frac{x^6}{6!} \cdots$$

Derivatives

$$d\ \sin x = \cos x\, dx$$

$$d\ \cos x = -\sin x\, dx$$

$$d\ \tan x = \sec^2 x\, dx$$

$$d\ \sinh x = \cosh x\, dx$$

$$d\ \cosh x = \sinh x\, dx$$

$$d\ \tanh x = \mathrm{sech}^2 x\, dx$$

Integrals

$$\int x^n dx = \frac{x^{n+1}}{n+1}$$

$$\int \frac{dx}{x} = \ln x$$

$$\int \frac{dx}{\sqrt{a+bx}} = 2\frac{\sqrt{a+bx}}{b}$$

$$\int \frac{x\,dx}{\sqrt{a+bx}} = \frac{1}{b^2}\left[a+bx - a\,\ln(a+bx)\right]$$

$$\int \frac{dx}{a+bx^2} = \frac{1}{\sqrt{ab}}\tan^{-1}x\frac{\sqrt{ab}}{a}$$

$$\int \frac{x\,dx}{a+bx^2} = \frac{1}{2b}\ln\left(a+bx^2\right)$$

$$\int \ln x\,dx = x\,\ln x - x$$

$$\int e^{ax}\,dx = \frac{e^{ax}}{a}$$

$$\int x e^{ax}\,dx = \frac{e^{ax}}{a^2}(ax-1)$$

$$\int \sin x\,dx = -\cos x$$

$$\int \cos x\,dx = \sin x$$

$$\int_0^{\pi/2} \sin x\,dx = \int_0^{\pi/2} \cos x\,dx = 1$$

$$\int \sin^2 x\,dx = \frac{x}{2} - \frac{\sin 2x}{4}$$

$$\int \cos^2 x\,dx = \frac{x}{2} + \frac{\sin 2x}{4}$$

$$\int_0^{\pi/2} \sin^2 x\,dx = \int_0^{\pi/2} \cos^2 x\,dx = \frac{\pi}{4}$$

$$\int \sin x \cos x\,dx = \frac{\sin^2 x}{2}$$

$$\int_0^{\pi/2} \sin x \cos x\,dx = \frac{1}{2}$$

$$\int \sin^3 x\,dx = -\frac{\cos x}{3}\left(2 + \sin^2 x\right)$$

$$\int \cos^3 x \, dx = \frac{\sin x}{3} \left(2 + \cos^2 x \right)$$

$$\int_0^{\pi/2} \sin^3 x \, dx = \int_0^{\pi/2} \cos^3 x \, dx = \frac{2}{3}$$

$$\int_0^{\pi/2} \sin^4 x \, dx = \frac{3\pi}{16}$$

$$\int \cos^5 x \, dx = \sin x - \frac{2}{3} \sin^3 x + \frac{1}{5} \sin^5 x$$

Bibliography

Ackers, P. and W. R. White (1973). Sediment transport: New approach and analysis. *J. Hyd. Div. ASCE*, **99**, no. HY11, 2041–60.

Adriaanse, M. (1986). De ruwheid van de Bergsche Maas bij hoge afvoeren. Rijkswaterstaat, RIZA, Nota 86.19. Dordrecht.

Akalin, S. (2002). Water temperature effect on sand transport by size fraction in the lower Mississippi River. Ph.D. dissertation, Colorado State University, 243 p.

Alonso, C. V., W. H. Neibling, and G. R. Foster (1991). Estimating sediment transport capacity in watershed modeling. *Trans. ASAE*, **24**, no. **5**, 1211–20.

American Society of Civil Engineers (1975). Sedimentation Engineering. ASCE Manuals and Reports on Engineering Practice no. 54. New York.

American Society of Civil Engineers (2007). "Sedimentation Engineering – Processes, Measurement, Modeling and Practice", Manual and Reports on Engineering Practice No. 110.

Anderson, A. G. (1942). Distribution of suspended sediment in natural streams *Trans. Am. Geoph. Union*, 678–83.

Athaullah, M. (1968). Prediction of bedforms in erodible channels. Ph.D. thesis, Colorado State University.

Atkinson, E. (1994). Vortex-tube sediment extractors I: trapping efficiency. *J. Hyd. Engr. ASCE*, **120**, no. 10, 1110–25.

Bagnold, R. A. (1954). Experiments on a gravity-free dispersion of large solid spheres in a Newtonian fluid under shear. *Proc. Roy. Soc. Lond.* **A225**, 49–63.

—— (1956). Flow of cohesionless grains in fluids. *Phil. Trans. Roy. Soc. Lond.*, **249**, no. 964, 235–97.

—— (1966). An approach to the sediment transport problem from general physics. Professional Paper 422-I. U.S. Geological Survey, Washington, D.C.

Baird, D. C. (2004). Turbulent flow and suspended sediment transport in a mobile, sand-bed channel with riprap side slopes. Ph.D. dissertation, University of New Mexico, 260 p.

Bakke, P. D., P. O. Basdekas, D. R. Dawdy, and P. C. Klingeman (1999). Calibrated Parker-Klingeman model for gravel transport. *J. Hyd. Div. ASCE*, **25** no. 6, 657–60.

Barekyan, A. S. (1962). Discharge of channel forming sediments and elements of sand waves. *Soviet Hydrol. Selected Papers*, 128–30.

Beverage, J. P. and J. K. Culbertson (1964). Hyperconcentrations of suspended sediment. *J. Hyd. Div. ASCE*, **90**, no. HY6, 117–28.

Bingham, E. C. (1922). *Fluidity and Plasticity*. New York: McGraw-Hill.

Bird, R. B., W. E. Stewart, and E. N. Lightfoot (1960). *Transport Phenomena*. New York: Wiley.

Bishop, A. A., D. B. Simons, and E. V. Richardson (1965). Total bed material transport. *J. Hyd. Div. ASCE*, **91**, no. HY2, 175–91.

Bogardi, J. (1974). *Sediment Transport in Alluvial Streams*. Budapest: Akademiai Kiado.

Borland, W. M. (1971). Reservoir sedimentation. In *River Mechanics*, ed. H. W. Shen. Ft. Collins, Colo.: Water Resources Pub.

Borland, W. M. and C. R. Miller (1958). Distribution of sediment in large reservoirs. *J. Hyd. Div. ASCE*, **84**, no. HY2, 1–18.

Bormann, N. E. and P. Y. Julien (1991). Scour downstream of grade-control structures. *J. Hyd. Engr., ASCE*, Vol. **115**, No.5, 579–94.

Bounvilay, B. (2003). Transport velocities of bedload particles in rough open channel flows. Ph.D. dissertation, Colorado State University, 155p.

Boyce, R. (1975). Sediment routing and sediment-delivery ratios. In *Present and Prospective Technology for Predicting Sediment Yields and Sources*, USDA-ARS-S-40, pp. 61–5.

Bray, D. I. (1982). Flow resistance in gravel-bed rivers. In *Gravel-Bed Rivers, Fluvial Processes, Engineering and Management*, ed. R. D. Hey, J. C. Bathurst, and C. R. Thorne. New York: Wiley. pp. 109–37.

Bridgman, P. W. (1922). *Dimensional Analysis*. New Haven, Conn.: Yale University Press.

Brooks, N. H. (1958). Mechanics of streams with movable beds of fine sediment. *Trans. ASCE*, **123**: 525–49.

Brown, C. B. (1950). Sediment transportation. In *Engineering Hydraulics*, ed. H. Rouse. New York: Wiley. pp. 769–857.

Brownlie, W. R. (1981). Prediction of flow depth and sediment discharge in open-channels. Report no. KH-R-43A. Pasadena: California Institute of Technology, W. M. Keck Laboratory.

Buckingham, E. (1914). On physical similar systems: Illustration of the use of dimensionless equations. *Phys. Rev.*, **4**, no. 4.

Bunte, K. and S. R. Abt (2001). Sampling surface and subsurface particle-size distributions in wadable gravel- and cobble-bed streams for analyses in sediment transport, hydraulics, and streambed monitoring. General Technical Report RMRS-GTR-74, U.S.D.A. Forest Service, Rocky Mountain Research Station, 428 p.

Bunte, K. and S. R. Abt (2005). Effect of sampling time on measured gravel bed-load transport rates in a coarse-bedded stream. *Water Resources Research*, **41**, W11405.

Bunte, K., S. R. Abt, J. P. Potyondy, and S. E. Ryan. (2004). Measurement of coarse gravel and cobble transport using a portable bedload trap. *J. Hyd. Engr., ASCE*, **130** No. 9: 879–93.

Bunte, K., S. R. Abt, J. P. Potyondy, and K. W. Swingle. (2008). A comparison of coarse bed-load transport measured with bedload traps and Helley-Smith samplers. *Geodinamica Acta*, **21**/1–2: 53–66.

Burkham, D. E. and Dawdy, D. R. (1980). General study of the Modified Einstein Method of computing total sediment discharge. Water-Supply Paper 2066, U.S. Geological Survey, Washington.

Carstens, M. R. (1952). Accelerated motion of a spherical particle. *Trans. Am. Geophy. Union*, **33**, no. 5.

Cecen, K., M. Bayazit and M. Sumer (1969). Distribution of suspended matter and similarity criteria in settling basins. Proc. 13th Congress of the IAHR, Kyoto, Japan Vol. 4: 215–25.

Chabert, J. and J. L. Chauvin (1963). Formation de dunes et de rides dans les modèles fluviaux. *Bull. Cen. Rech. Ess. Chatou*, no. **4**.

Chang, F. M., D. B. Simons, and E. V. Richardson (1965). Total bed material discharge in alluvial channels. Water Supply Paper 1498-I. Washington, D.C.: U.S. Geological Survey.

Chien, N. (1956). The present status of research on sediment transport. *Trans. ASCE*, **121**, 833–68.

Chow, V. T. (1964). *Open Channel Hydraulics*. New York: McGraw-Hill.

Colby, B. R. (1960). Discontinuous rating curves for Pigeon Roost and Cuffawa Creeks in northern Mississippi. ARS 41–36. Washington, D.C.: U.S. Department of Agriculture.

—— (1964a). Discharge of sands and mean velocity relationships in sand-bed streams. Professional Paper 462-A. Washington, D.C.: U.S. Geological Survey.

—— (1964b). Practical computations of bed material discharge. *J. Hyd. Eng.*, **90**, no. HY2, 217–46.

Colby, B. R. and C. H. Hembree (1955). Computations of total sediment discharge Niobrara River near Cody, Nebraska. Water Supply Paper 1357. Washington, D.C.: U.S. Geological Survey.

Colby, B. R. and Hubbell, D. W. (1961). Simplified method for computing total sediment discharge with the modified Einstein procedure. Water-Supply Paper 1593., U.S. Geological Survey, Washington.

Coleman, N. L. (1981). Velocity profiles with suspended sediment. *J. Hyd. Res.*, **19**, no. 3, 211–29.

—— (1986). Effects of suspended sediment on the open-channel velocity distribution. *Water Resources Res.*, **22**, no. 10, 1377–84.

Coles, D. E. (1956). The law of the wake in the turbulent boundary layer. *J. Fluid Mech.*, **1**, 191–226.

Culbertson, J. K., C. H. Scott and J. P. Bennett (1972). Summary of alluvial channel data from the Rio Grande Conveyance Channel, New Mexico, 1965–69. USGS Prof. Paper 562-J.

Dawdy, D. R. (1961). Studies of flow in alluvial channels, depth, height and discharge relation for alluvial streams. Water Supply Paper 1798c. Washington, D.C.: U.S. Geological Survey.

De Cesare, G., J. L. Boillat and A. J. Schleiss (2006). Circulation in stratified lakes due to flood-induced turbidity currents. *J. Env. Eng. ASCE* Vol. **132** No. 11, 1508–17.

DuBoys, M. P. (1879). Etudes du régime du Rhône et de l'action exercée par les eaux sur un lit à fond de graviers indéfiniment affouillable. *Ann. Ponts et Chaussées, ser.* **5**, 18, 141–95.

Edwards, T. K. and E. D. Glysson (1988). Field methods for measurement of fluvial sediment. Open-File Report 86–531. Reston, Va: U.S. Geological Survey.

Egiazaroff, I. V. (1965). Calculation of non-uniform sediment concentration. *J. Hyd. Div. ASCE*, **91**, no. HY4, 225–48.

Einstein, H. A. (1942). Formulas for the transportation of bed load. *Trans. ASCE*, **107**, 561–73.

—— (1950). The bed load function for sediment transport in open channel flows. Technical Bulletin no. 1026. Washington, D.C.: U.S. Department of Agriculture, Soil Conservation Service.

—— (1964). Sedimentation, Part II: River sedimentation. In *Handbook of Applied Hydrology*, ed. V. T. Chow. New York: McGraw-Hill, sec. 17.

Einstein, H. A. and N. Chien (1955). Effects of heavy sediment concentration near the bed on velocity and sediment distribution. MRD Series no. 8. Berkeley: University of

California: Institute of Engineering Research; Omaha, Neb.: U.S. Army Engineering Div., Missouri River, Corps of Engineers.

Emmett, W. W. (1979). A field calibration of the sediment trapping characteristics of the Helley-Smith bedload sampler. Open File Report 79–411. Denver, Colo.: U.S. Geological Survey.

Engelund, F. (1966). Hydraulic resistance of alluvial streams. *J. Hyd. Div. ASCE*, **92**, no. HY2, 315–26.

—— (1977). Hydraulic resistance for flow over dunes. Prog. Rep. 44, Inst. Hydrodynamics and Hydraulic Engineering, Tech University, Denmark, Dec. 19–20.

Engelund, F. and E. Hansen (1967). *A Monograph on Sediment Transport to Alluvial Streams*. Copenhagen: Teknik Vorlag.

Etcheverry, B. A. (1916). Irrigation practice and engineering: The conveyance of water. *Trans. ASCE*, **11**.

Fischer, H. B., E. J. List, R. C. Koh, J. Imberger, and N. H. Brooks (1979). *Mixing in Inland and Coastal Waters*. New York: Academic Press.

Fortier, A. (1967). *Mécanique des Suspensions. Monographie de Mécanique des Fluides et Thermique*. Paris: Masson.

Fortier, S. and F. C. Scobey (1926). Permissible canal velocities. *Trans. ASCE*, **89**, paper no. 1588, 940–84.

Frenette, M. and P. Y. Julien (1985). Computer modeling of soil erosion and sediment yield from large watersheds. Proc. 27th Annual Convention of the Institute of Engineers, Peshawar, Pakistan.

—— (1986a). Advances in predicting reservoir sedimentation. General lecture, Third International Symposium on River Sedimentation, ISRS-111, Jackson, Mississippi, March 31 to April 4, pp. 26–46.

—— (1986b). LAVSED I: A model for predicting suspended load in northern streams. *Can. J. Civ. Eng.*, **13**, no. 2, 150–61.

—— (1987). Computer modeling of soil erosion and sediment yield from large watersheds. *Int. J. Sed. Res.* **1**, 39–68.

Frenette, M., J. C. Souriac, and J. P. Tournier (1982). Modélisation de l'alluvionnement de la retenue de Péligre, Haiti. Proc. 14th International Congress on Large Dams, Rio de Janeiro, May 3–7, pp. 93–115.

Frenette, M., J. P. Tournier, and T. J. Nzakimuena (1982). Cas historique des sédimentation du barrage Péligre, Haiti. *Can. J. Civ. Eng.*, **9**, no. 2, 206–23.

Garde, R. J. and K. G. Ranga Raju (1985). *Mechanics of Sediment Transportation and Alluvial Stream Problems*, 2d ed. New York: Wiley.

Gessler, J. (1971). Beginning and ceasing of sediment motion. In *River Mechanics*, ed. H. W. Shen. Littleton, Colo.: Water Resources Pub., Chap. 7.

Gilbert, G. K. (1914). The transport of debris by running water. Professional Paper 86. Washington, D.C.: U.S. Geological Survey.

Goldstein, S. (1929). The steady flow of viscous fluid past a fixed spherical obstacle at small Reynolds number. *Proc. Roy. Soc. Lond.*, **123A**, 225–35.

Gordon, L. (1992). *Mississippi River Discharge*. San Diego, CA: RD Instruments.

Govier, G. W., C. A. Shook, and E. O. Lilge (1957). The properties of water suspension of finely subdivided magnetite, galena, and ferrosilicon. *Trans. Can. Inst. Mining Met.*, **60**, 147–54.

Graf, W. H. (1971). *Hydraulics of Sediment Transport*. New York: McGraw-Hill.

Guo, J. (1998). Turbulent velocity profiles in clear water and sediment laden flows. Ph.D. dissertation, Colorado State University.

Guo, J. and P. Y. Julien (2003). Modified log-wake law for turbulent flow in smooth pipes. *J. Hyd. Res, IAHR*, **41**, no. 5, 493–501.

Guo, J. and P. Y. Julien (2004). An efficient algorithm for computing Einstein integrals, Technical Note, *J. Hyd. Engr., ASCE*, **130**, no.12, 1198–1201.

Guo, J. and P. Y. Julien (2007). Buffer law and roughness effect in turbulent boundary layers, Fifth International Symposium on Environmental Hydraulics, Tempe, Arizona, 6p.

Guo, J. and P. Y. Julien (2008). Applications of the modified log-wake law in open channels. *J. Applied Fluid Mechanics*, **1**, no. 2, 17–23.

Guy, H. P. and V. W. Norman (1970). USGS techniques of water resources investigations. In *Fluid Methods for Measurement of Fluvial Sediments*. Washington, D. C.: U.S. Government Printing Office, Book 3, Chap. C-2.

Guy, H. P., D. B. Simons, and E. V. Richardson (1966). Summary of alluvial channel data from flume environments, 1956–1961. Professional Paper 462-I. Washington, D.C.: U.S. Geological Survey.

Happel, J. and H. Brenner (1965). *Low Reynolds Number Hydrodynamics*. Englewood Cliffs, N.J.: Prentice-Hall.

Hartley, D. M. and P. Y. Julien (1992). Boundary shear stress induced by raindrop impact. *J. Hyd. Res. IAHR*, **30**, no. 3, 351–9.

Hayashi, T., S. Ozaki and T. Ichibashi (1980). Study on bed load transport of sediment mixture. Proc. 24th Japanese Conference on Hydraulics.

Henderson, F. M. (1966). *Open Channel Flow*. New York: Macmillan.

Highway Research Board (1970). Tentative design procedure for riprap-lined channels. Report no. 108. Washington, D.C.: National Academy of Sciences, National Cooperative Highway Research Program.

Holmquist-Johnson, C. and Raff, D.A (2006). Bureau of Reclamation automated modified Einstein procedure (BORAMEP) program for computing total sediment discharge. U.S. Department of the Interior, Bureau of Reclamation, Denver, CO.

Hubbel, D. W. and D. Q. Matejka (1959). Investigation of sediment transportation, Middle Loup River at Dunning, Nebraska. USGS Water Supply Paper 1476.

Hunsaker, J. C. and B. G. Rightmire (1947). *Engineering Applications of Fluid Mechanics*. New York: McGraw-Hill.

Interagency Committee (1952). The design of improved types of suspended sediment samplers. Report no. 6. Iowa City: Subcommittee on Sedimentation, Federal Interagency River Basin Committee, Hydrology Laboratory of the Iowa Institute of Hydraulic Research.

Jenson, V. J. (1959). Viscous flow round a sphere at low Reynolds numbers. *Proc. Roy. Soc. Lond.*, **A249**, 346–66.

Ji, U. (2006). Numerical model for sediment flushing at the Nakdong River Estuary Barrage. Ph.D. dissertation, Colorado State University, 195 p.

Johnson, B. E., P. Y. Julien, D. K. Molnar and C. C. Watson (2000). The two-dimensional upland erosion model CASC2D-SED, *JAWRA*, AWRA, **36**, no. 1, 31–42.

Julien, P. Y. (1979). Erosion de bassin et apport solide en suspension dans les cours d'eau nordiques. M.Sc. thesis, Laval University.

—— (1982). Prédiction d'apport solide pluvial et nival dans les cours d'eau nordiques à partir du ruissellement superficiel. Ph.D. dissertation, Laval University.

—— (1988). Downstream hydraulic geometry of noncohesive alluvial channels. *Proc. International Conference on River Regime*, Wallingford, England, 9–16.

—— (1992). Study of bedform geometry in large rivers. Delft Hydraulics, Report Q1386. Emmeloord.

—— (2002). *River Mechanics*. Cambridge: Cambridge University Press, 434 p.

Julien, P. Y. and M. Frenette. (1986a) Predicting washload from rainfall and snowmelt runoff. Proc. Third International Symposium on River Sedimentation, ISRS-111, Jackson, Mississippi, pp. 1259–65.

—— (1986b). LAVSED II: A model for predicting suspended load in northern streams. *Can. J. Civ. Eng.*, **13**, no. 2, 162–70.

—— (1987). Macroscale analysis of upland erosion. *Hyd. Sci. J. IAHS*, **32**, no. 3, 347–58.

Julien, P. Y. and M. Gonzalez del Tanago (1991). Spatially-varied soil erosion under different climates. *Hyd. Sci. J. IAHS*, **36**, no. 6, 511–24.

Julien, P. Y. and G. J. Klaassen (1995). Sand-dune geometry in flooded rivers, *J. Hyd. Engr., ASCE*, **121**, no. 9, 657–63.

Julien, P. Y. and Y. Q. Lan (1991). Rheology of hyperconcentrations. *J. Hydr. Engr. ASCE*, **115**, no. 3, 346–53.

Julien, P. Y. and C. A. Leon (2000). Mudfloods, mudflows and debris flows: classification, rheology and structural design, International Seminar on the Debris Flow Disaster of December 1999. Caracas, Venezuala.

Julien, P. Y. and Y. Raslan (1998). Upper-regime plane bed. *J. Hydr. Engr.*, ASCE, **124**, no. 11, 1086–96.

Julien, P. Y., G. J. Klaassen, W. T. M. ten Brinke, and A. W. E. Wilbers (2002). Case study: bed resistance of the Rhine River during the 1998 flood, *J. Hydr. Engr., ASCE*, **128**, no. 12, 1042–50.

Julien, P. Y. and Rojas, R (2002). Upland erosion modeling with CASC2D-SED. *Intl. J. Sed. Res.*, **17**, no. 4, 265–74.

Julien, P. Y. and D. B. Simons (1985a). Sediment transport capacity of overland flow. *Trans. ASAE*, **28**, no. 3, 755–62.

—— (1985b). Sedimnt transport capacity equations for rainfall erosion. Proc. Fourth International Hydrology Symposium, Fort Collins, CO, 988–1002.

Julien, P. Y., G. Richard, and J. Albert (2005). Stream restoration and environmental river mechanics, Special Issue of the J. River Basin Management, IAHR & INBO, **3**, no. 3, 191–202.

Kalinske, A.A (1942). Criteria for determining sand transport by surface creep and saltation. *Trans. AGU*, **23**, part 2, 639–43.

—— (1947). Movement of sediment as bed load in rivers. *Trans. AGU*, **28**, no. 4, 615–20.

Kamphuis, J.W (1974). Determination of sand roughness for fixed beds. *J. Hyd. Res. IAHR*, **12**, no. 2, 193–203.

Kane, B. and P. Y. Julien (2007). Specific degradation of watersheds. *Intl. J. Sed. Res.*, **22**, no. 2, 114–19.

Karim, F. and J. F. Kennedy (1983). Computer-based predictors for sediment discharge and friction factor of alluvial streams. Report no. 242. Iowa Institute of Hydraulic Research, University of Iowa.

Keulegan, G. H. (1949). Interfacial instability and mixing in stratified flows. *J. Res. Nat. Bur. Stand.*, **43**, no. 487, RP 2040.

Kikkawa, H. and T. Ishikawa (1978). Total load of bed materials in open channels. *J. Hyd. Div. ASCE*, **104**, no. HY7, 1045–60.

Kilinc, M. Y. (1972). Mechanics of soil erosion from overland flow generated by simulated rainfall. Ph.D. dissertation, Colorado State University.

Kim, H. S. (2006). Soil erosion modeling using RUSLE and GIS on the Imba Watershed, South Korea. M.S. Thesis, Colorado State University, 118 p.

Kim, H. S. and P. Y. Julien (2006). Soil erosion modeling using RUSLE and GIS on the Imha Watershed, South Korea. Water Engineering Research, J. Korean Water Resources Association, 7, no. 1, 29–41.

Kircher, J. E. (1981). Sediment analyses for selected sites on the South Platte River in Colorado and Nebraska, and the North Platte and Platte Rivers in Nebraska; suspended sediment, bedload, and bed material. U.S. Geological Survey Open-File Report 81–207, United States Department of the Interior Geological Survey, Denver, CO.

Klaassen, G. J., J. S. Ribberink, and J. C. C. de Ruiter (1988). On sediment transport of mixtures. Pub. 394. Delft Hydraulics, Emmeloord.

Kodoatie, R. J. (1999). Sediment transport relationing in alluvial channels. Ph.D. dissertation, Colorado State University, 287 p.

Kopaliani, Z. D. (2002). Fluvial processes in rivers. Proc. State Hydrological Institute #361. Hydrometeoizdat, St. Petersburg, 146–84. (in Russian).

—— (2007). Problem of bed load discharge assessment in rivers. Proc. 10th Symposium of the International Research and Training Center on Erosion and Sedimentation, Moscow, 175–81.

Lacey, G. (1929). Stable channels in alluvium. *Proc. Inst. Civ. Eng.*, **229**.

Lambe, T. W. and R. V. Whitman, (1969). *Soil Mechanics*. Wiley and Sons, 553 p.

Lane, E. W. (1953). Progress report on studies on the design of stable channels by the Bureau of Reclamation. *Proc. ASCE*, **79**, no. 280.

Langhaar, H. L. (1951). *Dimensional Analysis and Theory of Models*. New York: Wiley.

Lara, J. M. (1966). *Computation of "Z's" for use in the Modified Einstein Procedure*. U.S. Department of the Interior, Bureau of Reclamation.

Lee, J. S. and P. Y. Julien (2006). Electromagnetic wave surface velocimeter. *J. Hyd. Engr., ASCE*, **132**, no. 2, 146–53.

Liu, H. K. (1957). Mechanics of sediment-ripple formation. *J. Hyd. Div. ASCE*, **183**, no. HY2, 1–23.

Long, Y. and G. Liang (1994). Database of sediment transport in the Yellow River. Tech. Rpt. No. 94001. Institute of Hydr. Research, Yellow River Conservation Comm., Zhengzhou, P. R. China, 15 p. (in Chinese).

Lowe, J., III and I. H. R. Fox (1982). Sedimentation in Tarbela Reservoir. Proc. 14th International Congress on Large Dams, Rio de Janeiro, 317–40.

Lucker, T. and V. Zanke (2007). An analytical solution for calculating the initiation of sediment motion. *Intl. J. Sed. Res.*, **22**, no. 2, 87–102.

Mantz, P. A. (1977). Incipient transport of fine grains and flakes by fluids-extended Shields diagram. *J. Hyd. Eng.*, **103**, no. 6, 601–16.

Meyer-Peter, E. and R. Müller (1948). Formulas for bed-load transport. Proc. 2nd Meeting IAHR, Stockholm, 39–64.

Migniot, C. (1989). Tassement et rhéologie des vases. *La Houille Blanche*, no. 1, 11–29.

Miller, C. R. (1953). Determination of the unit weight of sediment for use in sediment volume computations. Memorandum, U.S. Bureau of Reclamation, Department of the Interior, Denver, CO.

Mirtskhoulava, T. E. (1988). *Basic Physics and Mechanics of Channel Erosion*, transl. R. B. Zeidler. Leningrad: Gideometeoizdat.

Molnar, D. K. and P. Y. Julien (1998). Estimation of upland erosion using GIS. *J. Computers & Geosciences*, **24**, no. 2, 183–192.

Murphy, P. J. and E. J. Aguirre (1985). Bed load or suspended load. J. Hyd. Div. ASCE, **111**, no. 1, 93–107.

Neill, C. R. (1969). Bed forms in the lower Red Deer River, Alberta. *J. Hydro., The Netherlands* no. **7**, 58–85.

Nezu, I. and W. Rodi (1986). Open channel flow measurements with a laser Doppler anemometer. *J. Hyd. Engr., ASCE,* **112**, no. 5, 335–55.

Nordin, C. F. (1963). A preliminary study of sediment transport parameters, Rio Puerco near Bernardo, New Mexico. Professional Paper 462-C. Washington, D.C.: U.S. Geological Survey.

—— (1964). Aspects of flow resistance and sediment transport: Rio Grande near Bernalillo, NM. Water Supply Paper 1498-H. Washington, D.C.: U.S. Geological Survey.

Nordin, C. F. and Dempster, G. R. (1963). Vertical distribution of velocity and suspended sediment, Middle Rio Grande, New Mexico. Professional Paper 462-B, Washington: U.S. Geological Survey.

O'Brien, J. S. and P. Y. Julien (1985). Physical properties and mechanics of hyperconcentrated sediment flows. Proc. ASCE Specialty Conference on the Delineation of Landslide, Flashflood and Debris Flow Hazards, Utah Water Research Lab, Series UWRL/G-85/03, 260–79.

—— (1988). Laboratory analysis of mudflow properties. *J. Hyd. Engr. ASCE,* **114**, no. 8, 877–87.

Oseen, C. (1927). Hydrodynamik. In *Akademische Verlags Gesellschaft.* Leipzig: AVG, chap. 10.

Parker, G., P. C. Klingeman and D. G. McLean (1982). Bedload and size distribution in paved gravel-bed streams. *J. Hyd. Engr.,* **108**, no. 4, 544–71.

Patel, P. L. and K. G. Ranga Raju (1999). Critical tractive stress for non uniform sediment. *J. Hyd. Res., IAHR,* **37** no. 1, 39–58.

Peters, J. J. (1978). Discharge and sand transport in the braided zone of the Zaïre estuary. *J. Sea Res. The Netherlands* **12**(3/4), 273–92.

Petersen, M. (1986). *River Engineering.* Englewood Cliffs, N.J.: Prentice-Hall.

Piest, R. F. and C. R. Miller (1977). Sediment sources and sediment yields. In *Sedimentation Engineering,* ASCE Manual and Reports on Engineering Practice no. 54. New York, Chaps. 4 and 5.

Pitlick, J. (1992). Flow resistance under conditions of intense gravel transport. In *Water Resource Research,* AGU, 891–903.

Posada-Garcia, L. (1995). Transport of sands in deep rivers. Ph.D. dissertation, Colorado State University, 157 p.

Qian, N. (1982). Reservoir sedimentation and slope stability, technical and environmental effects. General Report Q54, 14th International Congress on Large Dams, Rio de Janeiro, May 1982.

Qian, N. and Z. Wan (1986). A critical review of the research on the hyperconcentrated flow in China. Beijing: International Research and Training Centre on Erosion and Sedimentation.

Ranga Raju, K. G., R. J. Garde and R. C. Bhardwaj (1981). Total load transport in alluvial channels. *J. Hyd. Div. ASCE,* **107**, no. HY2, 179–92.

Raudkivi, A. J. (1990). *Loose Boundary Hydraulics,* 3rd ed. New York: Pergamon Press.

Rayleigh, Lord. (1915). The principle of similitude. *Nature,* **95**.

Richardson, E. V. and P. Y. Julien (1986). Bedforms, fine sediments, washload and sediment transport. Third International Symposium on River Sedimentation, ISRS-III, Jackson, Mississippi, 854–74.

Richardson, E. V., Simons, D. B., and P. Y. Julien (1990). Highways in the River Environment. Design and training manual for US Department of Transportation, Federal Highway Administration, publication no. FHWA-HI-90–016. Washington, D.C.

Richardson, E. V., Woo, H. S., and P. Y. Julien (1986). Research trends in sedimentation. In *Proceedings of the Symposium on Megatrends in Hydraulic Engineering,* ed. M. L. Albertson and C. N. Papadakis. Fort Collins: Colorado State University, 299–330.

Rojas, R., Julien, P. Y., Velleux, M., and B. E. Johnson (2008). Grid size effect on watershed soil erosion models. *JHE, ASCE*, **134**, no. 9, 793–802.

Rouse, H. (1937). Modern conception of the mechanics of turbulence. *Trans. ASCE*, **102**, 463–543.

Rouse, H. (ed.) (1959). *Advanced Mechanics of Fluids*. New York: Wiley.

Rubey, W. (1933). Settling velocities of gravel, sand and silt particles. *Am. J. Sci.*, **25**, no. 148, 325–38.

Ryan, S. E. and W. W. Emmett (2002). The nature of flow and sediment movement in Little Granite Creek near Bondurant, Wyoming, Report RMRS-GTR-90, 48 p.

Savage, S. B. and S. McKeown (1983). Shear stresses developed during rapid shear of concentrated suspension of large spherical particles between concentric cylinders. *J. Fluid Mech.*, **137**, 453–72.

Schlichting, H. (1968). *Boundary Layer Theory*. New York: McGraw-Hill.

Schoklitsch, A. (1934). Geschiebetreib und die geschiebefracht. Wasser Kraft und Wasser Wirtschaft. Jgg. 39, no. 4.

Schultz, E. F., R. H. Wilde, and M. L. Albertson (1954). Influence of shape on the fall velocity of sedimentary particles. Report of the Colorado Agricultural and Mechanical College, Fort Collins, CO MRD Sediment Series, no. 5.

Schwab, G. O., R. K. Frevert, T. W. Edminster, and K. K. Barnes (1981). *Soil and Water Conservation Engineering*, 3rd ed. New York: Wiley.

Sedov, L. I. (1959). *Similarity and Dimensional Methods in Mechanics*, transl. M. Hold. New York: Academic Press.

Shah-Fairbank, S. C. (2009). Series expansion of the Modified Einstein Method. Ph.D. dissertation, Colorado State University.

Shen, H. W. and C. S. Hung (1983). Remodified Einstein procedure for sediment load. *J. Hyd. Div. ASCE*, **109**, no. 4, 565–78.

Shen, H. W. and P. Y. Julien (1993). Erosion and sediment transport. In *Handbook of Hydrology*. New York: McGraw-Hill, Chap. 12.

Shields, A. (1936). Anwendung der Aehnlichkeitsmechanik und der Turbulenzforschung auf die Geschiebebewegung. Berlin: Mitteilungen der Preussische Versuchanstalt für Wasserbau und Schiffbau.

Simons, D. B. (1957). Theory and design of stable channels in alluvial material. Ph.D. dissertation, Colorado State University.

Simons, D. B., and E. V. Richardson (1963). Form of bed roughness in alluvial channels. *Trans. ASCE*, **128**, 284–323.

—— (1966). Resistance to flow in alluvial channels. Professional Paper 422-J. Washington, D.C.: U.S. Geological Survey.

Simons, D. B. and F. Senturk (1977). *Sediment Transport Technology*. Littleton, CO: Water Resources Pub., 1977, revised, 1992.

Simons, D. B., R. M. Li, and W. Fullerton (1981). Theoretically derived sediment transport equations for Pima County, Arizona. Prepared for Pima County DOT and Flood Control District, Tucson, Ariz. Ft. Collins, Colo.: Simons, Li and Assoc.

Singamsetti, S. R. (1966). Diffusion of sediment in a submerged jet. *J. Hyd. Div. ASCE*, **92**, no. 2, 153–68.

Smerdon, E. T. and R. P. Beasley (1961). Critical cohesive forces in cohesive soils. *Agr. Eng.*, **42**, 26–9.

Stevens, M. A. and D. B. Simons (1971). Stability analysis for coarse granular material on slopes. In *River Mechanics*. Littleton, Colo. Water Resources Pub., Chap. 17.

Streeter, V. L. (1971). *Fluid Mechanics*, 5th ed. New York: McGraw-Hill.

Thomas, W. A. (1976). Scour and deposition in rivers and reservoirs. Report no. 723-G2-L2470. Vicksburg, Miss.: US Army Corps of Engineers, Hydrology Engineering Center.

Toffaleti, F. B. (1968). A procedure for computation of the total river and sand discharge and detailed distribution, bed to surface. Technical Report no. 5. Vicksburg, Miss.: US Army Corps of Engineers, Committee on Channel Stabilization.

Tsujimoto, T. (1992). Fractional transport rate and fluvial sorting. Proc. Grain Sorting Seminar, Versuchanstalt für Wasserbau, Mitteilungen 117, ETH. Zurich, 227–49.

US Bureau of Reclamation (2006). *Erosion and Sedimentation Manual*. U.S. Govt. Printing Office, Washington, D.C. 20402–0001, 616 p.

van Rijn, L. C. (1984a). Sediment transport, Part II: Suspended load transport. *J. Hyd. Div. ASCE*, **110**, no. 11, 1613–41.

—— (1984b). Sediment transport, Part III: Bedforms and alluvial roughness. *J. Hyd. Div. ASCE*, **110**, no. 12, 1733–54.

Vanoni, V. A. (1946). Transportation of suspended sediment by water. *Trans. ASCE*, **3**, paper no. 2267, 67–133.

Vanoni, V. A. and N. H. Brooks (1957). Laboratory studies of the roughness and suspended load of alluvial streams. Report E-68. Pasadena: California Institute of Technology, Sedimentation Laboratory.

Vanoni, V. A., N. H. Brooks, and J. F. Kennedy (1960). Lecture notes on sediment transportation and channel stability. Report KH-RI. Pasadena: California Institute of Technology, W. M. Keck Laboratory.

Velleux, M., Julien, P. Y., Rojas-Sanchez, R., Clements, W., and J. England (2006). Simulation of metals transport and toxicity at a mine-impacted watershed: California Gulch, Colorado. *Environmental Science and Technology*, **40**, no. 22, 6996–7004.

Velleux, M., England Jr., J. F., and P. Y. Julien (2008). TREX: Spatially distributed model to assess watershed contaminant transport and fate. *J. Science in the Total Environment*, **404**, 113–28.

Wargadalam, J. (1993) Hydraulic geometry equations of alluvial channels. Ph.D. dissertation, Colorado State University.

Waterways Experiment Station (1935). Studies of river bed materials and their movement with special reference to the lower Mississippi River. USWES, paper no. 17. Vicksburg, Miss.

Weinhold, M. (2002). Application of a site-calibrated Parker-Klingeman bedload transport model, Little Granite Creek, Wyoming. M.Sc. Colorado State University, 99 p.

White, F. M. (1974). *Viscous Fluid Flow*. McGraw-Hill, 725 p.

Whitehouse, J. S., R. L. Soulsby, and J. S. Damgaard (2000). Discussion Technical Note 17149, *J. Hyd. Engr., ASCE*, **126**, no. 7, 553–5.

Wijbenga, J. H. A. (1991). Analyse prototype-metingen (niet-) permanente ruwheid. Report Q1302. Emmeloord: Delft Hydraulics.

Williams, D. T. and P. Y. Julien (1989). On the selection of sediment transport equations. *J. Hyd. Engr. ASCE*, **115**, no. 11, 1578–81.

Williams, G. P. and Rosgen, D. L. (1989). Measured total sediment loads (suspended load and bed loads) for 93 United States Streams. 89–67, United States Geological Survey, Denver, CO.

Wilson, K. C. (1966). Bed-load transport at high shear stress. *J. Hyd. Div. ASCE*, **92**, no. 6, 44–59.

Winterwerp, H. (1999). On the dynamics of high-concentrated mud suspensions. Report 99–3, Dept. of Civil Engineering and Geosciences. Delft University of Technology.

Wischmeier, W. H. (1972). Upslope erosion analysis. In *Environmental Impact on Rivers*, ed. H. W. Shen. Littleton, Colo.: Water Resources Pub. Chap. 15.

Wischmeier, W. H. and D. D. Smith. (1978). *Predicting Rainfall Erosion Losses: A Guide to Conservation Planning*. USDA Agr. Handbook 537. Washington, D.C.: U.S. Department of Agriculture.

Woo, H. S. and P. Y. Julien (1990). Turbulent shear stress in heterogeneous sediment-laden flows. *J. Hyd. Eng. ASCE*, **116**, no. 11, 1416–21.

Woo, H. S., Julien, P. Y., and E. V. Richardson (1986). Washload and fine sediment load. *J. Hyd. Eng. ASCE*, **112**, no. 6, 541–5.

—— (1988). Suspension of large concentrations of sands. *J. Hydr. Engr. ASCE*, **114**, no. 8, 888–98.

Wu, B. (1999). Fractional transport of bed-material load in sand-bed channels. Ph.D. dissertation, Colorado State University, 193 p.

Wu, B., A. Molinas, and P. Y. Julien (2004). Bed-material load computations for sediment mixtures. *J. Hydr. Engr., ASCE*, **130**, no. 10, 1002–12.

Wu, W. and S. S. Y. Wang (2006). Formulas for sediment porosity and settling velocity. *J. Hydr. Engr. ASCE*, **132**, no. 8, 858–62.

Yalin, M. S. (1964). Geometrical properties of sand waves. *J. Hyd. Div. ASCE*, **90**, no. HY5, 105–19.

—— (1977). *Mechanics of Sediment Transport*, 2nd ed. London: Pergamon Press.

Yalin, M. S. and E. Karahan (1979). Inception of sediment transport. *J. Hyd. Div. ASCE*, **105**, no. HY11, 1433–43.

Yalin, M. S. and A. M. daSilva (2001). *Fluvial Processes*. IAHR Monograph, Balkema, 197 p.

Yang, C. T. (1973). Incipient motion and sediment transport. *J. Hyd. Div. ASCE*, **99**, no. HY10, 1679–1704.

Yih, C. S. (1979). *Fluid Mechanics*. Ann Arbor, Mich.: West River Press.

Young, R. A. and C. K. Mutchler (1967). Soil movement on irregular slopes. *Water Resources Res.*, **5**, no. 5, 1084–5.

Yücel, O. and W. H. Graf (1973). Bed load deposition in reservoirs. Proc. 15th Congress of IAHR, Istanbul.

Zhou, D. and C. Mendoza (2005). Growth model for sand wavelets, *J. Hydraul. Eng., ASCE*, **131**, no. 10, 866–76.

Zingg, A. W. (1940). Degree and length of land slope as it affects soil loss by runoff. *Agricultural Engineering*, **21**, no. 2, 59–64.

Zyserman, J. A. and J. Fredsøe (1994). Data analysis of bed concentration of suspended sediment. *J. Hyd. Div. ASCE*, **128**, no. HY9, 1021–42.

Index